Template and info pages for

"Mapping the Body with Art"

A video e-course to be accessed via: www.ellenjmchenry.com

Please note that samples of the final drawings
can be found at the back of this book.

1: THE WATER MOLECULE

Water is essential to life. Our bodies are 60 to 70 percent water. (Younger people tend to have more water, and older people have less.) If you removed all the water from our bodies, we'd be nothing but a small pile of dry, dusty minerals.

Water is made of one oxygen atom and two hydrogen atoms. The hydrogens stay attached to the oxygen because they are sharing electrons. Oxygen would like to gain two electrons (it has six in its outer shell but would like to have eight) so it works out very well if two hydrogens come over and share their electrons. This type of bond (sharing electrons) is called a **covalent bond**. In general, non-metal atoms (such as carbon, nitrogen, and oxygen) form covalent bonds.

Atoms at the top of the Periodic Table obey what is called the "Octet Rule." They want to have 8 electrons in their outer shell. (If you would like a review activity about electrons shells and orbitals, try the "Quick and Easy Atom-izer" posted at www.ellenjmchenry.com; click on free downloads, then on chemistry.) Ideally, an atom with only 2 electrons in its outer shell will try to pair up with an atom that has 6 electrons in its outer shell. Between the two of them they will have 8, fulfilling the Octet Rule. It doesn't always work out this perfectly. Sometimes atoms have to "double share" to make things work out. For example, the oxygen in the air we breathe forms O_2 molecules. Each oxygen atom has 6 electrons in its outer shell and wants 2 more. They each give the other a pair of electrons, forming a **double bond**. It's not a covalent bond where everything adds up to eight, but it seems to work well enough for the O_2 molecule. Oxygen molecules can even take on a third oxygen temporarily, forming ozone, O_3.

Chemists have actually been able to measure the angle at which the hydrogen atoms sit on the oxygen atom, and found it to be about 104 degrees (meaning geometry degrees, not heat). The placement of the hydrogens makes the molecule look a bit comical to us; it has been compared to a teddy bear, or to Mickey Mouse or Kermit the Frog.

Water molecules are constantly vibrating. The hydrogens go in and out, closer to and further from the oxygen. They also tumble around, bumping into each other like bumper cars at an amusement park. The faster they move, the more heat there is. Slower activity goes along with decreased temperatures. (Try the online interactive demo posted as an activity for this lesson. It lets you increase and decrease the movement of the molecules and see what happens.)

The oxygen atom is bigger than the hydrogen atoms. The oxygen atom has 8 protons, whereas the hydrogens have just one. This gives the oxygen a huge advantage when it comes to controlling the 8 electrons they are all sharing. The oxygen's large nucleus attracts those electrons strongly and keeps them circulating around it most of the time. Those poor hydrogens get an electron whizzing around them once in a while, but the electrons spend more time around the oxygen atom. This causes the entire molecule to become slightly unbalanced, so to speak. If you imagine the molecule to look like a teddy bear head, the chin of the bear becomes more negative because of the constant presence of the electrons, which carry a negative charge. The bear's ears, those hydrogens, are essentially made of nothing but a proton, and since protons carry a positive charge, the ear side of the molecule is more positive. This difference in charge between the two sides makes the molecule **polar**. Polarity is when a molecule is more negative on one side and more positive on the other. Polarity is what gives water so many of its amazing qualities, such as being able to dissolve so many substances. We'll meet the concept of polarity again in future lessons. It is very important.

Hydrogen atoms sometimes leave their water molecules. This means that at any given time, there are some H's and some OH's floating around. We call these partial molecules **ions**. An ion is an electrically unbalanced atom or molecule, with more electrons than protons or more protons than electrons. When a hydrogen atom leaves its water molecule, it leaves its electron behind. Without its electron, hydrogen becomes nothing but a proton. In a normal hydrogen atom there is one proton and one electron, so the positive and negative charges are equal. But a hydrogen ion has only one proton and zero electrons, so its overall electrical charge is +1. We can use the abbreviation H^+ to represent a hydrogen ion. (You must always remember that the words "hydrogen ion" and "proton" can be used interchangeably since they are the same thing.)

In the case of OH, it has lost a proton (H+), so now it has one more electron than it does protons. Overall, it carries a negative (-) charge. We will therefore write it like this: OH^-. The proper name of this molecule is the **hydroxide ion**.

The strangest molecule in a drop of water has to be the **hydronium ion**, H_3O. A hydrogen ion (a proton) goes and attaches itself to a regular molecule, turning H_2O into H_3O. This does not work out so well, and one of the hydrogens immediately leaves, returning the molecule to normal water.

structural formula

ball-and-stick model   space-filling model

Water molecules are covalent. Electrons are shared.   Water molecules are polar.

Hydrogen ions are protons, H+.
Hydroxide ions are OH-.
Hydronium ions are H_3O.

Out of 55,000,000 (55 million) water
molecules, you will find only one ion.

(There are about 470,000,000,000,000,000,000
water molecules in one drop of water.)

2: CARBON ATOMS and FATTY ACIDS

The carbon atom is like no other atom. It is so flexible in the ways that it can bond with other atoms, that it is the basis for thousands of molecules, many of which are found in living organisms. Starting with carbon, we can build sugars, proteins, fats, and nucleic acids —the essential molecules for life.

Carbon is element number 6 on the Periodic Table. This means it has 6 protons. Because it has 6 protons it also has 6 electrons. Of these 6 electrons, 2 fill the small inner shell and the remaining 4 occupy the outer shell. As you will remember, the outer shell would like to have 8 electrons in it, so carbon tries to bond with other atoms in order to gain 4 electrons. In other words, carbon has 4 places that it can bond to another atom.

The most basic carbon molecule is *methane*: 1 carbon atom bonded to 4 hydrogen atoms. (The root "meth" means "one." The way you count carbon atoms isn't, "One, two, three, four," it's, "Meth, eth, prope, bute.") When we look at the electrical situation in this molecule, we see that the molecule is evenly balanced. The hydrogen atoms move around so that they are the maximum distance apart, evenly spaced around the carbon. We need to remember that hydrogens don't want to be right next to each other. Their positively charged protons don't like to be next to other positive charges. As they say, "Like charges repel, opposite charges attract." So the hydrogens space out evenly. This means that unlike water, methane does not have a positive and a negative side. All sides are the same. Methane is *nonpolar*.

Another important feature of carbon is that it likes to bond to itself, making chains or rings. A 2-carbon chain is called ethane, a 3-carbon chain makes propane, and 4 is butane. You will perhaps recognize the words propane and butane and know them to be fuels we use in things like lighters and outdoor grills.

Methane is a very light molecule so it is a gas (at standard temperature and pressure). As the carbon chains get longer, the molecules get bigger and heavier. By the time we get to the 8-carbon chain called octane, we have a liquid. Octane is found in the gasoline (petrol) we put into cars. When carbon chains get to be really long they form solids such as wax. Plastics are also made of extremely long chains of carbon atoms, though sometimes other types of atoms are mixed in, too, such as chlorine in PVC (polyvinyl chloride).

Chemists use a short cut when drawing carbon chains. They draw only a zig-zag line. They know that at the point of each V there is a carbon atom. Since they all know this, they don't bother drawing it. They also assume, unless otherwise indicated, that there are hydrogens attached to the carbons in order to give each one its required 4 bonds. A carbon at the end of a chain will have 3 hydrogens and those in the middle will have 2.

The carbon chains found in our bodies are called *fatty acids*. The chains are usually 12 to 18 carbons long, and they have a special group of atoms on one end. This group is called the *carboxyl group*, or carboxylic acid, **COOH**. In the last lesson we briefly mentioned that a liquid with a lot of protons (hydrogen ions) in it will be acidic. In COOH, the H is not attached very well and can "fall off." Anything that creates loose H's is an acid. As we will see in the next lesson, this group has a special feature that will let us connect the carbon chain to other molecules, making larger structures. Just to confuse you, it will be the OH part of COOH that will fall off. However, this will not make the environment alkaline. The OH will immediately get picked up and joined to another atom to make a harmless substance.

The fatty part of a fatty acid is the carbon chain. We know that fats and oils are greasy and don't mix with water. Carbon chains are nonpolar so they don't have positive or negative sides that can attract water molecules. Other substances can dissolve fats, but not water. Water can only dissolve polar substances.

Carbon wants to make 4 bonds.

ball and stick model of methane

ethane and butane

Simplyfing how we draw carbon chains

Carbon can make double bonds (and even triple bonds).

A fatty acid is a chain of carbons with -COOH at one end.

3: LIPIDS (part 1)

A fatty acid is a carbon chain with the COOH carboxyl group attached to one end. In this lesson we will see how fatty acids can be attached to a larger molecule, called glycerol.

Glycerol is a carbon-based structure, with a chain of three carbons at the center. The presence of three oxygen atoms makes it different from the carbon structures we saw in the previous lesson. Think of glycerol as a hanger, from which one, two, or three fatty acids can hang. The fatty acids hanging down from glycerol usually have from 12 to 20 carbon atoms.

To attach a fatty acid to a glycerol "hanger" you must take the OH off the COOH, leaving just CO. On the glycerol end, a hydrogen, H, is pulled off. The unhappy atoms who have an empty bond site dangling are the carbon on the fatty acid and the oxygen on the glycerol. The carbon and the oxygen and matched up and they bond to each other.

What happens to the OH and the H that were pulled off? They join together quite happily to form a molecule of water, H_2O. This does not happen randomly, on its own, however. A little "machine" (an enzyme) does the attaching. As we will see in future lessons, enzymes are like little robots who do only one job. The enzyme at work here was designed to do this one task: join fatty acids to glycerol by popping off OH's and H's. This enzyme has an extremely long name that is almost impossible to remember even for an biochemist, so just knowing that an enzyme in involved is enough information for us.

This process has a name you will need to know; it's called **dehydration synthesis**. (The word "synthesis" is from the Greek word for "make.") As it turns out, this is a common way of joining two molecules. Pull off an OH and an H and join the "ragged edges" you leave behind, making a water molecule in the process. The name might seem backwards. When we think of dehydration it's usually in the context of water evaporating and disappearing, not being made. Think of the word "dehydration" in this way: "de" means "from," and "hydro" means "water." You are using water to synthesize something. The end product is coming <u>from</u> the process of making <u>water</u>. This process can be reversed, too. Water can be used to separate two molecules. The water is broken into OH and H, and those ions are used to "plug" the ends of the two broken pieces.

A glycerol that has 3 fatty acids hanging on it is called a **triglyceride**. ("Tri" is Greek for "three.") A triglyceride's fatty acids can be all the same, or they can be very different. Sometimes a glycerol will have only one fatty acid, so it will be called a **monogylceride**. ("Mono" is Greek for "one.") A glycerol with 2 fatty acids would be a **diglyceride**. ("Di" is Greek for "two.") Fatty acids that are not attached to a glycerol are called **free fatty acids**.

Where does glycerol come from? The body can make it from glucuse sugar or it can use glycerols that come from the things we eat (fats from plants or animals). If glycerol needs to be manufactured from glucose, there are specialed enzyme "robots" that will perform this task. If glycerol is taken out of food, there are enzymes for that, too.

If a fatty acid has all the hydrogens it can possible hold, it is called **saturated**. Saturated fats tend to be solid at room temperature and are most often found in animal products such as meat and butter. An unsaturated fat is a carbon chain that is missing some hydrogens. A place where two hydrogens are missing creates a double bond, which then causes as a slight bend in the chain. If the chain has only one double bond, it is called **monounsaturated**. If it has many double bonds and therefore many bends, it is called **polyunsaturated**. ("Poly" is Greek for "many.")

Nutritionists are still debating whether it is better to consume saturated or unsatured fats. To complicate matters, there are **trans fats** where hydrogens have been artificially added, to break those double bonds, but the hydrogens end up on opposite sides of the molecule. ("Trans" means "across.") The trans fats tend to stick to the insides of our blood vessels and clog them up. Transfat are most often found in desserts and snack foods.

Fatty acids often have strange-sounding names such as myristoleic acid, sapeinic acids, vaccenic acids and caprulic acid. We won't be learning those. Fatty acids that are very common and also have decent names are lauric acid, (a short one with only 12 carbons and found in coconuts and palms), and palmitic acid with its 16 carbon chain (found in palms). There is a whole group of fatty acids that all have 18 carbons, though the number of double bonds varies. Stearic acid has no double bonds so it is completely saturated. It is found in great aboundance in animal fat, particularly in the fat inside bones (which is where it gets its name). Oleic acid has one double bond and is found in great aboundance in olive oil. The "ol" at the beginning of each word (olive and oleic) helps us to remember the connection. Linoleic acid has two double bonds in its chain and is found in flax (from which **linen** is made). Alpha-linolenic acid has three double bonds and is found in nuts and seeds. Why all the fuss about double bonds? Those double bonds cause the chain to have bends or "kinks" in it. This alteration of the shape causes it to behave differently in cells.

Omega-3 fatty acids have a double bond starting at the third carbon from the end. ("Omega" means "last"). Omega-6 fatty acids have a double bond starting at the carbon sixth from the end. Our diets should contain more omega-3 fats and less omega-6 fats. Unfortunately, the modern diet has these reversed. Omega-3 fats are found abundantly in fish oil.

Our bodies can make some fatty acids. The ones we must get from food are called Essential Fatty Acids (EFAs).

LIPIDS (part 1: triglycerides)   3

FATTY ACID

GLYCEROL	TRIGLYCERIDE
TRIGLYCERIDES DIGLYCERIDES MONOGLYCERIDES	

What kind of fatty acids might be attached to glycerol?

_____ acid has ___ carbon atoms. It is found in coconuts and palms. Palmitic acid has ___ carbon atoms.

These fatty acids all have ___ carbons:
1) _____ acid has ___ double bonds and is found in abundance in _____.
2) _____ acid has ___ double bond and is found in abundance in _____.
3) _____ acid has ___ double bonds and is found in abundance in _____ and _____.
4) _____ (ALA) has ___ double bonds and is found in _____ and _____.

Our bodies can make some fatty acids. Others must come from our diet and are called _____.

4: LIPIDS (part 2: phospholipids)

When lipids are joined to phosphate molecules they form a large molecule called a **phospholipid**. Phospholipids are one of the key building blocks of cells. As we will see in lesson 5, they will form the outside layer of cells, and also of many smaller cell parts called organelles.

A phospholipid molecule is made of a glycerol hanger that is holding on to two fatty acids and a phosphate. A phosphate is made of 1 atom of phosphorus and 4 atoms of oxygen.

Phosphorus can make 5 bonds. (The Periodic Table can help you determine how many bonds an atom wants to make. The atoms in the first column (Li, Na, etc.) all have one extra electron in their outer shell, so they want to make one bond. The second column atoms all have 2 electrons in their outer shell so they are good for 2 bonds. The third column has 3, and so on. Phosphorus is in the fifth column, so it can make 5 bonds.) Since there are only 4 oxygen atoms in this molecule, one lucky oxygen atom will get a double bond. The other oxygen atoms have one bond with phosphorus, but are also holding on to an extra electron, which will give the molecule an overall electrical charge of negative 3. We can write PO_4^{3-}. This molecule is a **polyatomic ion**. Simple ions are made of one atom. Polyatomic ions are made of more than one atom. (By the way, don't let the word "ion" confuse you. It is common for students to have trouble remembering what an ion is. If you think of an ion as broken molecule, that's sort of right. Ions have had electrons added or taken away, ruining their original neutrality.)

The phosphate ion is best explained by looking at phosphoric acid, H_3PO_4. In this molecule, the oxygens have their two bonds, one with phosphorus and one with a hydrogen. Hydrogen atoms, as we have seen, have the bad habit of easily wandering off. When they leave, however, the oxygen atoms insist on keeping the electrons. You'd think that would make a hydrogen want to stay, but no, it goes off as nothing but a proton. (This makes the surrounding environment acidic. That's what acids do — they donate protons.) If all 3 hydrogens are gone, you have a phosphate ion.

One oxygen of the phosphate ion is bonded to the third hanger on the glycerol. One oxygen has a double bond and just sits there. One oxygen hangs off unbonded, and the fourth oxygen is usually attached to another molecule which can be labeled R. An R group is the variable part of a molecule. R could be any one of a number of different options. We don't need to know any specific molecules for R in this case. We will just write an R and leave it at that.

Two of glycerol's hangers are connected to fatty acids. There are many possibilities for what type of fatty acids these might be, but we are going to draw two special fatty acids: **EPA** and **DHA**. EPA stands for <u>e</u>icosa<u>p</u>entaenoic <u>a</u>cid, and DHA stands for <u>d</u>ocosa<u>h</u>exaenoic <u>a</u>cid. (That's why we calle them EPA and DHA. EPA has 20 carbon atoms and 5 double bonds. DHA has 22 carbons and 6 double bonds. The double bonds cause bends that make the molecules look almost like circles. EPA and DHA are especially valuable in some types of cells, such as nerve cells. They seem to have the ability to keep other molecules from getting too tangled. (This will make more sense a few lessons from now when we make membranes.) These fatty acids are found in other places, too, not just in phospholipids. They are important ingredients in messenger molecules, for instance. They help to stop inflammation.

The most important thing to know about a phospholipid molecule is that the head portion is **hydrophilic,** or "water loving," and the tails are **hydrophobic**, or "water hating." This is a hugely important fact in biochemistry.

Carbon is a very flexible atom; besides being able to form long chains, it can also make rings. One of the most basic rings is called benzene. This is not a molecule found in your body (well, hopefully not), but it is a good example of a carbon ring. Carbon rings are often in the shape of a hexagon or a pentagon.

Cholesterol is made of four carbon rings with some extra atoms attached. It is a natural body substance, and as we will soon see, it is found in and around phospholipid molecules. It can tuck in amongst the fatty acid tails. Cholesterol is not a harmful substance; it is a natural and necessary component found in all body cells. Too much of anything can be bad, so, indeed, a very high level of cholesterol is not good for you. On the other hand, too little is not healthy, either.

PHOSPHATE is PO$_4$

It carries an electrical charge
of -3 because it has three
unhappy oxygen atoms.
Oxygen always wants two
bonds. One of these oxygens is
lucky— it has a double bond. The
others have an empty place where
they need a bond.

Additional fact: Carbon can form rings.

A PHOSPHOLIPID is made of a phosphate
group, a glycerol "hanger" and two fatty acids.

CHOLESTEROL is a natural body substance,
very necessary for proper functioning of cells.

5: MEMBRANES (part 1)

If you could throw a whole bunch of phospholipid molecules into a bucket of water, the water-hating tails would go into a panic and try to find a way to get away from the water molecules. The tails would all congregate together, to make an area that was water-free. More and more molecules would join in until a sphere had formed. A sphere is the most efficient, compact shape they could form. A simple. single-layer sphere of phospholipids is called a *micelle*. We will see these in a future lesson. (A micelle is shown and discussed in the video lesson, but not drawn as part of this lesson.)

Another way the phospholipids could arrange themselves is into a double layer, with all the tails turned inward. A double layer of phospholipid molecules is called a *phospholipid bilayer*. This term is used quite frequently in cell biology. The phospholipid bilayer is the basic structure of a cell membrane. It separates the inside from the outside. Most cell organelles are also surrounded by a phospholipid bilayer.

A simple phospholipid sphere can be called a *liposome*. ("Lipo" is Greek for "fat," and "soma" is Greek for "body.") We will look at three cell parts that have the structure of a liposome.

An empty liposome can be called a *vacuole*. "Vacuus" is Latin for "empty." Vacuoles can contain water or air.

A *vesicle* is basically a vacuole that has stuff inside of it. Cells use vesicles like we use plastic or paper bags. Anything that needs to be stored, or transported across the cell, can be put into a vesicle "bag." Sometimes vesicles filled with certain substances (food, fats, chemicals) are given fancy names, rather than just being called a vesicle. We'll see some of these in future lessons.

A *lysosome* is a very special kind of liposome. It contains digestive enzymes, so it is a bit like a stomach. Things that need to be broken down and recycled (old cell parts, or even bacteria) are put into a lysosome. The digestive enzymes are able to tear them apart and turn them into simple proteins, sugars, and other molecules that cells can use. It's like destroying a Lego sculpture and using the parts to build something else. The word "lys" means "to dissolve or break apart." Lysosomes can contain as many as 50 different kinds of enzymes. Each enzyme does only one job. (This is a very important fact to remember. Enzymes are highly specific and can only do ONE thing. In fact, if the enzyme's substrates or the enzymes' active site are altered even a tiny bit, the enzyme will not be able to function. Many medicines work in this way. A molecule is given to the patient that prevents an enzyme from doing its job.)

Scientists used to think that if a lysosome burst, the cell would die because the enzymes would get out and go around digesting all the cell parts. (Enzymes are not smart. They don't know what they are supposed to digest. They digest anything in their path.) Then they discovered that the environment inside a lysosome prevents this from happening. The digestive enzymes need an acidic environment to be able to function. Little proton pumps in the membrane of the lysosome bring protons inside. An environment with lots of protons floating around will be acidic. When the lysosome bursts, the enzymes do escape, but they suddenly find themselves in an environment that is neutral, not acidic. Therefore they stop functioning (or at least slow way down).

NOTE: Lysosomes do participate in cell death (apoptosis or necrosis, which we will discuss in future lessons) and are sometimes "given instructions" to digest cell parts, but under normal circumstances lysosomes don't endanger their cells.

In a polar substance (like water) phospholipids form a bilayer with the hydrophobic tails pointing to the inside.

Phospholipids will form a sphere. This is the basic structure of all membranes.

Liposomes form basic cell parts such as:

VACUOLES: empty "bubbles" filled with water or air

VESICLES: "storage bags" used to transport things around the cell

LYSOSOMES: vesicles that are acidic inside and contain digestive enyzmes

An empty phospholipid sphere is called a liposome.
"Lipo" is Greek for "fat," and "soma" is Greek for "body."

6: MEMBRANES (part 2)

ALL cells are surrounded by a membrane. Even plants cells, with their thick cell wall made of cellulose, still have a membrane under that wall. And what is that membrane made of? Phospholipid bilayer, of course. The membrane separates the inside from the outside; kind of obvious, perhaps, but a concept often emphasized in textbooks. Once "inside" and "outside" have been established, the cell now needs a way to bring things in and send things out. The cell will need to bring in nutrients and other helpful molecules, and will need to send manufactured products to other cells and to get rid of wastes. There are a number of different methods of getting things across the membrane, depending on the size and chemical properties of the materials being transported. The first thing to consider is whether these methods use energy or not.

We use the word "transport" to describe the process of crossing the membrane. We add the word "passive" to describe transport that does not require any energy. So *passive transport* is crossing the membrane without using energy.

PASSIVE TRANSPORT: There are two types of passive transport. They both use the principle of *diffusion*, so we need to discuss that first. Diffusion comes from a Greek word meaning "to spread out." Diffusion is what happens when you open a bottle of something very smelly in a closed room. At first, people standing at the far end of the room can't smell anything. Then, as time goes on, the smell spreads and fills the room. Soon all areas of the room are equally smelly. Diffusion is the movement of molecules from where there are more of them to where there are less of them. The "goal" of the molecules is to end up equally distributed everywhere. The molecules will keep moving after this, but they will remain equally distributed. In our example, the smelly molecules did not have anything blocking them from moving around so we were not too surprised that they could fill the room.

If we had put a paper wall across the middle of the room and sealed all the edges, this would have made it more difficult for the smelly molecules to fill the room, but not impossible if the molecules were small enough to go through the microscopic holes in the paper. A paper wall isn't going to stop a smell like gasoline, for instance. This shows us that diffusion can still happen across a barrier if the chemistry is right. Water would also be able to get across our paper wall but not quite as quickly. Barriers that allow things to pass right through are called *permeable*. (So why have them in the first place, right?) Barriers that allow only some things to pass through are *semi-permeable*. Cell membranes fall into the category of semi-permeable, but an even better word would be *selectively permeable* because cells can select some things to come in and other things to stay outside. Again, these words mean exactly what they say, so it should not be hard to remember them.

Now that we know what diffusion is, we can learn the two types of passive transport. They are **1)** *simple diffusion* and **2)** *facilitated diffusion*. Simple diffusion is often just called diffusion without the word simple in front of it. However, it might be easier to remember these terms if their format is the same. The brain likes pairs and groups and always looks for similarities, so we can help out brains remember if we cooperate and try to group things logically. The visual layout of the drawing will help you, also. Passive on the left, active on the right, with the types listed below.

Simple diffusion: Some things can diffuse right through the phospholipid membrane. If the concentration of that type of molecule is greater outside the cell than inside, the molecule will diffuse in. What kind of molecule will be able to do this? As you might guess, it would have to be small. Size is important. What about chemical properties? Look at the phospholipid bilayer. Which area is thicker—the head area or the tail area? The tails are very long, so the lipid layer in the middle is much thicker. This means that any molecule passing through will have most of its journey be through the fatty, non-polar area. Overall, this means that **fat-friendly, non-polar molecules** stand a better chance of getting through the bilayer than water-friendly, polar ones. Examples of small, fat-friendly molecules that can diffuse through the membrane are **vitamins A, D, E and K, and steroids**. These vitamins are often called the "fat-soluble" vitamins. Their molecular structure includes rings of carbons, like we saw in cholesterol, so they get along very well with the cholesterol molecules that sit in and among the fatty tails. They have no trouble slipping through. Steroids are also based on rings of carbon. In fact, your body turns cholesterol into steroid molecules (and also into vitamin D). Cholesterol isn't a poison; it is an essential molecule you can't live without. The steroids made from cholesterol include estrogen, testosterone, and anti-inflammatory steroids.

Oxygen and carbon dioxide are small and are non-polar so they can use simple diffusion, too. They are very numerous and must get across quickly, so it is good that they can just cross on their own. If they had to wait for a molecular gate to open, this would cause a chemical traffic jam for sure!

Water used to be on the list of molecules that use simple diffusion, but now that has been called into question. It is true that even though **water** is polar, the molecules are very small and they do indeed often slip through the membrane to the other side. The "pull" of diffusion (wanting to be where there are less of them) will sometimes be enough to get them through. However, we now know that even though water does sometimes diffuse through, the molecules prefer to go through a channel that we will discuss on the next page.

Facilitated diffusion: The word "facilitate" means "to make easier." In faciliated diffusion we find some molecules that can diffuse if they have just a little bit of help. Molecules that are not-so-small, or are polar, can't slip through that fatty middle layer of tails. There's just no way they are getting through that "We hate water and polarity" zone. They need some kind of tunnel that they can go through. The tunnels are often called "channel proteins." (There isn't one officially correct name for these channels. Some authors call them more complicated names, like "transmembrane integral proteins." We're going to stick with "channel proteins.") You don't know exactly what a protein is yet, but we'll go ahead and use that word since you are familiar with it. The key word is "channel." These channels provide a way for small polar or electrically charged molecules to diffuse in and out of the cell. Now we can go back to discussing water.

A) **Aquaporins:** Until 1992, it was thought that water simply diffused into cells. This is true to some degree. It's not impossible for a water molecule to get through the membrane. However, the journey is not easy. Scientists began to suspect that water was getting through another way but they did not know how. Then an American scientist named Peter Agre discovered a protein channel that he named **aquaporin**. It's a pore that lets "aqua" (water) through. This discovery was so important that he received the nobel prize in 2003. (Just think of all the textbooks that had to be rewritten!) Aquaporins are channels that let water, and only water, diffuse in and out of the membrane. As Agre's team studied aquaporins they found out that there are many different kinds, depending on the cell type. Some aquaporins are found only in brain cells, other in eye or skin cells. You'll learn a little more about aquaporin in one of the activities that goes with this lesson.

At this point, we need throw another vocabulary word at you. You may already know this word; it shows up in botany and physics books, too. When water diffuses through a membrane, we don't call it diffusion — we call it **osmosis**. We might guess that "osmos" means "water" in another language, but it doesn't. "Osmos" is Greek for "push." Water looks like it is pushing through the membrane? The person who named it apparently seemed to think so. Even though it doesn't mean "water," thinking of water when you see "osmo" is still a good idea. Most science words that start with "osmo" have something to do with water.

As long as we are throwing vocabulary words at you, we might as well go ahead and cover the last major diffusion term that you need to know. The action of going from where there are more of them (a higher concentration) to where there are less of them (a lesser concentration) is often called "following the **concentration gradient**." Sometimes biologists will say that the molecules go "**down the concentration gradient**." You know what concentration means. Let's look at the word "gradient." The word "grade" is used by landscapers to describe the steepness of a hill. The grade of a hill is given in degrees (of geometry, not temperature). A gentle slope might be 3 degrees; a steep one would be 30 degrees. The word "grade" can also be used as a verb. Landscapers will "grade" the dirt around a house so that it goes downhill, away from the house. (This helps to keep rain water out of the basement.) So when you see the term "concentration gradient," don't panic. Just think of grading dirt so that water flows DOWN the slope and away from the house. The key word is DOWN. Molecules go DOWN their concentration gradient, from places of high concentration to places of lower concentration. When you see the word "gradient," think of a slope where things role from high places to low places. Think of rolling DOWN a hill. High to low, high to low.

It's actually not a hard a concept, despite the fact that the term "concentration gradient" sounds like it might be difficult. You will see this term used all the time in biology books, so it is best not to be scared of it!

B) **Ion channels:** Another example of facilitated diffusion is the **ion channel**. An ion channel is designed to let only one type of ion get through. The most common types of ion channels are for sodium (Na^+), potassium (K^+), chlorine (Cl^-) and calcium (Ca^{2+}). Ion channels tend to look like two funnels stuck together at their narrow ends. The wide top and bottom are described as "water-filled." The narrow part in the middle is very small indeed, perhaps only one or two atoms wide. The channels can be constantly open or they can be "gated," meaning that they only allow ions through when certain conditions exist. Gates can act as a triggering mechanisms, allowing a sudden influx of ions that will cause a whole series cellular events.

Gated channels come in many kinds. Some respond to light (cells in the eye), temperature (skin cells), or pressure (skin cells). Some are triggered by messenger molecules coming from outside. The gated channels in your nerve cells are triggered by differences in electrical voltage. There are about 300 different types of ion channels present in most cell membranes. (That's 300 types, not 300 total.) The most important thing to know about ion channels is that they do not require energy to operate. They are passive, not using energy.

One particular type of channel needs special mention—the channel that transports **glucose** sugar molecules into cells. This channel relies on shape. Glucose molecules look like a hexagonal ring. When a glucose molecule goes into the channel, it is like a key fitting into a lock. The shape clicks in place. When it clicks in, this automatically causes the channel to change shape so that the glucose drops out the bottom and into the cell. With glucose gone, the shape returns to normal.

ACTIVE TRANSPORT: Active transport requires energy. Your body is full of tiny molecular re-chargeable batteries. The most well-known is called **ATP**. We'll take a closer look at ATP in a future lesson. Chances are good that you've heard of ATP already, so you won't be in too much suspense till then. Right now all you need to know is that ATP stands for "adenosine triphosphate" and that it is like a rechargeable battery for your cell. The molecule has three phosphates, and the third one can be popped on and off. When it is popped off, energy is released. Phosphates have all kinds of uses in the cell, but this is the most famous one. ATP provides energy of cellular work.

Two other (slightly less famous) rechargeable molecules are **NADH** and **FADH**. You don't need to know what those letters stand for. (But for those of you who are curious, NADH stands for Nicotinamide Adenine Dinucleotide.) These molecules carry electrons to ion pumps. There are some very important ion pumps in your cells that depend on these molecules to deliver high-energy electrons. We will see them again during the lesson on what goes on inside a mitochondria.

We are going to look at three kinds of active transport: ***ion pumps, endocytosis and exocytosis***.

Ion pumps: These look similar in structure to ion channels. They cross the entire bilayer, and have an entry or exit at each end. Recently, researchers have found even more similarities and in future years you might be reading about their discoveries. The most important difference between them is that the pumps are going "***against the concentration gradient***." Thinking of ordinary water pumps, the water goes downhill, with gravity, but pumps can force it to go up, against gravity. Pumps need the energy of your arm or a motor to make them go. Ion pumps in cell membranes use ATP energy. There are places at the bottom of the pumps where ATP molecules attach and release their third phosphate. Details about this energy and how it is used by the pump are beyond the scope of this course. (Very quickly you get into quantum physics!) Ion pumps are found in plant cells, too, not just animal cells. The process of photosynthesis relies heavily on ion pumps. We'll also see ion pumps as a major feature of cellular respiration in a future lesson.

The most famous ion pump is the sodium-potassium pump located in the membrane of nerve cells. It pumps sodium out of the cell and potassium into the cell. The ions would not naturally go in this direction, because this is against their concentration gradient. Huge numbers of the ions build up on one side of the membrane, then they are let back in very suddenly when the gates of their ion channels open. (There are separate channels for sodium and potassium.) A pump and two channels work together, so to speak.

Endocytosis: "Endo" means "in," and "cyto" means "cell." So endocytosis just means going inside the cell. Endocytosis is used to take in very large things. In a future lesson, we'll meet immune system cells that take in all kinds of things, including bacteria. There is a special name for when a cell takes in just a tiny amount of something. This is called ***pinocytosis***. It is often described as cells "drinking or sipping." We will see cells doing this when we study the cells that line the insides of blood vessels and capillaries.

The best way to learn about endocytosis is to watch an animation of it. An indent starts to form in the membrane, and it becomes deeper and deeper. The particle gets trapped inside this pocket. The pocket goes so deep that it starts to close off and pull away from the membrane. The pocket breaks away from the membrane and becomes a vesicle with the particle inside. This vesicle can be transported to where it is needed in the cell.

Exocytosis: "Ex" means "out" so exocytosis means things moving to outside the cell. Exocytosis is the opposite of endocytosis. When the cell wants to send something out, it puts it into a vesicle then send the vesicle to merge with the membrane. When the vesicle touches the membrane, it become part of it. The vesicle turns into a deep pocket which becomes more and more shallow until there is nothing left. The particle that was inside the vesicle now finds itself outside the membrane. Some cells of your body produce molecules that are needed by cells all over the body. These molecules exit the cell by exocytosis, and go into the blood so that they can circulate throughout the body.

The phospholipid bilayer membrane separates the inside from the outside. However, cells need to bring things in and send things out. There are many methods for getting things in and out, depending on the size and the chemical properties of those things. Some ways require energy and some don't.

PASSIVE TRANSPORT Does not require energy.	**ACTIVE TRANSPORT** Uses energy (often from ATP or NADH)

1) SIMPLE DIFFUSION
   Very small, non-polar molecules, such as oxygen and carbon dioxide can go right through the membrane.
   Small lipids can also diffuse because they get along so well with the fatty acid tails in the middle layer.
	1) PUMPS
	   Use energy to push molecules across the membrane against the "concentration gradient."
	Types of energy: ATP, NADH carrying high-energy electrons

2) FACILITATED DIFFUSION (facil = to make easier)
   Molecules that are polar or electrically charged can't use simple diffusion; they must use channel proteins.

A) Aquaporins (for water)

Center of tube is tiny and will allow only 1 water molecule through at a time. However, 1 million water molecules get through every second!

B) Ion channels

2) ENDOCYTOSIS   (when particles are brought inside)

3) EXOCYTOSIS   (when particles are sent out)

Types of gates: light, temperature, pressure, voltage, binding of messenger molecules

7: MEMBRANES (part 3)

The plasma membrane of a cell is a busy place. Not only does it have many channels and pumps, it also is able to send and receive many kinds of molecular messages. The arrangement of these items in and around the plasma membrane is called the "fluid mosaic model." The word "fluid" tells us that it is not hard or solid, but flowing and changeable. All the phospholipids are able to move around, so the embedded proteins are also able to move. (Imagine a bathtub full of ping pong balls, with other objects floating among the balls.) The word "mosaic" refers to the art form where small colored tiles are used to make a picture. Artists' pictures of membranes often remind us of a colorful mosaic image.

We've already learned about some of the things you will find in a plasma membrane: channels and pumps. Aquaporins are also there, but we're going to be non-specific in our drawing and just put in a generic channel and pump (they could be anything). Proteins like channels that go all the way through the membrane are called **transmembrane proteins**. "Trans" means "across." These proteins go across the membrane. (Don't worry about what the word "protein" means exactly—we'll get to that in the next lesson.) There are also proteins that go only half way in, and proteins that stick to the inside or outside and don't go in at all. Proteins that go in, either halfway or all the way across are called **integral proteins**. (So all transmembrane proteins are integral proteins, but not all integrals are transmembrane.) Proteins that don't go inside at all are called **peripheral**. "Peri" means "around or outside." Our integral protein here looks like it has a hook attached to it. There are proteins that do function like little hooks.

One of the most interesting transmembrane proteins is **Flippase**. Finally, a name that is easy and makes sense! Flippases can flip phospholipids from the top to the bottom, or vice versa. (The one that flips bottom to top is often called "Floppase.") Why would this be necessary? Several reasons, but the easiest to understand is that during the process of endocytosis a new circle of membrane is being formed. Imagine a vesicle as a running track. The inside lane is shorter than the outside lane. If you put balls along the lanes, you would need more balls for the outside lane because it is longer. In the same way, there will need to be more phospholipids on the outside layer of the vesicle than the inside because the outer layer is larger. As soon as endocytosis starts, Flippase begins flipping phospholipids over, to prepare for this new geometry. Brilliant!

There are many **receptors** on the outside of plasma membranes. Receptors receive messages in the form of special molecules. Each receptors has a unique shape so only one kind of message molecule will fit. If a message comes along and snaps into place, then chemical changes will take place inside the cell.

Some proteins have sugar chains attached to them. Because they have sugars as part of their structure, they are called **glycoproteins**. "Glyco" is Greek for "sugar." Cells use glucose molecules for more than just energy. Short strings of sugars are used as tags or labels. These chains are called **oligosaccharides**. "Oligo" is Greek for "few," and "sacchar" is Latin for "sugar." (The English word "sugar" came from the Latin "sacchar.") Oligosaccharides are used inside the cell, also, often as "mailing labels" directing where things such as vesicles should be taken. The arrangement of the sugar molecules contains information.

Every cell the body has an "ID tag" called **MHC1**. We will meet this molecule again when we study the immune system. For now, you just need to know it exists, and that its job is to label that cell as belonging to the body. When roving immune cells come to inspect, they will look for that tag. If they don't find MHC I on the cell, they will probably destroy it.

Transmembrane proteins have a middle section that is hydrophobic. This is how they stay in the membrane. In our drawing, the protein's hydrophobic region looks like a spring.

Some floating proteins must work together to do a job. If they drifted apart in the fluid mosaic, they would not be able to work together, so they need to be kept near each other. This is accomplished by a **lipid raft**. The raft is an area with extra cholesterol and other fatty molecules. The lipids stick everything in that area together, so the proteins can't float away. The raft can drift around, but the proteins will all stay put inside the raft. Lipid rafts are especially important in nerve cells and in immune system cells.

(the "Fluid Mosaic Model")

The phospholipid bilayer membrane is a busy place. Not only does it have many channels and pumps, it also has lots of gadgets for sending and receiving messages. Also present are "ID tags" called MHC 1. Every cell in the body has these MHC 1 tags on them so that immune system cells know they belong to the body.

This model is called "fluid" because everything can move around. The word "mosaic" is an art term used to describe a picture made from many small colored tiles. Recent research has revealed that some things in the membrane stay in place more than expected, but this model still seems to be valid, nonetheless.

Vocab to know:
1) **Integral membrane proteins** are attached to the membrane. They can be on one side, or all the way through.
2) **Transmembrane proteins** go all the way through from one side to the other.
3) There are two words that both means "sugar." **Glyco** is from Greek, and **sacchar** is from Latin.
4) **Oligo** means "few."

8: PROTEINS (part 1)

We've already looked at a number of things that are made of proteins. All the "gadgets" in the plasma membrane, such as ion pumps, aquaporin channels, and receptors, are made of proteins. Now it is time to find out what a protein is.

The basic unit of all proteins is the ***amino acid***. The word "amino" refers to this group of atoms: NH_2. The presence of **nitrogen** is primarily what makes proteins different from lipids. Nitrogen is number 7 on the Periodic Table, so this means it has 5 electrons in its outer shell. The first two are hidden in the small inner shell, so only the last 5 are in the outer shell. This means that nitrogen wants to make 3 bonds. (5+3=8)

The word "acid" refers to a group we've already met: COOH. If we combine these two groups, we get an amino acid. Like lipids, amino acids are built around carbon. At the core of the amino acid is a carbon atom. Attached to the four bonding sites of that carbon are: 1) the amine group, NH_2 2) the acid group, COOH   3) a hydrogen, H   4) the "wild card" R group Remember that R stands for the **R**est of the molecule. (It really stands for "radical," but "rest" works just as well and is much easier to understand.) There are several dozen possibilities for what R could be. We will look at just five of them in the next lesson.

We can't forget that molecules are not flat. They have a 3D shape. What would happen if we switched the positions of the amine group and the carboxyl group, "flip-flopping" them? The new molecule would contain the very same atoms, so its chemistry would be the same. The flip-flopped molecule, however, would be the mirror image of the original. We could call the original one the "left-handed" molecule and the mirror image would be "right-handed." To help us imagine the difference between left and right handed molecules, we should think of a pair of gloves. The gloves are mirror images of each other and seem identical. However, if you try to put a right-handed glove onto your left hand, it doesn't work so well. In biology, the shape of a molecule is vitally important. Many cellular processes are based on shape. It turns out that right-handed molecules will not make usable proteins for any form of life. All proteins in your body and in every living thing on the planet are made from left-handed amino acids. Life is left-handed. Left-handed proteins are designated by the letter "L" and right-handed ones by the letter "D." (You can use the fact that "dexter" is Latin for "right" to help you remember that D is the right-handed form.)

The correct name for "handedness" is ***chirality***. "Chiro" is Greek for "hand." (<u>Chiro</u>practors manipulate your spine with their hands. Bats are classified as <u>Chiro</u>ptera, meaning "hand wings.") The importance of chirality was first seriously recognized in the early 1960s when a drug called Thalidomide was given to pregnant women to prevent nausea. The medicine contained equal amounts of the left and right handed versions of the molecule. Unfortunately, one of them was harmful to developing fetuses and thousands of babies were born without arms and legs. (Recently it has been determined that even if you eliminate the harmful version, the human body will convert the harmless version into the harmful version.) Another example of a molecule where chirality is important is the sweetener Aspartame. One version of the molecule is sweet, the other is bitter. Asparatme is controversial, but not because of its chirality. Penicillin, ibuprofen and DNA are also chiral molecules, but their chirality is not an issue that causes problems.

The biggest difference between proteins and lipids is the presence of nitrogen, N. Nitrogen is number 7 on the Periodic Table, so it has 5 electrons in its outer shell. This means it wants to make 3 bonds (5+3=8).

The nitrogen atom: **The basic unit of all proteins is the AMINO ACID:**

LEFT-HANDED AMINO ACID   RIGHT-HANDED AMINO ACID

"Handedness" is called *chirality.* "Chiro" is Greek for "hand."
Living things are made of left-handed amino acids only!

9: PROTEINS (part 2)

Proteins are chains of amino acids. Think of a long bead necklace; amino acids are like the beads. The bond that holds amino acids together is called a **peptide bond**. The peptide bond is created using a process we are familiar with: **dehydration synthesis**. The OH on the COOH is removed on the carboxyl group, and one of the Hs is removed on the amine group. The OH is connected to the H to make H_2O. The C and the N are stuck together with a very firm peptide bond. ("Pep" means protein.) The peptide bond is not very flexible and can't rotate. The other parts of the molecule can rotate a bit, and will adjust their position depending upon what situation they find themselves in. More on this in the next lesson.

When two amino acids are joined, we call this a **dipeptide**. Three aminos make a tripeptide. Many aminos in a long chain are called a **polypeptide**. Polypeptides can be as short as 4 aminos or as long as 4,000. Amino acids are most often represented by circles, sometimes with a three-letter abbreviation written on them. The abbreviations are usually the first three letters of their names.

There are only 20 amino acids in the human body. (These aminos are also found in other forms of life.) Remember, they are all left-handed aminos. Some amino acids are said to be **essential**, meaning your body can't manufacture them. You have to get these from the foods you eat. The other are **non-essential** because your cells can make them (often from other aminos). There are 9 essential aminos and 11 non-essential.

The R group determines the chemical characteristics of each amino acid. Aminos can be hydrophilic or hydrophobic (polar or non-polar). They can carry an electric charge or they can be neutral. They can be acidic or basic. Some contain extra elements such as sulfur. The nature of each amino acids is what enables long polypeptides to fold into specific shapes.

Let's look at glycine, valine, lysine, glutamic acid and cysteine.

Glycine is the smallest and simplest of the aminos. The R group is simply an H. Because glycine is so small, it is useful in proteins that are wound tightly, such as collagen. Its main feature is its size.

Valine is an example of an amino that is hydrophobic. This is easy to predict when you see that the R group is made of carbons and hydrogens, looking very much like a short fatty acid. The difference between valine and a fatty acid is that valine's carbons branch apart, forming a V. Valine is very hydrophobic (non-polar) and will always be found on the inside of a large protein, trying to hide from water molecules. A group of hydrophobic aminos will bend the polypeptide chain so that they can be together, forming a hydrophobic region. This will help to determine the overall shape that the polypeptide makes. (More on this in lesson 10.) Valine is also the amino acid that your body can use to make glucose for energy.

Lysine is one of the essential amino acids that you must get from food. Good sources of lysine include milk, meat, fish, eggs and beans. Lysine is hydrophilic (polar), basic (alkaline), and also is ionized, meaning it carries an electrical charge. Lysine has (NH_3^{+1}) as its R group. Nitrogen only really needs two H's so the third one can easily be replaced with something else. Lysine is used in strategic places where tiny molecular switches are needed, such as on the spools around which DNA is wound. Molecular "tags" can be popped on and off lysine, signaling the spool to tighten or loosen. Two examples of molecular tags are the methyl group, CH_3, and the acetyl group, $COCH_3$.

Glutamic acid is the official name of this amino acid, although it is often known as **glutamate**. As you might guess from the word "acid" in its name, it has the carboxyl group as its R group. As we've mentioned before, that H on the end of COOH comes off very easily, so that the COOH is very often just COO^-. When this happens, glutamic acid is then called glutamate. So basically glutamic acid and glutamate are the same thing, but the acid version still has its COOH intact. However, the body's natural pH encourages the H to fall off, so for all practical purposes, this amino exists in the body as glutamate. So you can call it either one. Right now we will switch to calling it glutamate so that you will understand a video on the playlist. One of glutamate's most important jobs is to allow nerve cells to transfer electrical signals. Both glutamate and glycine must bind to the top of an ion channel in the nerve cell's plasma membrane. When they bind to the receptors on the ion channel, this allows a sudden influx of calcium ions. The sudden influx of ions is what helps to generate the electrical signal in the cell.

When glutamate is produced artificially and a sodium (Na) atom is added, you get **monosodium glutamate, MSG**. This substance is used as a food additive to make the food taste better. MSG affects your sensory nerve cells, not the food. Your cells are excited and told to send more signals to the brain. Wow, your brain gets the message that the food tastes better! Unfortunately, the glutamate in MSG is the right-handed D form, not the left-handed L form. This may or may not be the reason that MSG causes health problems for so many people. It could also be that artificial glutamate floods the body with too much glutamate all at one time. MSG is very controversial, just like Aspartame is. Some people don't have any problems with it; others have terrible reactions to it.

Cysteine has a sulfur atom as its R group. (Sulfur needs two bonds, so it has an H attached to it.) Sulfur is an element that allows substances to form cross links. Charles Goodyear discovered this when he discovered how to vulcanize rubber. His rubber recipe was horrible until he tried adding sulfur. The addition of sulfur allowed the chains of rubber molecules to form cross links, like rungs on a ladder, so that the substance became very sturdy, even in extreme heat and cold. In the body, there are also substances that need to be cross-linked and tough, so they use a lot of cysteine in their structure. An example of a protein high in cysteine is insulin.

A protein is a chain of amino acids. Amino acids are attached by a **peptide bond**. When you see the root word "pep," think "protein." (Ex: the enyzme "pepsin" digests protein)

To form a peptide bond, we can use the same attachment method that we used for making triglycerides. We will use **dehydration synthesis**.

amino acid   dipeptide   tripeptide   polypeptide   (poly=many)

There are only 20 amino acids (all left-handed, of course).   The R group determines the chemical characteristics. R groups can be hydrophililc or hydrophobic (or neutral), and they can carry an electrical charge (or not).

10: PROTEINS (part 3)

The two most important things to remember about proteins are:
    1) Proteins are made of amino acids.
    2) The shape of a protein is what allows it to do its job.

This lesson will show you how a polypeptide chain will coil, bend, and fold to create a unique 3D shape. There are three levels of organization in this folding process, plus a fourth that some large proteins have.

1) *Primary structure*: This is the sequence of amino acids. Since amino acids have different "personalities" the order in which the aminos occur will determine how it folds up.

2) *Secondary structure*: There are basically two secondary shapes: the alpha helix and the beta sheet. In the alpha helix, the polypeptide coils tightly to look like an old-fashioned telephone cord. In the beta sheet, the chain bends back and forth like a paper fan, sometimes with multiple strips lining up to make a wider sheet. (Other formations include turns, loops and random coils.) Hydrogen bonding between the H that is bonded to the N, and the O that is double-bonded to the C, allows these formations to hold their shapes.
    We saw hydrogen bonding between water molecules when the positive and negative sides were attracted. Here we have something similar, with oxygen acting like an "electron hog," holding the electrons a little bit longer than carbon does. (The correct term is "electro-negativity." Oxygen is more "electronegative.") The poor H has a nucleus consisting of only one proton, so it is no match for nitrogen when it comes to attracting the electrons. Therefore the H nucleus (a proton) sits by itself a fair amount of the time, making that side of the molecule more positive than negative. The negative and positive areas are attracted to each other and they form an association that is strong enough to let the helices and beta sheets hold their shape.

3) *Tertiary structure*: Polypeptides can have sections curled into helices or bent into beta sheets, but the entire chain also folds up into a 3D shape. A simple and memorable tertiary shape made of nothing but beta sheets is the beta barrel. These proteins make good channels and portals in membranes. Aquaporin is made of nothing but helices. Another 3D shape that has a memorable name is the "zinc finger." Zinc fingers are used in proteins that grab and hold DNA. (For those of you who love trivia, the aminos that hold the zinc atom are 2 cysteines and 2 histidines.)
    Amino acids that are hydrophobic will all try to gather together on the inside as the protein folds, creating a hydrophobic zone. The hydrophilic aminos like to be on the outside edges. Various other interactions occur between positive and negative areas. Besides alpha helices and beta sheets, you will find loops, turns, and random coils.

4) *Quaternary structure*: This is when several individual proteins get together and form a large shape. Hemoglobin is the most well-known example of a quaterny structure. Four smaller proteins combine to make the final shape.

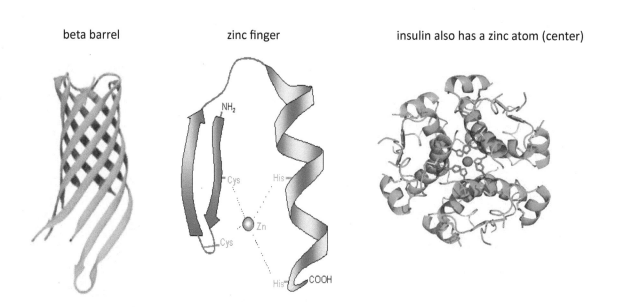

beta barrel   zinc finger   insulin also has a zinc atom (center)

The PRIMARY structure of a protein is the sequence of amino acids in the chain.

The SECONDARY structure of a protein is either an alpha helix or a beta sheet.

The TERTIARY structure of a protein is the way it folds up into a unique 3D shape.

The QUATERNARY structure is when two or more proteins bond together to form a final protein shape. Hemoglobin is made of four tertiary structures.

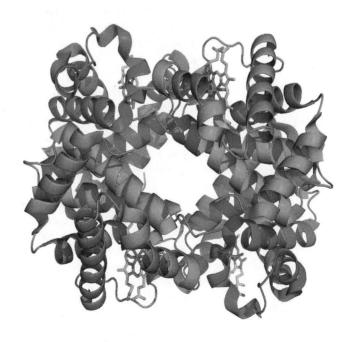

Hydrophobic aminos try to hide in the center. Sometimes atoms of other elements (like zinc) are incorporated into the shape.

11: PROTEINS (part 4)

Many cell parts and body parts are made of protein. It's like wood or steel—a building material that is strong and can be used in a variety of ways. This page is a brief visual catalog of cellular (and *extracellular*, meaning outside of cells) "gadgets" that are made of protein. It is very helpful to think of these cells parts as biological equivalents to non-biolgocial items we are already familiar with.

The cell's plasma membrane has lots of embedded gadgets, as we've already seen. We've got *channels* (or we could call them tunnels or portals), *pumps*, *hooks*, *anchors*, *receptors* ("mailboxes"), *messages* ("letters"), and *motors.* We'll meet this particular motor in a future lesson.

Inside the cell we will meet cables and ropes. The largest cable is called a *microtubule*. Even though it is the largest, it is still small, so it still qualifies as "micro." Microtubules are hollow and they criss-cross the inside of the cell, attached to each end. This network of cables is called the *cytoskeleton* and it allows the cell to maintain its shape. In the case of *motile* cells (those that can move on their own, like amebas), the cytoskeleton is what causes the movement. The microtubule is made of small units of a protein called *tubulin*. The microtubules can be made longer by the addition of tubulin units on the end. The tubulin units "self assemble," that is, they don't need an enzyme robot to do the assembly. The units just snap into place automatically. There are trigger molecules that get the process started by taking off an "end cap." When the end cap is removed, tubulins will start snapping into place and will continue to do so until an end cap is put back on again. This process happens very quickly. Millions of units can snap on in a few seconds. If you watch an ameba crawl along, you can see how fast the microtubules can lengthen.

This network of microtubules, the cytoskeleton, also functions as a road system. There are proteins that act like vehicles, driving along these microtubule highways. Or perhaps a better analogy would be to imagine the microtubules as tightropes in a circus, with acrobats walking on them, sometimes while carrying objects larger than themselves. The proteins that travel along microtubules look like they have legs and feet, and their motion looks very much like walking. In video animations of these walking proteins, the microtubules underneath their "feet" often look very thin, more like tightropes than like roads or sidewalks.

Microtubules also form the basic structure of *flagella* and *cilia*. Flagella are whip-like tails found on motile (moving) cells such as bacteria, protozoans, and sperm cells in both plants and animals . Cilia look more like tiny hairs than like tails, thus, the name "cilia" which means "hairs". Cilia are found in great numbers on some cells, where they beat back and forth, moving in unison. Some cells, such as bacteria and protozoans use cilia to move, and other cells, such as those in your trachea, use cilia to sweep their surface clean. We even find ciliated cells in the fallopian tubes of females, sweeping egg cells along. For both flagella and cilia, tiny "motors" at their bases pull on the protein cables, causing them to move.

The medium-sized cables are called *intermediate filaments*. They look a bit like a braided rope. There are many different types of intermediate filaments, the most well-known being *keratin*. One type of keratin, *hard keratin*, is tough and relatively stretchy and is used to form hair and nails. In animals, scales and horns are also made of hard keratin. *Soft keratin* is found in skin cells and helps to make skin waterproof. Other types of intermediate filaments (not keratins) occur as part of the cytoskeleton and act as strengthening cables. Unlike microtubules, intermediate filaments can't help the cell move; they simply provide structure.

The smallest cables are called *microfilaments*. They are made of only two strands, so they are super thin. The strands are made of a protein called *actin*, which is found in great abundance in all cells, but especially in muscle cells where they interact with other proteins in such a way that your muscles contract. Microfilaments can't help the cell move, like microtubules can, but they are responsible for changing the shape of the cell. For example, when a cell reproduces by splitting in half, the microfilaments are what carry out the action of splitting.

Enzymes are proteins that act as "scissors" or "staplers." The things they join or separate are called *substrates*. The scissor-like enzymes are little robot molecules whose job it is to tear things apart. Our digestive processes rely heavily on these destructive enzymes which tear apart food molecules and reduce them to individual molecules that our bodies can then use. Other enzymes are designed for joining things together. We already met one of these when we talked about dehydration synthesis. Constructive ("stapler") enzymes are required for most cellular processes. And don't forget that they are specific to their jobs—an enzyme can only do ONE thing.   An enzyme that can build a particular type of protein can't also build other proteins, and it certainly can't build lipids or sugars.

Often, an enzyme is described as something that *speeds up reactions*. The reaction might, just maybe, take place given enough time, but the enzyme goes in and does the job quickly and efficiently. Enzymes, like all workers, work best in certain environmental conditions. Some people like to keep their office toasty warm, others like their rooms on the cool side. Some people like to have music playing while they work, others do not. Enzymes have environmental preferences, too. Some like it hot, some like it cold. Some like to be in an acidic environment; others like alkaline. And just like some people need help (secretaries, support staff), some enzymes are dependent on other molecules to help them do their job. These helper molecules are called *coenzymes*. ("Co" means "with.") Usually the coenzymes fits into a special slot in the main enzyme, to help create the correct shape needed for the job.

Proteins can be used to make "vehicles" that can carry other molecules. We've already mentioned the **motor proteins**. The correct name for this protein is **kinesin**. *(keh-NEE-sin)* "Kine" is Greek for "motion," and also appears in the word "cinema.") They look like they have little legs and feet and they do something very similar to walking as they travel along the microtubule cables. (Note, however, that their "feet" are actually called "heads.") Their job is to carry loads, often heavy and large loads, from one place to another. We've talked about vesicles traveling around the cell or moving out to the membrane, but we've not said how. This is how it's done, via motor proteins. Motor proteins can also drag entire organelles across the cell. If the load is too large, several kinesins will join together and cooperate.

Kinesins (motor proteins) travel in one direction, usually starting from the center and going toward the outside. They know what direction to travel because microtubules are polar. Kinesins will go toward the positive end. A different type of motor protein is required to go the other way. These "reverse direction" motor proteins are called **dyneins**. *(DIE-nins)* Dyneins are also responsible for the motion of cilia. Motor proteins that can go both ways do exist but are rare. When a motor protein reaches the end of the microtubule, it releases its cargo then gets off the microtubule. It jumps off and floats back to the starting point. Then it is ready to reattach to another microtubule, take on new cargo, and begin a new journey.

Some cells use proteins to manufacture transportation devices that can be likened to boats because they float along in the bloodstream. Why do we need boats in our blood? Blood is mostly water, which is polar. When a hydrophobic molecule needs to be transported via the blood, it will refuse to get in unless it is safely enclosed in a "vehicle" of some kind. Just think— if you are afraid of water but need to cross a river, would you swim or ride in a boat? Same concept applies to molecules. The most common protein boat you'll find in blood is called **albumin**. (This word looks similar to albumen, which you find in egg whites. Notice that the egg word has an "e" instead of an "i.") Albumins account for half of the proteins found in blood. Albumins can carry free fatty acids, hormones, ions, calcium, broken pieces of hemoglobin that are in the midst of being recycled, and also some prescription drugs such as the blood thinner **warafin**, which we will meet again in our lesson on blood clotting. Albumins do another job, besides transport things. Their very presence in the blood works to regulate osmosis as water diffuses in and out of the blood. Albumins help to maintain proper blood pressure.

The last major category is "tags." We've seen how sugars can also be used as tags, both inside the cell and at the surface. Protein tags are specifically used as "flags" to mark foreign invaders. These tags can be called **gamma globulins**, **immunoglobulins**, or **antibodies**. All three words are correct and can be used interchangeably. They are Y-shaped, with the upper part designed to stick to the foreign invader, and the base part designed to stick to a cell membrane if need be. Some float around freely and others stick to the outside of a cell. Each Y has a unique shape and can stick to only one type of invader. Your body makes millions of differently shaped antibodies, hoping that a small percentage of them will actually be useful.

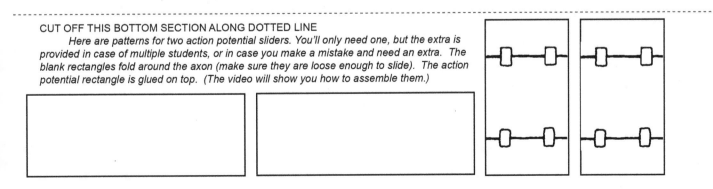

CUT OFF THIS BOTTOM SECTION ALONG DOTTED LINE
Here are patterns for two action potential sliders. You'll only need one, but the extra is provided in case of multiple students, or in case you make a mistake and need an extra. The blank rectangles fold around the axon (make sure they are loose enough to slide). The action potential rectangle is glued on top. (The video will show you how to assemble them.)

Proteins are the building blocks of numerous "gadgets" in and around cells.

EMBEDDED GADGETS in the plasma membrane

CABLES

Microtubules

Intermediate
filaments

Microfilaments

"SCISSORS" and "STAPLERS" (enzymes)	"VEHICLES" transportation around the body	TAGS mark invaders for destruction by immune cells
	MOTOR ALBUMIN PROTEINS: Acts like boat --Kinesins (or taxi) --Dyneins	They go by three names: --gamma globulins --immunoglobulins --antibodies All three are correct-- you choose.
A and B are called "substrates."		

12: CARBOHYDRATES

Carbohydrates, as the name implies, are a mixture of carbon and water ("hydra"). This means that their chemical formula will include the elements carbon, hydrogen and oxygen. Just like proteins have a basic building block (amino acids), carbohydrates also have basic unit: monosaccharides. *Monosaccharide* is a fancy word for "single sugar." (Remember, "mono" means "one," and "sacchar" means "sugar.") However, they are usually called <u>simple</u> sugars instead of single sugars. The three simple sugars we will look at in this lesson are *glucose, fructose and galactose*. All three have the same chemical formula: $C_6H_{12}O_6$. The only thing that is different is the arrangement of the atoms.

Glucose (from "gluco" or "glyco" both meaning "sugar") is the most abundant and most useful sugar molecule in your body. It's the one we'll see again and again in this course. Glucose is the molecule that is "burned" as fuel in the energy-producing processes inside your cells. Glucose also shows up as a structural element in glycoproteins and glycolipids and in these forms is often found attached to the outside of cells. Strings of glucose are used by cells almost like mailing labels, tagging proteins for delivery.

Fructose is found in fruits. It tastes much sweeter than glucose. Your liver has to work hard to break it down, though, and can only do so at a certain rate. High-fructose corn syrup, a sweetener used in candy and carbonated drinks, overloads the liver with too much fructose and can cause long-term health problems. The amount of fructose found in natural fruits is fine. It would be hard to overwork your liver by eating too many apples and bananas.

Galactose got its name from the Greek word "galaxias," meaning "milky one," referring to the Milky Way in the night sky. So the words "galaxy" and "galactose" are related. The first person in history to study galactose was Louis Pasteur in 1856. Galactose is most abundantly found in milk, where is it attached to a molecule of fructose. Like glucose, galactose can also be used by the body as a structural component. For example, you find galactose in the "tags" on the outsides of red blood cells, marking them as type A, B or O.

The arrangement of the atoms in a molecule is extremely important, as we saw in the case of amino acids where the mirror image molecule is useless (or harmful) to the body even though the atoms are identical. Sugars can also be right or left-handed, and though our bodies are made exclusively of left-handed aminos, they use right-handed sugars. Left-handed sugars can't be digested. The little enzyme robots that take apart right-handed sugars can't operate on left-handed ones. In the last years of the 20th century, food chemists discovered a way to convert galactose to its left-handed form, which they named *tagatose*. The patent for making tagatose was passed around among several European food companies until someone found a way to make it inexpensively in great volumes. Tagatose began to be marketed in the USA in 2003. Brand names include Naturlose, PreSweet, and Tagatesse.

When you join two simple sugars together, using *dehydration synthesis*, you get a *disaccharide*. The most well-known disaccharide is *sucrose*, or "table sugar." The enzyme that cuts the bond between the glucose and the fructose is called *sucrase*. Notice how the sugar ends in "-ose" and the enzyme ends in "-ase." (*Sucralose* is an adpated form of sucrose. Not only are the atoms rearranged, but three chlorine atoms are also added. Food chemists say that it tastes 600 times sweeter than sucrose. The brand name "Splenda" is sucralose. Though the FDA says that it is perfectly safe, YouTube is full of videos claiming otherwise.)

Another common disaccharide is *lactose*. This is the sugar found in milk. We already learned that galactose is one of milk's simple sugars; the other is glucose. Lactose is a galactose bonded to a glucose. The enzyme that can tear apart this bond is called *lactase*. Some people lose the ability to make this enzyme as they get older, and they become "lactose intolerant."

When you join many simple sugars together you get a *polysaccharide*. Two types of polysaccharides that you are very familiar with are *starch and cellulose*. A third, *glycogen*, is less familiar because it is inside your body where you can't see it.

Starch is what we call a long string of glucose molecules. These strings can have several thousand glucose molecules in them. Starches are made by plants and are stored in seeds and roots. Food items high in starch include wheat, oats, rice, corn, beans, potatoes and carrots. Notice how the glucose molecules are arranged in a starch. The "flags" on the glucose molecules are all pointing the same direction. This configuration is easy for your digestive system to deal with because you have enzymes designed to tear apart those chemical bonds.

Cellulose is similar to starch, but has the glucose "flags" in alternating directions. This small difference in molecular structure makes a big difference in physical characteristics. Cellulose is made by plants to be used in their tough cell walls. It is fibrous and dense and is the stuff that leaves and stems are made of. Most animal bodies don't make the right enzymes for breaking down cellulose. So why can cows live on nothing but grass? They have a gut full of bacteria that digest cellulose. Cows (and all herbivores) rely on bacteria to digest their food for them. Without the bacteria, they would starve. In humans, most of the cellulose we eat (leaves, stems) passes through our system undigested. This is not really such a bad thing, though. We need "bulk" in our intestines to keep things moving along. We do have some bacteria in our gut, though, and they do munch on our spinach and broccoli a bit. Keeping the friendly, vegetarian bacteria happy helps to keep the population of bad bacteria low.

Glycogen is a type of polysaccharide made by our bodies as a way to store glucose. After we eat, the glucose level in our blood goes way up and if it stays high this causes major health problems. The hormone *insulin* is released by the pancreas and it signals the enzyme *glycogen synthase* to start collecting glucose molecules out of the blood and begin stringing them together. These strings then get stored mainly in the liver and in the muscles. When the level of blood glucose begins to fall too low, these strings come out of storage and the glucose molecules are clipped off and put back into the blood.

CARBOHYDRATES

MONOSACCHARIDES are simple sugars

GLUCOSE   FRUCTOSE   GALACTOSE

DISACCHARIDES are "double" sugars made of two simple sugars

POLYSACCHARIDES are long strings of simple sugars (up to 4,000 units long!)

1) Starch

Plants make starches and store them in seeds and roots: wheat, rice, corn, beans, potatoes, beets, carrots, etc.

2) Cellulose

Plants make cellulose and use it for their tough cell walls. We eat it as fiber (leaves and stems).

3) Glycogen

Our bodies make glycogen as a way to store glucose in the liver and in muscles.

The hormone Insulin signals the body to turn glucose into glycogen. Adrenaline does the opposite.

13: HOW PROTEINS ARE MADE (part 1: TRANSLATION)

We've been fairly thorough in our discussion of proteins so far. We've gone into detail about the atomic structure of amino acids, what R groups are made of, how polypeptides fold into helices and beta sheets, and how proteins fold into complex shapes. We skipped over one very important detail, however. The arrangement of the amino acids on the polypeptide chain is critically important. The order in which the amino acids occur will determine every quality that protein will have. If even one amino acid were to be added or deleted, it might cause the protein to be useless or even hazardous. The sequence of the amino acids is vitally important. So how do the amino acids get lined up in the correct order?

We are going to present the answer by working backwards. First, we'll see a polypeptide being assembled by a tiny "factory" called a ribosome. Then we'll find out where the ribosome's instructions came from. We'll eventually end up at the source of all cellular information: the DNA In the nucleus.

Ribosomes are the "factories" that make polypeptides. Though they might look like it, they are NOT protein gadgets. Well, they do have a little bit of protein in them, but not enough to call them protein gadgets. Ribosomes are mostly made of **RNA,** or **ribonucleic acid**. (RNA is sort of like one half of DNA. We'll study RNA in more depth in the next lesson.) The RNA strands are twisted into a complex shape (similar to a protein's tertiary shape). Ribosomes are made of two sections: a **large subunit** and a **small subunit**. These subunits float around separately if they are not actively engaged in making a polypeptide.

Ribosomes "read" a strip of coded information that is also made of RNA. This strip is called **messenger RNA** (**mRNA**). The mRNA tells the sequence of amino acids for a particular polypeptide. There are "letters" in this code that correspond to the different types of amino acids. (We will see this in the next lesson.) The mRNA "tape" goes in one side, is read in the middle, and comes out the other side. When it is finished, the mRNA can be used again to make another identical polypeptide. The ends of the tape are different, and biochemists keep track of which end is which by using the numbers 5 and 3. They use"prime" marks that look like apostrophes after the numbers, so that the leading end, the one that is read first, is called 5' (five prime), and the end that gets read last is 3' (three prime).

Amino acids are brought to the ribosome by little "taxis" that are also made of RNA. The taxis are called **transfer RNA** (**tRNA**). It is fairly easy to remember what tRNA does, because tRNA actually looks a bit like the letter T, which is the first letter in the words "taxi" and "transfer." The tRNAs carry amino acids very much like taxis carry passengers. On the other end of the tRNA (the end that does not attach to an amino acid) is a special site called the **anticodon**. Each anticodon will match a corresponding site on the mRNA called a **codon**. (We'll learn about codons in the next lesson. A codon is a secret code for an amino acid.) As the mRNA slides through the ribosome, the codons on the mRNA will determine which aminos will be brought by the tRNAs.

The ribosomes have "parking places" for the tRNA taxis. The first two are called A and P, and the last one is E for Exit. The mRNA coded tape slides along underneath these sites, exposing the codes, one by one, under the A site. The mRNA waits until a matching tRNA comes along and parks in the A site. That tRNA then moves to the P site, allowing a new tRNA to occupy the A site. As a tRNA moves from the A to the P site, it transfers its amino acid to the bottom of the growing peptide chain. After successfully attaching its amino to the chain, it sits briefly on the Exit site, then leaves. This continues until the end of the mRNA tape is reached. When finished, all the parts separate.

The new polypeptide then folds up, often with the help of chaperone proteins. Some chaperones are designed to "spell check" or "shape check" and if the protein does not pass inspection, it will be sent to a shredding machine and the amino acids will be recycled.

It's interesting to think about the fact that the chaperone proteins are made of the same stuff they are folding (protein) and were therefore also made by ribosomes. So who folded the chaperones—other chaperones? And who folded those other chaperons? And then who folded the chaperones that folded the chaperones? We see a cycle that must continue, unbroken, with no room for mutations or gradual change.

All cells have ribosomes, even bacteria cells. The antibiotic called **erythromycin** kills bacteria by blocking the active sites of their ribosomes. The erythromycin molecules stick in the P site so that the tRNAs can't move out of the A site. This means that the bacteria will not be able to manufacture the proteins that it needs to survive. Thus, the bacteria dies.

HOW PROTEINS ARE MADE (part 1: Translation)

RIBOSOMES are the "factories" that make proteins. They are the smallest organelle in a cell.

(NOTE: They are <u>NOT</u> protein gadgets. They are made primarily of RNA with only a few bits of protein mixed in.)

ribosome

Image of ribosome from the Center for Molecular Biology of RNA, Univ. of CA Santa Cruz, ma.ucsc.edu/rnacenter

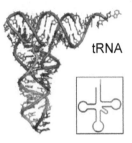

tRNA

"TRNA-Phe yeast 1ehz" by Yikrazuul - https://commons.wikimedia.org/wiki/File:TRNA-Phe_yeast_1ehz.png#/media/File:TRNA-Phe_yeast_1ehz.png

1) All pieces are separate until a mRNA snaps onto a small subunit. (There are special molecules that cause the mRNA to stick.)

Floating nearby are millions of transfer RNAs (tRNAs) holding amino acids.

2) The two subunits snap together.

3) tRNAs start bringing amino acids to the ribosome, and they are added one by one to the growing protein chain.

4) When the protein is finished, all parts separate.

5) After the polypeptide is done it gets folded, often with help from chaperone proteins.

14: HOW PROTEINS ARE MADE (part 2: RNA)

Now we are going to find out exactly what RNA is. RNA is an abbreviation for ribonucleic acid. "Ribo" is short for "ribose." **Ribose** is classified as a simple sugar, but has only 5 carbon atoms, so its formula is $C_5H_{10}O_5$. The other simple sugars we learned about all had 6 carbons ($C_6H_{12}O_6$). The basic structure of ribose is a pentagon, so it looks similar to fructose. The "nucleic" part of the name refers to the nucleus, which is mainly where RNA is made. (DNA is also a nucleic acid.) The "acid" part of the name refers to the phosphate group that is attached to the molecule. (You'll remember that the phosphate ion used to be H_3PO_4, phosphoric acid. The H's went running off as protons, leaving their electrons behind.)

When you combine a ribose sugar, a phosphate ion, and something called a **base** (more on this in a minute) you get a molecule called a **nucleotide**. Nucleotides are the individual units that string together to make RNA and DNA.

The phosphate ion functions sort of like glue, or maybe a paper clip or a staple, and holds the ribose sugars together. The pattern is: ribose, phosphate, ribose, phosphate, ribose, phosphate, etc. Since ribose is a simple sugar, this string is often called the "sugar-phosphate backbone" of the molecule.

The ribose's job is to hold on to a molecule called a **base**. There are five different types of bases, three of which are common to both RNA and DNA, and two that are found in just one or the other. The common bases are: **adenine, cytosine and guanine**. **Thymine** is found only in DNA and **uracil** is found only in RNA. Thymine and uracil are very similar, with thymine being basically a uracil with a CH_3 added to it. CH_3 has a special name: the **methyl** group. (Remember, "meth" means "one" when you count carbons.) This methyl group must be added when the base is used in DNA in order to make DNA less likely to tear apart at that point. In RNA, it is better not to have that methyl group there.

The bases contain several nitrogen atoms, so sometimes they are called **nitrogenous** (nie-TRODGE-en-us) bases, meaning "nitrogen-containing." The bases pair up with each other in a predictable way: A with T (or U), and C with G. In DNA they are always paired up in this way. In RNA they only pair up at certain times, such as when DNA is being copied or when tRNAs are matching up with mRNA.

RNA is a long chain of nucleotides. The order of the nucleotides is critically important, because they code for amino acids. Each set of three nucleotides forms a trio called a **codon**. It's easy to remember what a codon does because the word "codon" looks like the word "code."

A codon codes for an amino acid. The bases A, C, G and U provide 64 possible codons, which is more than enough since there are only 20 amino acids. Some amino acids have more than one code. Lysine has two codes: AAA and AAG. Valine, one of the hydrophobic aminos, has four codes: GUU, GUC, GUA, and GUG.

Some codons don't code for any amino acid. For example, UAG, UGA and UAA mean "stop." To signal "start" the codon AUG is used. This codon codes for the amino acid **methionine**. Therefore, methionine is always the first amino acid in every polypeptide.

Amino Acids			Codons
Alanine	Ala	A	GCA GCC GCG GCU
Cysteine	Cys	C	UGC UGU
Aspartic acid	Asp	D	GAC GAU
Glutamic acid	Glu	E	GAA GAG
Phenylalanine	Phe	F	UUC UUU
Glycine	Gly	G	GGA GGC GGG GGU
Histidine	His	H	CAC CAU
Isoleucine	Ile	I	AUA AUC AUU
Lysine	Lys	K	AAA AAG
Leucine	Leu	L	UUA UUG CUA CUC CUG CUU
Methionine	Met	M	AUG
Asparagine	Asn	N	AAC AAU
Proline	Pro	P	CCA CCC CCG CCU
Glutamine	Gln	Q	CAA CAG
Arginine	Arg	R	AGA AGG CGA CGC CGG CGU
Serine	Ser	S	AGC AGU UCA UCC UCG UCU
Threonine	Thr	T	ACA ACC ACG ACU
Valine	Val	V	GUA GUC GUG GUU
Tryptophan	Trp	W	UGG
Tyrosine	Tyr	Y	UAC UAU

Transfer RNA (tRNA) needs to be mentioned again at this point. One end of tRNA holds the amino acids. The other end has an **anticodon**. "Anti" can mean "against" or "opposite." In this case, "opposite" is a good translation because the anticodon contains the opposite match for each of mRNA's codons. The bases come in pairs. C and G are matches, and A matches with U in RNA. The anticodon for AAA would be UUU; for AUG it would be UAC; for GCA it would be CGU. So as the mRNA slides through the ribosome, a codon is exposed right under that A parking space. When a codon is right under the A site, that is when a tRNA with the correct anticodon will come over and match up. If the mRNA codon AUG is exposed, then a tRNA that has the anticodon UAC will come over and match up. The UAC tRNA will be carrying the amino acid methionine.

RNA is found in several places in a cell. We've seen most of these already. Messenger RNA (mRNA) is used by ribosomes to make proteins. Ribosomal RNA (rRNA) is the RNA that ribosomes are made of. Transfer RNA (tRNA) is used as a taxi for amino acids, bringing them to the ribosomes to be hooked together to form polypeptides. MicroRNA (miRNA) comes in very short pieces and is used to control which parts of DNA get used. miRNA can "shut down" the production of a protein by binding to the area of DNA that has the instructions for how to make it. This is part of epigenetics and we'll learn more about it in a future lesson.

HOW PROTEINS ARE MADE (part 2: RNA)

RNA is **ribonucleic acid**. The individual units of RNA are called **nucleotides**.
A nucleotide is made of three pieces: a ribose, a phosphate, and a "base."

Ribose is a simple sugar
but has only 5 carbons.

Phosphate is PO_4.
This is the "acid."

Bases contain
nitrogen. There
are 5 types.

A NUCLEOTIDE
is a ribose, a phosphate and a base.

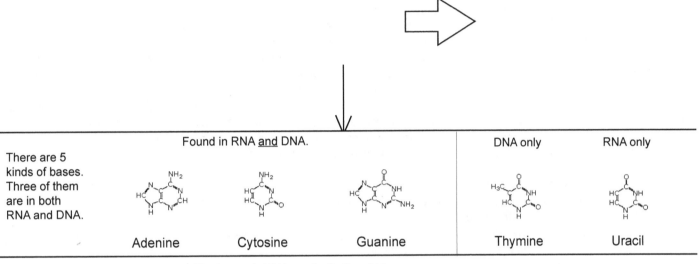

	Found in RNA <u>and</u> DNA.		DNA only	RNA only
There are 5 kinds of bases. Three of them are in both RNA and DNA.	Adenine	Cytosine	Guanine	Thymine	Uracil

RNA is a long chain of nucleotides. The bases in RNA are A, C, G and U.

We could draw RNA
showing the riboses, the
phosphates and the bases.

Or we can short-cut and
draw it like this:

To make it even simpler
we can draw it like this:

Each set of 3 nucleotides forms a unit called a CODON. (64 possible codons)	tRNA has an anticodon that matches up with a codon during translation in a ribosome.

RNA is found in more than one place in a cell.

1) _____ (messenger RNA)
is a copy of a section of
DNA and is used to make

_____.

2) _____ (transfer RNA)
is used as a taxi
for _____ _____.

3) _____(ribosomal RNA)
is folded up to make

_____.

4) _____ (microRNA)
are very short pieces of
RNA used to regulate
gene expression.

15: HOW PROTEINS ARE MADE (part 3: Transcription)

We have approached this topic in the opposite direction from most book and video presentations. We looked at proteins first, then began working backwards, step by step, to see where they came from. You'll need to follow up this lesson with some video animations showing the process in the "forward" direction. Animations usually start in the nucleus with DNA, then show mRNA being made, then show translation in a ribosome. However, it's good to look at this process from many directions, so that your learning is not a matter of mere memorization but of understanding.

Messenger RNA is a copy of one small section of DNA. DNA is often described as the "library" of the cell, containing all the information a cell will ever need. In fact, it contains a lot of information the cell will <u>never</u> need. Every cell in the body has a complete set of DNA. Therefore, a skin cell has information about how to be a lung cell or a liver cell, but this extra information is permanently locked up, never to be used. The DNA library is so vital to the health of the cell that it never leaves the safety of the nucleus. The membrane around the nucleus is twice as thick as the outer membrane of the cell. This **nuclear envelope** is made of two phospholipid bilayers. The envelope has holes in it, to let things come in and out. These holes are specialized **pores** designed to regulate the traffic of molecules in and out of the nucleus, including mRNA.

If you look at a cell under a microscopic, and the cell has been stained so that the nucleus shows up very well, you will see that there is a dark spot inside the nucleus. This is called the **nucleolus**. This is a place where the DNA is particularly dense. In this area you will also find lots of RNA, because the nucleolus is where ribosomes are made. There are also some proteins being manufactured in this area, mostly associated with the ribosomes. You'd have a lot of loose nucleotides floating around, too. We'll soon see why.

DNA is similar to RNA, but there are four main differences:
1) DNA is a double helix, having two strands. RNA is a single helix, having one strand.
2) DNA has **T**hymine; RNA has **U**racil.
3) DNA has **deoxyribose**; RNA has **ribose**. ("Deoxy" means "missing an oxygen.")
4) DNA is found only in the nucleus; RNA is found in all parts of cell.

Since DNA cannot leave the nucleus, whenever a piece of information is needed outside the nucleus, a messenger must be sent out. Messenger RNA is an exact copy of a small section of DNA. (mRNA can be thousands of nucleotides long, but since DNA has billions of nucleotides, a few thousand seems like a small number.) You might be able to guess that a special gadget is required to copy this information from the DNA. This gadget is called **RNA polymerase**. Polymerase can be broken into "poly," which means "many," and "mer," which means "individual unit," and "ase," which means "enzyme." All molecules ending in "ase" are protein gadgets. This one acts like both a scissor (or zipper) and a stapler. It makes a "polymer" from nucleotides.

RNA polymerase snaps on to a section of DNA, reading some special codes (written with A, T, C and G) that allow it to identify the correct section to copy. Then the DNA goes into the polymerase machine where it meets a splitter that separates the two strands. DNA is easy to split because the two helices are held together only by hydrogen bonding, which is fairly weak. In our picture, the top strand is the one that is being copied. If the RNA was copied directly from the correct strand, it would then be a reverse copy, with A's instead of T's and C's instead of G's. This would mess up the protein manufacturing. So to get a correct copy that makes sense to the ribosome, the opposite strand of DNA must be copied. Sometimes this strand is called the **antisense** (or nonsense) strand. It is the reverse of the strand you want to copy. When you copy it with RNA, you get a correct, "sense" strand. (Don't forget, when the mRNA copies it, uracil is used to replace thymine. Uracil is thymine without a CH_3 group.)

Nucleotides go streaming into the polymerase through a special funnel-like opening. When they meet the antisense strand of DNA, they begin to match themselves up. A strand of RNA begins to form. As it gets longer, it exits the polymerase. This new strand of RNA is not ready to be used yet. It is called **pre-mRNA** because it must be processed before it can leave the nucleus.

The first end to come out is called the 5' ("five prime") end. Scientists have to keep track of the ends of RNA and DNA because they are not the same. There is a front and back, so to speak. The front end is called 5' and the back end is called 3'. As soon as the 5' end comes out, a special cap is put onto it by a little enzyme robot designed to do that job. The cap ends with a G and a **methyl tag**, CH_3. (Remember, "meth" means "one" when you count carbons. "Meth-, eth-, prop-, but-" means "1, 2, 3, 4.") You will see methyl tags again soon. They keep sections of DNA zipped shut, and are therefore a key to cell differentiation and epigenetics. The lagging end that comes out last (the 3' end) gets a tail put onto it. The tail is called the **poly(A) tail** because it is made of a long string of A's. AAAAAAAA... etc.

There is one last step in the preparation of mRNA. There are sections that must be snipped out. The information does not come in one long piece, but in little pieces with "filler" sections in between. The necessary pieces are called **exons** and the extra filler pieces are called **introns**. An RNA gadget called a **spliceosome** comes along and pinches the introns into a loop, then snips them off. The snipped ends are glued back together, so that, when finished, you have a long line of exons with no introns left. (Don't think of the introns as "junk." They play a necessary role. If you got rid of them, DNA would get really messed up.) After the introns are gone and the cap and tail are stuck on, the mRNA leaves the nucleus and is ready to meet a ribosome.

DNA contains all the information a cell will ever need. DNA is so important that it can never leave the protection of the nucleus. Therefore, when information is needed outside the nucleus, a mRNA copy of that section of DNA is created and sent out.

DNA

DNA is similar to RNA. It is made of nucleotides.

These are the DIFFERENCES:

1)

2)

3)

4)

The nucleus is made of TWO bilayers of phospholipids. It has a double-thick membrane.

A protein gadget called **RNA polymerase** acts as both a scissor (or maybe zipper) and a stapler. It unzips the DNA and then staples nucleotides together to match one of the strands of the DNA. (Notice which strand is being copied.)

The mRNA that leaves the polymerase is not ready for immediate use. There are "filler" pieces called *introns* that must be snipped out. Also, a special cap must be put on the 5' end, and a poly(A) tail on the 3'.

The introns are cut out by a machine made of RNA called a *spliceosome*.

After the introns are gone, the mRNA is ready to leave the nucleus.

16: PROKARYOTES (part 1)

We've got enough cell parts now to make a cell—but not a human cell. The only type of cell we can draw is a bacterium; but that fits in well with our study of the human body because humans are full of bacteria. For every one human cell in your body, there are ten bacterial cells. Estimates on the number of cells in a human body range from 15 trillion to 70 trillion. Many sources say about 37 trillion. Using the 37 trillion figure, that means you have 370 trillion bacterial cells inside of you, or on your skin. As long as the many species stay in balance, you remain healthy.

Bacteria are classified as **prokaryotes.** ("Karyo" means "nut or kernel," referring to the nucleus.) Prokaryotes **do not have a true nucleus**. They have DNA, but they do not have a nuclear envelope around it. Another related group of organisms, the **archaea**, are also classified as prokaryotes. Archaea used to be classified as bacteria. When scientists began examining bacterial DNA, they found that some types of bacteria had their DNA wound on little spools and some did not. The ones with the little spools were moved to a new group and given a new name: "archaea," meaning "old." You have both kinds of prokaryotes in your intestines. Some types of archaea produce methane gas, a partial explanation of where all that air and gas in your intestines comes from.

There are many types of bacteria. We will start by drawing something that represents what all bacteria have in common. It's labeled as a "basic" bacteria. This is a no-frills model, without any of the special features that many bacteria have. We will look at some of the special features after we see what they all have in common.

Bacteria are cells, and therefore must have a plasma membrane. However, unlike our cells, bacterial cells also have a cell wall outside of the membrane. The cell wall is made of a substance called **peptidoglycan**. "Peptid" means "protein," and "glyc" means "sugar." It was mentioned in a previous lesson that sugars can be used as structural elements, not just as food. The cell wall is made of long strings of sugar molecules held together by protein cables. The cables that hang down are made of just four amino acids: alanine, glutamate, lysine and another alanine. We've already met glutamate and lysine.

Alanine has been called the most boring amino acid because it doesn't have a strong "personality." Its R group is the methyl group, CH_3. (Yes, the methyl group is showing up again already!) Although CH_3 is non-polar and therefore technically hydrophobic, it's small enough in comparison to the whole alanine molecule that it does not affect the amino's "personality" very much. Alanine does not act either hydrophobic or hydrophilic. It also does not carry an electrical charge It basically minds its own business and gets along well with everyone. Perhaps you know someone with alanine's personality? In some situations that's exactly what you want. Alanine shows up at some point in most proteins. About 8% of the amino acids in your body are alanine.

In this situation, alanine's job is to hang on to a sugar molecule or a glycine cable. The cables that run horizontally, parallel to the plasma membrane, are made of five gylcine amino acids. (Glycine is the smallest amino, with only an H as its R group.) There are also long molecules that act like molecular ropes, attaching the sugar layers from top to bottom and anchoring it to the plasma membrane. These long molecules have phosphate "hooks" at intervals. Phosphates, [PO_4's] are good at hanging onto sugar molecules, as we can see in DNA and RNA.

Bacterial cells have little enzyme robots that do jobs, just like our cells do. One particular enzyme attaches the glycine cables to the alanine cables. This enzyme can be prevented from doing its job by **penicillin** and other "cillin" antibiotics. The penicillin molecule interferes with the active site on the enzyme so it can't do its job. Without those cross-linking glycine cables, the bacteria's cell wall falls apart and the cell dies.

Bacteria do not have a true nucleus, but they do have DNA. Their DNA is circular, but the circle is so large that it folds and crumples into a lump. This central clump of DNA is called its **genomic DNA**. A genome is a complete set of DNA— all the instructions the cell will ever need. We need to use the word "genomic" in order to distinguish this DNA from other DNA in the cell. Bacteria also have small circles of DNA called **plasmids**. These are just bits of "bonus" information that might help the cell survive in certain situations. For example, resistance to antibiotics is often found as DNA code on plasmids. Bacteria can share this "bonus" information with other cells of it species by a process called **conjugation**. In this process, a plasmid is transferred from one bacteria to another via a small tunnel that grows between the two cells. The donor cell makes a copy of the plasmid DNA and sends this strip through the tunnel to the recipient cell. The strip of DNA can then form a circle, and the recipient cell now has a new plasmid.

Bacteria must have **ribosomes**. They've got enzymes, such as that enzyme that attaches the glycine cables, and enzymes (as well as all other protein gadgets) are made by ribosomes. They also have a **cytoskeleton**, though it is not exactly the same as the cytoskeleton of an animal cell. Textbooks used to say that bacteria did not have a cytoskeleton, but this has been corrected as more research has revealed that they do indeed have their own bacterial version of a cytoskeleton. Some bacteria have gas-filled **vacuoles** that help them float or move. The watery fluid that fills the cell is called **cytosol**. Cytosol is made of water, minerals, salts, sugars and enzymes. Sometimes there are particles such as oil droplets or mineral crystals floating around in the cytosol. These particles are called **inclusions**. The word **cytoplasm** can be easily confused with the word cytosol. The cytoplasm is the cytosol plus the organelles floating in it, such as the ribosomes. In a human cell, the cytoplasm is basically everything outside of the nucleus.

The word *morphology* means "shape." Or, more correctly, "study of shape," because "-ology" means "study of." Fortunately, bacterial shapes are easy to draw.

- The *cocci (cock-eye, or cocks-eye)* are little spheres.
- *Diplococci* are two spheres stuck together.
- The *bacilli (ba-sill-eye)* look like little rods or sticks, though some are stubby and rounded on the ends.
- *Vibrio* are C-shaped. (The most well-known vibrio is *cholera*, an intestinal disease that often strikes after natural disasters like earthquakes, as drinking water is contaminated.)
- *Streptococcus* looks like a long chain. The illness we call strep throat is caused by one species of streptococcus, but there are many other species, too.
- *Staphylococcus* looks like a bunch of grapes. There are many species of staphylococci, but the one that gets most of the public's attention is *Staphylocuccus aureus*, the one that causes "staph" infections on the skin. (S. aureus is part of our natural bacterial population and does not cause harm as long as it is kept in check by other species of bacteria.)
- The *spirilli (spir-ill-eye)* are spiral-shaped and often have a flagella at each end of their bodies. They can look squiggly as they move, but their bodies are not actually flexible. Spirilli are less common that the cocci and bacilli, but can be found in watery places like sewers.
- Unlike the spirilli, the *spirochetes (spi-ro-keets)* are flexible and can really bend around as they move. This is mainly due to their inner flagellum. They can spiral their way right through soft tissue (such as the connective tissue in our joints). The most famous spiriochete is the one that causes Lyme Disease.

Some bacteria have *fimbriae (fim-bree-eye)*. These hair-like structures help the bacteria to hold on to surfaces, including our cells. Some bacteria have additional hair-like things called *pili (pie-lie)*. These have several functions. They help the bacteria stick to surfaces. They can help the bacteria by acting like grappling hooks. The bacteria can toss out a few pili and the ends will stick to the surface out in front of the bacteria; then it retracts the pili so that its cell body is moved forward. It then repeats the process, throwing the pili out again and then pulling its body to close the gap. A special pilus can also be used for conjugation. This type is often called a conjugation pilus or a sex pilus. The pilus is extended in order to grab a passing bacteria. The trapped bacteria is then "reeled in" until it is almost touching. The pilus then acts as a transfer tube, allowing the plasmid DNA to be passed from donor to recipient, as we discussed on the previous page.

NOTE: Scientists are not in agreement as to whether fimbriae and pili are the same thing. Some insist that they are different and give good reasons for thinking so. Other scientists don't see any difference and use the words interchangeably. This is good to know as you read books and websites on this topic. Be aware that the word "pili" might be used to describe what we have defined as fimbriae.

Bacteria are not part of human anatomy, but they are part of our physiology. Bacteria form a vital portion of our internal ecosystem and they affect our digestion and nutrition, our immune responses, and sometimes even our mental health. In the past, doctors and medical researchers did not realize the importance of bacteria in our bodies. Now they are beginning to understand how critically important it is, and some researchers have begun to catalog the *human microbiome*, a list of all the microorganisms (including fungi, protozoans and viruses, too) normally found in and on humans. Medical care of the 21st century will increasingly be focused on maintaining a healthy microbiome.

PROKARYOTES (part 1)

- For every 1 human cell in your body, you have 10 bacteria cells.
- There are two types of prokaryotes: BACTERIA and ARCHAEA. (Archaea used to be classified as bacteria. They tend to be the "extremophiles" who survive in harsh conditions.)
- Prokaryotes do not have a true nucleus. They have a clump of DNA but it does not have an envelope around it.

ANATOMY of a "basic" BACTERIA

NOTE: The "cillin" antibiotics interfere with the enzyme robot that does the cross-linking. Without these links, the cell wall is too weak and the bacteria falls apart.

MORPHOLOGY (means "shape")   (SIZE: 1 to 3 μm)

coccus (cocci)   diplococcus   bacillus (bacilli)   vibrio

streptococcus   staphylococcus   spirillum (spirilli)   spirochete

FIMBRIAE (fim-bree-eye) are little "hairs" that allow the bacteria to stick to surfaces. They are more common in Gram negatives.

PILI (pie-lie) are similar to fimbriae (some people think the names are interchangeable) but are longer, and thicker. They are used to move (like rock climbers use grappling hooks). A "sex plius" is very long and can grab another bacteria and "reel it in" until they touch. Then DNA can be shared.

17: PROKARYOTES (part 2)

The word *motility* means movement. Bacteria motility is most commonly the result of one or more flagella, but there are also a few other ways they can move, such as making internal air bubbles that float them to the surface if they are in water. Some bacteria, that don't live in water, can glide using a slippery slime make of sugars (polysaccharides). Scientists speculate that the slime is extruded out the back and acts as a very slow jet engine causing an equal and opposite reaction that propels the bacteria forward, albeit extremely slowly.

Most often, bacteria rely on flagella for movement. Flagella are made of microtubules. At the base, where it attaches to the cell wall, there is a tiny protein gadget motor that spins the flagella around at speeds up to 200 rotations per second. Flagella are similar to cilia, a feature often found on unicellular protozoans but also found on some human cells. Cilia beat back and forth, creating a coordinated waving motion. Flagella beat in circles, acting more like an outboard motor.

Some bacteria have only one flagellum. The cell can still move backward and forward, however, because the direction of the flagellum's rotation can be reversed. Other bacteria have a clump of flagella instead of just one. Some have a flagella at each end. Perhaps the strangest arrangement is the all-over look, where there are so many flagella that the bacteria starts to look hairy. Bacteria with multiple flagella can spread them out and vibrate them in such a way that they do a "tumbling" motion, which helps them to turn and go in a different direction. Spirochetes have an inner flagellum, set between an inner and outer membrane. This geometry allows the spirochete to flex and twist, allowing it to "swim" through soft body tissue such as skin and cartilage. Whatever type of motility a bacteria has, it is used to go toward sources of food and away from things that would harm them.

Bacteria can be classified several ways. One is by shape, as we have seen. This is important for some purposes, but not for applying antibiotic medicines. When you need to get rid of harmful bacteria in your body, you must know which medicines will be most effective. *Pathologists* (doctors who study diseases) want to know about the bacteria's biochemistry. The most important testing procedure is called the *Gram stain*. The results of this test will give you information about a bacteria's outer coating. Some bacteria have a thick layer of peptidoglycan, as we saw in the previous drawing. These bacteria will turn purple when the Gram stain method is applied. The peptidoglycan layer will soak up a lot of purple stain. Bacteria that turn purple with Gram staining are called *Gram positive*.

Other bacteria have a much thinner peptidoglycan layer and have an additional outer membrane. These bacteria don't have enough peptidoglycan to turn purple, but will hold a reddish-pink stain. Bacteria that turn pink with Gram staining are called Gram negative. The names "Gram positive" and "Gram negative" don't mean "good" and "bad." There are good and bad bacteria in both categories. And from the bacteria's point of view, they are just trying to survive, which to them is always good.

Gram negative bacteria also usually have an additional feature that causes problems for their hosts. They have toxic sugars sticking up from their outer membrane. These toxins can be a problem even if the bacteria has died. In fact, if you kill off these bacteria too rapidly your body can be overwhelmed with trying to get rid of all the toxins. Your liver can only process toxins at a certain rate. The sick feeling (fever, aches, nausea) you get from an overload of these toxins is called the *Herxheimer reaction*.

Some bacteria have yet another outer layer called a *capsule*. The capsule is made of sugars (polysaccharides) and is very soft and sticky; it is often called the *slime layer*. Having a slime layer is a big advantage for a bacteria. It prevents the cell from drying out, it helps it to stick to surfaces, and it makes it much harder for our immune cells to eat and digest it.

The archaea used to be classified as bacteria. Basically they are bacteria. They look and act like bacteria. The only reason they now get their own kingdom is because of some minor differences. Regular bacteria are now called *eubacteria*, *(yu-bacteria)* meaning "true" bacteria. Here are some differences between archaea and eubacteria:

1) The protein structure of archaea RNA polymerase (the thing that makes mRNA) is slightly different from eubacteria.

2) The archaea ribosome has a slightly different structure and can't read eubacteria DNA code that tells where to start.

3) The phospholipids in the archaea plasma membrane have their fatty acids attached differently. This difference makes it possible to have a molecule with heads on both ends, so that there is no need for two molecules to make a bilayer. (Imagine joining the tails of all those phospholipids in our bilayer pictures.) This makes a very tough plasma membrane—one that can survive extreme conditions such as boiling hot or extremely salty.

4) Archaea cell walls are not made of peptidoglycan. Their walls are made of sugars, proteins and various combinations of those, sometimes looking very much like peptidoglycan, but not close enough for chemists to use the term "peptidoglycan."

5) Archaea DNA is wound around little spools made of protein balls called *histones*. Plant and animal cells also have their DNA wound on histone spools. Eubacteria do not have histones.

For better or for worse, depending upon the situation, archaea will react differently to antibiotics than eubacteria will. We've seen how two types of antibiotics work: erythromycin and penicillin. Erythromycin attacks bacterial ribosomes at their P site, and penicillin attacks peptidoglycan cell walls. Archaea are not susceptible to either of these medicines because their ribosomes and cell walls are different from those of eubacteria.

PROKARYOTES (part 2)

MOTILITY (means "movement")

Bacteria sometimes have "tails" called flagella that are made of microtubules. ("Flagella" means "whip.")

Some bacteria can glide using a sugary slime.
They can also use pili as grappling hooks to pull forward.

Spirochetes have an inner flagellum.

CLASSIFICATION by pathologists ("Patho" means "disease.")
The Gram stain is used to find out what kind of cell wall the bacteria has so right kind of antibiotic can be prescribed.

GRAM positive (+)   GRAM negative (-)

(Gram staining is named after Danish scientist Hans Christian Gram.)

CAPSULES (the "slime layer")

Capsules keep them from drying out
and from being eaten by immune cells.

ARCHAEA

Archaea look and act very much like bacteria.

- Used to be classified as bacteria
  (They have their own kingdom now.)

- Most of them live in extreme environments like the bottom of the ocean or in hot mineral springs. However, a few species live in our intestines where they produce methane gas.

- None of them are pathogens.
  (They won't make you sick.)

- Coccus, bacillus and spirillum shapes

- Some have flagella

Differences between bacteria and archaea:

1)

2)

3)

4)

5)

18: ATP and GLYCOLYSIS

Living things need energy. In this lesson we will begin to learn about cellular energy and see the first step in breaking down glucose to release its energy. But first, let's take a look at something that is not alive and therefore does not need energy. Viruses are often confused with bacteria. We use the word "germ" to describe anything that makes us sick. Many people are not aware of how different bacteria and viruses are.

A *virus* is basically nothing more than a piece of DNA or RNA wrapped in a protein shell. Sometimes there is an extra layer made of phospholipid membrane. Protein hooks are often attached to this lipid layer. The hooks have a particular shape that will match receptors on the outside of the cells they attack. This means that a virus that invades skin cells probably won't be able to get into stomach cells. Viral diseases of fish or reptiles are not a threat to humans because the viral protein hooks only match the receptors on fish cells or reptile cells. Viruses called bacteriophages can only attack bacteria cells. (It's funny to think that bacteria can come down with viruses!)

A virus does not have ribosomes so it cannot make proteins. In order to manufacture its protein coat and its protein hooks, it needs to borrow the ribosomes of a living cell. The cell's ribosomes follow the instructions written in the viral DNA or RNA and they manufacture viral proteins. These parts assemble into thousands of new viruses and they fill the cell until it bursts. Then each new virus goes and attacks another cell.

Viruses are not considered to be living organisms because they do not have the basic characteristics of all living things. Living things can grow, move, reproduce, respond to their environment, and use energy. Cellular energy will be the theme of the rest of this lesson.

ATP is a tiny molecule that can store energy. *ATP* stands for *adenosine triphosphate*. Adenosine is an adenine connected to a ribose. If you add a phosphate to adenosine you get adenosine monophosphate, *AMP*. (Notice how similar this is to the nucleotide that includes adenine.) If you add another phosphate you get adenosine diphosphate, *ADP*. A third phosphate brings you to adenosine triphosphate. This third phosphate is what stores and releases energy. The phosphates carry a negative charge so they don't want to be next to each other. They act like a compressed spring, storing energy. When the third phosphate pops off, energy is released and you are left with ADP plus a single phosphate. That phosphate can be put back on so that ADP gets recharged and goes back to being ATP. Recharging ATP requires energy, just like recharging a battery does. In lesson 20 we will meet an enzyme gadget that is very efficient at recharging ATPs.

Glucose is the primary molecule from which our bodies derive energy. If all the energy in glucose were released at once, it would be too much for our cells to handle. They'd explode. Glucose must be torn apart very slowly, with a series of very small steps. The process of *glycolysis* is the first part of releasing the energy locked up in the chemical bonds of the molecule. "Glyco" means "glucose," and "lysis" means "break apart." Glycolysis will break glucose in half.

Glycolysis happens in the cytosol (cytoplasm) of the cell. No special organelle is needed. This is not true of many cell processes. Many of them only happen inside a particular location. Glycolysis happens anywhere and everywhere outside the nucleus.

There are ten steps in glycolysis. The first and third steps require a phosphate. The phosphates are used as part of the process of snipping the hexagon and turning it into a line. Then this chain of 6 carbons (with all their associated oxygens and hydrogens) is cut in half. In steps 6-10, 4 ATPs are recharged as well as 2 NADH molecules.

NADH is a "taxi" for high-energy electrons. It carries 2 electrons (and 1 proton) from glycolysis over to an assembly line that includes 3 proton pumps. The electrons will power the pumps and push protons across a membrane. (We'll look at this in lesson 20.) When the NADH taxi is empty it is called NAD+. When it is carrying 2 electrons and 1 proton, it is called NADH. Like ATP, NADH can be used over and over again.

At the end of step 5, we have two 3-carbon molecules. Steps 6- 10 make small alterations to the molecules so that by the end of step 10 we have two identical 3-carbon molecules of *pyruvate*. The pyruvates will go on to the next step where they'll get one of their carbons snipped off, leaving them as 2-carbon molecules. Then the 2-carbon molecules will begin a complicated process called the Krebs Cycle (or Citric Acid Cycle).

As we said, in steps 6-10 we have 4 ATPs being recharged, as well as 2 NADHs. Since we used 2 ATPs in the beginning of the process, our net gain of ATPs is 2. The most important thing that most books/teachers/ texts/tests want you to know about glycolysis is that it produces 2 ATPs and 2 NADHs. (Often, they only ask about ATPs.)

.

What does it mean to be alive? We just looked at bacteria, which are definitely alive.
But what about viruses? Are they alive? Viruses are also part of our microbiome.

To be alive, an organism must:

1)

2)

3)

4)

5)

A virus is basically a piece of
DNA or RNA inside a protein shell
sometimes with a lipid membrane.

HOW DO LIVING CELLS USE ENERGY?
Most cellular processes are powered by **ATP** (adenosine triphosphate). ATP acts like a rechargeable battery.

GLYCOLYSIS means "breaking glucose"

Glucose is a like a stick of dynamite, full of potential energy.
It must be disassembled little by little so the energy
is released slowly, not all at once.

What is NADH?

19: SPERM and MITOCHONDRIA

Male *gametes* (reproductive cells) are called *sperm*. It's not just animals that make sperm — plants make them as well. Mosses and ferns make sperm that swim. Most plants, however, make non-motile sperm that must be carried by pollen grains. In animals, sperm are made continually and only have a life span of a few days. Mature male mammals make tens of millions of sperm every day (about 1000 per second). The male reproductive tract also makes fluid for the sperm to swim in, but that is outside the scope of this lesson.

A sperm has three body sections: a head, a midpiece and a tail called a *flagellum*. The head is about 5 *microns* long, making sperm the smallest human cells. (Compare this to a coccus bacteria which is 1 micron in diameter. A micron (or micrometer) is 1/1000 of a millimeter.) The midpiece contains the *centrioles* and the *mitochondria*, which are discussed below. The flagellum is about 50 microns long and has a central core called the *axoneme*, which is made of microtubules. ATP energy is used at the base of the flagellum to make it go back and forth. The entire sperm, including the flagellum, is surrounded by plasma membrane.

The head contains the *acrosome*, which has enzymes that will dissolve the outer portion of the egg, and a nucleus containing tightly packed DNA. The DNA contains 23 *chromosomes,* very long pieces of DNA. This is half the number of chromosomes required for a human cell (46), so gametes are called *haploid* cells. The human genome contains about 3 billion base pairs ("rungs" on the DNA "ladder.") Chromosomes are often shown looking like little sticks. That's not what we find here. Inside this nucleus, the DNA strands have been specially packaged for traveling.

A swimming sperm has one goal: get to the egg first, and enter it. Each sperm has this goal, but there are usually about 100 million of them competing against each other, so the odds of winning are pretty small. A sperm with the slightest defect won't even get close to the egg. Some sperm have unfortunate mutations such as two tails, or a head that is too large. The sperm have to be so perfect that even small disadvantages will cause them to be losers in the race. The DNA packed into the nucleus is bundled very tightly using special protein spools called *protamines*. The protamines allow the DNA strands to get closer than they would ordinarily, and they also cause it to bundle into donut shapes called toroids, which are then stacked right next to each other. If the protamines aren't perfect, the DNA will be packed too loosely and this will cause the sperm to swim more slowly, losing the race. If you use Google image search with key words "sperm protamines" you will see many illustrations showing the spools and the toroids. (Regular DNA is wound on spools called histones. You might see these, as well.)

Since the sperm has a very particular purpose and a very short life span (less than a week), many organelles are not needed. For example, it is not going to need any ribosomes since it won't be making any proteins. The only organelle the sperm really needs are the mitochondria, which generate lots of ATP energy that can be used for making the flagellum whip back and forth. However, the sperm does have one other organelle: a pair of *centrioles*.

Centrioles do two things: 1) they form flagella (and cilia), and 2) they act as a focal point for the creation of a spindle of microtubles used in cell division. In other words, centrioles allow a cell to divide. It's good that the sperm have these because, as far as we know, egg cells don't. Researchers think that it is likely that the centrioles in every cell of our bodies can be traced back to those original centrioles that came from the sperm of our fathers. Centrioles make duplicates of themselves before each cell division. By the time a baby is born, those original centrioles from the sperm will have made copies of themselves trillions of times! Centrioles are easy to overlook in sperm anatomy and are often not even shown in diagrams, yet they are essential to the creation of a new life.

The mitochondria are often called the "powerhouses" of the cell. It is inside these organelles that those pyruvate molecules are completely "burned" and ATP energy is generated. Sperm have about 100 mitochondria, but they are fused together (unlike those found in regular cells) and then shaped into a coil. We will consider the structure of a single mitochondrion.

The mitochondrion has an oddly shaped "bag" in its interior. The bag has lots of folds and creases (called *cristae*) which allow a lot of surface area to be packed into a small space. The "fabric" that this bag is made of is something you are very familiar with: plasma membrane (phospholipid bilayer). The fluid inside this bag is called the *matrix*. The word matrix is very common in biology and means "a central area." The matrix is similar to the fluid of the cytoplasm, with lots of tiny molecules floating in it, such as oxygen, carbon dioxide, minerals, phosphates, pyruvates, tRNAs, nucleotides, and many enzymes that do various jobs. Also inside the matrix are rings of DNA called *mitochondrial DNA (mtDNA)*. These rings have 37 genes that code for proteins that the mitochondrion will need frequently. (A *gene* is a strip [or compilation of several strips] of DNA that codes for one protein.) The mtDNA also has codes for tRNAs and ribosome parts. The mitochondrion still needs to access the DNA in the nucleus, though, because the mtDNA does not hold all the necessary information, only some of it. Why these particular genes (the ones on the mtDNA) are found inside the mitochondria is still a mystery. The mtDNA is inherited predominantly from the mother, which makes it useful for tracing family ancestries. MtDNA mutates faster than the *genomic DNA* (in the nucleus), making it ideal for studying genetic changes in the human genome over time.

Pyruvates are taken into the matrix, where they go through the *Krebs Cycle* (named after its discoverer, Hans Krebs), also known as the *Citric Acid Cycle*. (You can use either name; it's your choice. I chose "Krebs" because it is easier to remember.) The Krebs Cycle recharges electron carriers that will be used in the *Electron Transport Chain*. The ETC is made of three pumps, several shuttles, and a motor, and is found embedded in the plasma membrane of the matrix.

The sperm cell is the smallest human cell. It has half the normal amount of DNA and has very few organelles.

MITOCHONDRIA are often called the "powerhouses" of the cell because they make lots of ATPs.

Mitochondrial DNA (mtDNA) is inherited primarily from the mother. The egg cell will donate thousands of mitochondria compared to the sperm's one hundred or so. This fact makes mtDNA useful for researching ancestry.

Most genes are for tRNA, some are for protein gadgets found in the Electron Transport Chain, and some are for ribosomal RNA.

A "gene" is a strip of DNA (or several strips spliced together) that code for a particular protein.

The Electron Transport Chain (ETC) is an assembly line of machines that will turn ADP back into ATP.

20: The KREBS CYCLE and the ELECTRON TRANSPORT CHAIN

Pyruvate molecules are transported into the mitochondrion through channel (portal) proteins. Once inside, they enter the matrix through a special channel that does two jobs. (The channel is not actually shown.) First, an enzyme "scissor" snips off one of the three carbons. This carbon has two oxygens attached to it, so it goes floating off as carbon dioxide, CO_2. The carbon dioxide will go to your lungs to be exhaled. Next, an enzyme "scissor" attaches a molecule called Coenzyme A to the remaining two carbon atoms. This makes a molecule called **acetyl-CoA** *(ah-SEE-till-co-A)*. Acetyl-CoA is a very important molecule, despite the fact that you have probably never heard of it. Just like a piece of wood can be used either to build something, or to be burned as fuel, acetyl-CoA can be "burned" in the Krebs Cycle, or it can be used in other places in the cell to build various parts. This step, forming acetyl-CoA, doesn't have a cool name like the Krebs Cycle does. It usually just gets called the "pre-step," since "pre" means "before." The technically correct name is "pyruvate oxidation." Though it is not one of the 8 steps in the cycle, it is still often considered to be part of the cycle.

NOTE: The Krebs Cycle isn't really a circle. It consists of various chemical processes, which are going on all over the place all the time. However, in order to understand the chemistry, it is very helpful to draw a diagram that looks like a circle.

Acetyl-CoA's are attached to a 4-carbon molecule to create a 6-carbon molecule called **citric acid**. Many scientists prefer to call the Krebs Cycle the **Citric Acid Cycle**, naming it after the molecule formed in the first step. The citric acid then has two carbons snipped off and releases (once again) a carbon dioxide. The most important thing to know about the cycle is what energy molecules it produces. Out of the Krebs Cycle come 3 NADH, 1 GTP (which is converted to ATP), and 1 **FADH$_2$**. The FADH$_2$ is another taxi for high-energy electrons; its function is similar to NADH. By the time the cycle is complete, you are back to having the same 4-carbon molecule you started with, and it is ready to accept another acetyl-CoA. (The two carbons that get snipped off during the cycle are not the ones that come in on the acetyl-CoA. Those two carbons go to the back of the molecule, so to speak, and get moved up each time the cycle goes around. So after two turns of the cycle, those carbons that came in on acetyl-CoA will have moved up enough so that they will be the ones that get snipped off during the next cycle.

At the end of the Krebs Cycle, the original glucose molecule is finally gone. NADH and FADH$_2$, now contain high-energy electrons that will be carried to the Electron Transport Chain.

The ETC is a series of protein gadgets, including three pumps, a few shuttles and a large motor. (The motor at the end is what will generate the ATPs.) NADH goes over to the first pump in the chain and donates its two electrons. The electrons go into the pump and cause 4 protons to be pumped up from the matrix, out into the inter-membrane space above. A shuttle then takes the electrons to the next pump. FADH$_2$ is a supplemental shuttle between the first two pumps. It brings its electrons to a protein between the first two pumps. The electrons go through the second and third pumps, causing each one to pump protons up and out. By the time the electrons exit the third pump their energy is used up and they need to be discarded. The two "tired" electrons are matched up with two protons to form two hydrogen atoms, which are then attached to an oxygen atom to make water, H_2O. It is important to remember that oxygen is final "electron acceptor" at the end of this chain. The oxygen comes from the air you breathe.

NOTE: You should also be familiar with the term **oxidative phophorylation**. "Oxidative" means that oxygen is used, and "phosphorylation" means "adding a phosphate." Phosphorylation means "adding a phosphate." This happens as the ATP synthase machine pops the third phosphate onto ADP. This is a mechanical process, with the ADP and the phosphate going into those beater-looking things, being pressed together, and then coming back out as ATP.

In the end, **each glucose molecule makes a maximum of 36 ATP's**. The count is often given as **32-36**, depending on how you count molecules in all the cycles, and depending on whether the pumps are operating at 100% efficiency.

Mitochondrial diseases are caused by defects in the molecular machinery inside the mitochondria. About 1 in 4000 babies will be born with a mitochondrial disease (more common than all childhood cancers combined). How many things could go wrong inside a mitochondrion? A lot, as you can see. Mistakes in coding for proteins could cause the shapes of the portals, pumps, motors or shuttles to change, such that they could not function properly. It would take only a small mistake in the DNA, perhaps even just one base (A, T, C, G) to mess up the shape of the protein. The mistakes could be in either the mtDNA or the genomic DNA found in the nucleus of the cell. Each mitochondrial disorder is named after the part that is broken. Since the problem originates in the DNA, there is no cure. Doctors just try to ease symptoms. Perhaps some day they'll figure out how to restore the correct sequence to the DNA.

Even without mitochondria, cells would still be able to make some ATP's using glycolysis, but this would not be enough to adequately power the cell. Without lots of ATPs, cells can't function properly because ATPs are needed for every cell activity. Cells would eventually die. Mitochondrial disorders can range from mild to very severe. In the most severe cases, babies die shortly after birth. Mitochondrial problems probably account for at least some of the cases of Sudden Infant Death Syndrome (SIDS). In moderate cases, the children live into their teen years, but with much suffering. They can have a range of problems, including seizures, blindness, deafness, diabetes, muscle weakness, liver and kidney disease, and brain problems. When mitochondrial disease is mild, the symptoms might not show up until the person is well into adulthood.

In the mitochondrial matrix, pyruvates are cut apart to make acetyl-CoA's, which are then sent to the Krebs Cycle.

The Krebs Cycle breaks apart the remaining carbon bonds and uses the energy to recharge "taxi" molecules NADH and FADH2. These taxis then go over to the E.T.C.

The end result of all these processes is that one molecule of glucose can yield up to 36 ATPs. CO_2 and H_2O are given off as wastes.

---

1) The PRE-KREBS step happens as the pyruvate crosses the inner membrane (into the matrix)

2) The KREBS CYCLE finishes the "burning" of glucose

An enzyme "scissor" snips off one carbon. This carbon has two oxygens attached to it, so the carbon goes sailing off as CO_2.

NADH is a "taxi" for high-energy electrons. One taxi gains two electrons in this step.

An enzyme "stapler" adds a piece called **coenzyme A** to the remaining 2-carbon molecule (which is called acetyl). CoA is actually a very large molecule.

This cycle is also called the Citric Acid Cycle

---

3) The Electron Transport Chain (ETC)
The ETC is an elaborate system of protein gadgets. There are 3 pumps, several shuttles, and a motor at the end. The goal? ATPs!

  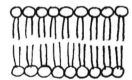

Since this process requires O_2 and it results in phosphates being put back on, we can call it **oxidative phosphorylation**.

DRAWING 21: THE OVUM and FERTILIZATION

In drawing 19 we met the smallest human cell. Now we meet the largest: the *ovum* (egg). The ovum is about 200 microns (.2 mm) which is so large you can almost see it without a microscope. You wouldn't be able to see any detail, of course. It would look like a tiny dot, smaller than the period at the end of this sentence.

The ovum starts out in the ovary as a cell called a primary oocyte. By the time a baby girl is born, she has already produced all the eggs she will ever have (on the order of a million). By the time she is a teenager, the number of egg cells has decreased by half that number, and then through the middle of her life, the number of eggs drops steadily. However, there are still plenty left, and one egg each month is released from the ovary. It travels down the fallopian tube to the uterus. A future drawing will give more details about this process. Here in this drawing we will just look at the ovum itself.

The ovum is surrounded by a layer of protective cells called the *corona radiata* (named for the sun's corona). These cells have been surrounding the ovum for a very long time. In fact, there used to be a lot more, but when the ovum left the ovary, only a few layers of these cells managed to stick close enough to be able to go along for the ride down the fallopian tube. These cells have nourished the egg for years. This is mainly to protect the ovum's DNA from damage. The normal metabolic processes that go on inside cells can produce dangerous by-products (free radicals) that can bounce around inside the cell and damage DNA if they happen to run into it. Cells have ways of dealing with this damage and are constantly making repairs, but if a few body cells die.. oh well, there are millions more like it. Not so with an egg cell; there's a lot at stake here! This DNA will rise to a whole new human being. Damage here can be fatal. So to keep the environment safe inside the ovum, those nourishing cells do all the metabolic work and the ovum sits there, pampered and lazy, letting the corona cells do all the work of feeding her and keeping the cell neat and clean.

Inside the corona radiata is the *zona pellucida*. The word "lucid" there in the middle means "clear." This layer is clear and jelly-like. Later is will harden, but right now it is soft. Then comes the plasma membrane. Inside the membrane we find cytoplasm, a watery goo in which many things float around: mitochondria, ribosomes, enzymes, fats, proteins, minerals, ions, and dissolved gases such as oxygen and carbon dioxide. An ovum has about a million mitochondria, ready and waiting to start churning out ATPs if and when the egg joins with a sperm and begins to form an embryo. The egg's mitochondria will use the fats in the cytoplasm, chopping them up into acetyl-CoA's and burning them in the Krebs Cycle to make energy for the ETC which will generate ATPs. The cytoplasm contains enough energy to keep the embryo going until it can implant itself into the mother's uterus and begin drawing upon her supply of glucose.

The ovum has a nucleus with 46 chromosomes. It will need to reduce down to 23 chromosomes before it joins with the sperm's 23. If the ovum's nucleus retained all 46 and then joined with the sperm's 23, it would have one-and-a-half times the normal amount and we would call this situation "triploid." Haploid (N) means half the correct number, diploid (2N) means the correct number, triploid (3N) is 1.5 times too many, and tetraploid (4N) means having twice the correct number. Some organisms are okay with having more or less than the diploid number. Moss plants and male bees are haploid (N), seedless watermelons are triploid (3N), goldfish and salmon are tetraploid (4N), and wheat plants can be tetraploid or hexaploid (6N). The Ugandan clawed frog is the record-holder at dodecaploid (12N)! However, in most cases, if an organism gets too many chromosomes, it does not survive. These are very unusual examples. If a human embryo gets more than 46 chromosomes, it either dies or is born with serious problems. (Note: Each species has its own correct number of chromosomes. The least number on record is a certain type of ant with only 6. The highest is a species of butterfly with 268 chromosomes. There is not a detectable pattern as to which species have how many.)

Gametes (eggs and sperm) are produced by a process called *meiosis*. In meiosis, a primary cell divides, then divides again. These two divisions yield four haploid cells, three of which will be discarded. These discards are called *polar bodies*. The ovum has already been through the first part of meiosis, and the first two discarded cells (polar bodies) are still visible at the outside rim. They will just dissolve and disappear. Now the nucleus is waiting for its final division. If a sperm enters, those 46 will split and 23 of them will be discarded as a third polar body. If no sperm enters, the entire egg will die. It will then be flushed out during menstruation.

Notice that the nucleus has a *nucleolus*. Remember, this is a section of dense DNA that manufactures ribosomes. It will become very active if the ovum in fertilized.

As the sperm approach the ovum, they must force their way into and through both the corona radiata and the zona pellucida. They are aided by the enzymes in the acrosome, which can dissolve the zona pellucida. The ovum secretes chemicals that aid this process, too. Some researchers say they've seen the ovum stay in this state (with sperm burrowing in) for several hours. The action of the sperm flagella cause the ovum to rotate slowly, like a Ferris wheel. It must be an amazing sight!

Right outside the ovum's membrane are little protein receptors that act like finish line buttons at the end of a swimming pool. Like swimmers in a race, the sperm will announce their arrival at the finish line by hitting one of these protein buttons. And like the buttons in a swimming pool, they are wired to allow only one winner, and once a button is hit, the others are disabled in a split second. When a sperm makes contact with a receptor protein, immediate changes take place. First, there is an instantaneous electrical change across the membrane, preventing the binding of any other sperm. However, this electrical change doesn't last very long, so the ovum triggers its secondary "security system." The vesicles that are waiting right inside the membrane fuse with the membrane (exocytosis), spilling their chemical contents into the zona pellucida, causing it to become hard and impenetrable. This hardened shell will stay around the embryo for about a week (at which point the embryo will have to "hatch" out of it!).

Meanwhile, the winning sperm's membrane is fusing with the ovum's membrane, allowing the sperm's nucleus, centrioles, mitochondria and even the axoneme to enter the ovum. The sperm's 23 chromosomes then join with the egg's 23, making a new cell called a *zygote*. The sperm's centrioles will allow the zygote to make its first division (mitosis).

The ovum (egg) is the largest human cell at about 200 microns (the length of a paramecium).

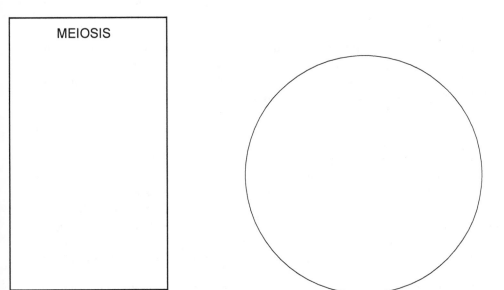

MEIOSIS

FERTILIZATION

As soon as one sperm touches a protein receptor, two things happen:

1) There is an immediate electrical reaction (called depolarization) across the membrane, preventing any other sperm from fusing.

2) The vesicles begin dumping their chemicals into the zona, causing it to harden. Later, the embryo will have to "hatch" out of this hardened shell.

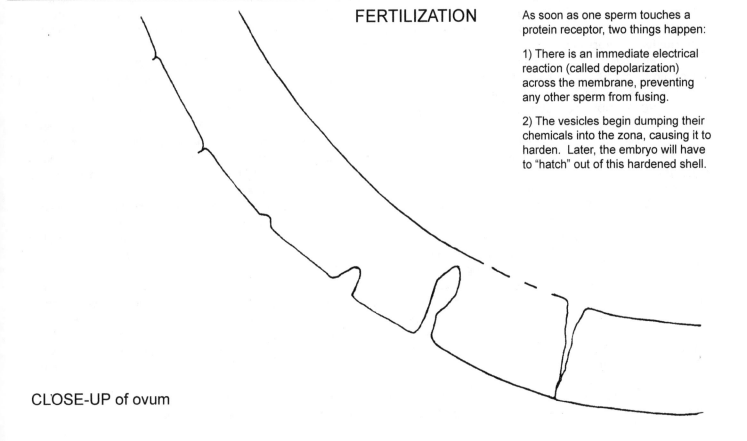

CLOSE-UP of ovum

22: THE ZYGOTE

In this drawing, we finally achieve our goal of making a human cell. As soon as the sperm nucleus fuses with the ovum nucleus, a complete cell is formed. This cell is ***totipotent***, meaning "all powerful." It can turn into ANY cell, not only body cells but placenta cells and umbilical cord cells, as well. We can't forget how important these supporting parts are. Although you discarded your placenta when you were born, it was once part of you. It's an odd thought, but every cell of the placenta and the umbilical cord contained a full set of your DNA.

In reality, a newly formed zygote would be a bit more complicated than this. As soon as the sperm nucleus enters, those polar bodies are expelled as the ovum completes meiosis. And while that is happening, the cell is also beginning its first division using a process called mitosis, which we will look at in the next lesson. Here we have a simplified zygote cell that has not yet started dividing. Actually, this diagram is more of a generic (average) cell, not specifically a zygote cell.

NUCLEUS: This is where the full set of DNA instructions is kept. Most of the time, when the cell is not dividing, the DNA looks like a pile of spaghetti and is called ***chromatin***. When the cell goes to divide, the chromatin will organize into bundles called ***chromosomes***. The nucleus is surrounded by a double layer of phospholipid membrane and has many pores.

NUCLEOLUS: This is a particularly dense part of the nuclear DNA and under the microscope it looks like a dark spot. This is where ribosomes are made. There are many copies of these instructions , so many ribosomes can be made quickly.

NUCLEAR PORES: The nucleus has over 3000 pores which control what goes in and out. The pore isn't just a hole; it is a complex protein gadget (called the ***pore complex***). The pore has several rings, some filaments, and a basket. The inner ring can open and close so that only the necessary molecules can move in and out. (One molecule that needs to get out is mRNA.)

RECEPTORS and PORTALS: The cell is covered with protein gadgets, many of them ***receptors*** that receive messages from other cells, and ***portals*** that control the flow of various molecules, including nutrients and ions.

MHC1: One of the proteins stuck to the outside of the cell is MHC1, which functions as an identification tag, telling other cells that this cell is part of the body and is not a foreign invader. (M= Major, H= Histocompatibility, C= Complex)

CENTROSOME: This is a blob of protein goo with two barrel-shaped centrioles inside. The most important thing to know is that the centrosome is responsible for making the spindle at the beginning of mitosis (as we shall see in lesson 23). Centrioles also seem to be responsible for making cilia and flagella. (Cilia are found on cells lining the trachea.)

SMOOTH ER: Smooth endoplasmic reticulum is a network of tubes right outside the nucleus. In fact, it is attached to the nucleus's outer membrane. They often say that the ER is "continuous" with the nuclear membrane, which means it is attached to it (or is part of it). Smooth ER does not have any ribosomes attached to it. Its primary function is to make phospholipids that can replenish the membrane. It can also make steroid hormones and it can store calcium (especially in muscle cells). It has other minor functions, depending on what type of cell it is in.

ROUGH ER: Rough endoplasmic reticulum is also a network of tubes. It is identical to smooth ER except that it has lots of ribosomes attached to it. The ribosomes are what make it look "rough." The ribosomes feed their polypeptide chains into the rough ER so that they can be wrapped in membrane, forming a vesicle. Often, these vesicles leave the rough ER and go to a Golgi body for further processing.

RIBOSOMES: These are the only organelles that are not "membrane bound" (surrounded by membrane). Ribosomes are made mostly of RNA with just a few small protein bits added in. They come in two halves: the large and small subunits. Ribsomes make proteins of all kinds using mRNA instructions.

GOLGI BODY: It can also be called the ***Golgi apparatus***. Its structure is similar to the ER, as it is made of hollow tubes and "pancakes" of phospholipid membrane. The side facing the ER (the "cis" side) receives vesicles from the rough ER. The proteins enter the Golgi and have various tags (often sugars) added to them. Many of the extra tags act like labels telling where the proteins are supposed to go. For this reason, the Golgi is often called the "post office" of the cell. Then the vesicles exit the far side (the "trans" side) and go off to wherever they are headed. No one is sure how the Golgi is able to keep its enzymes inside, because the "pancakes" are constantly shifting. (Videos of the Golgi can show this shifting process.)

MITOCHONDRIA: The site of the Krebs Cycle and the Electron Transport Chain (that make ATPs). The mitochondria contain a ring of DNA, mtDNA (mitochondrial DNA), that has information on how to make tRNA and parts for the E.T.C.

LYSOSOME: This is the recycler of the cell. It is a hollow ball made of membrane, very much like a vesicle. However, it is filled with enzymes that can break apart proteins and fats. It is often called the "stomach," the "trash can," or the "recycling center" of the cell. Without lysosomes, cellular garbage would pile up and poison the cell. ("Lys" means "break apart.")

PEROXISOMES: This organelle was not discovered until 1967. The peroxisomes contain enzymes that work to neutralize toxins that we take in (alcohol, for example) and also dangerous "free radicals" (molecules with extra electrons or oxygen atoms). A by-product of the neutralization process is ***hydrogen peroxide***, H_2O_2, which isn't exactly good for you, either. Therefore, they also contain an enzyme that neutralizes the H_2O_2. Liver cells contain a lot of peroxisomes. Peroxisomes also work to break down long chain fatty acids into medium length chains which then get sent over to the mitochondria to be chopped up into acetyl-CoAs. (You'll remember that acetly-CoAs are what the Krebs Cycle uses.) Peroxisomes have a few othe minor functions, but these two (dealing with toxins and chopping fatty acids) are the most important ones to remember.

CYTOSKELETON: This was discovered in the 1960s, but the motor proteins that travel along it were not discovered until the mid-1980s. The cytoskeleton is made of ***microtubules***, which are made of protein units called ***tubulin***.

THE ZYGOTE

When the sperm nucleus fuses with the ovum nucleus, a zyote is formed. This cell is **totipotent** and is capable of turning into any type of cell, including not only body cells, but also placenta cells or umbilical cord cells.

[ ] for receiving messages

[ ] for taking in large molecules such as proteins or fats.

[ ]
- Holds the cell's shape
- Provides "roads" for motor proteins.

[ ]
- The location of the Krebs Cycle and the Electron Transport Chain (which makes lots of ATPs).

[ ]
- Is an identification tag showing that the cell is "self" and not an invader.

[ ]
- Makes microtubules and will form the spindle during mitosis.

[ ]
- DNA when it looks like spaghetti

[ ]
- Is a thick clump of DNA that makes ribosomes

[ ]
- Makes phospholipids to be used in membrane and vesicles
- Stores calcium
- Makes steroid hormones

[ ]
- Contain several dozen types of digestive enzymes that can break down proteins, fats, sugars and nucleic acids into individual units that can be used by the cell to build cell parts. (Without lysosomes, cellular garbage builds up and poisons the cell.)

[ ]
- Make proteins

[ ]
- Has ribosomes stuck to it. The ribosomes feed polypeptides into it. The ER wraps them in membrane to form a vesicle. The vesicle will likely go to a Golgi body to be labeled for delivery.

[ ]
- Break long chain fatty acids into medium length chains that get delivered to the mitochondria (where they will then be turned into acetyl-CoA's).
- Neutralize toxins (ex: alcohol) and free radicals by transferring dangerous H atoms to O_2 molecules, making H_2O_2 (hydrogen peroxide). Another enzyme neutralizes the H_2O_2. Liver cells contain many peroxisomes.

NUCLEAR PORES

GOLGI BODIES

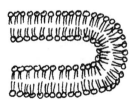

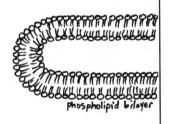

phospholipid bilayer

Golgi bodies add sugar tags that act like address labels so vesicles get delivered to the correct place in or out of the cell.

23: MITOSIS

Mitosis is a word that has two definitions, one formal and the other informal. Informally, the word mitosis is used to mean the ordinary process of cell division. Cells are constantly dying and being replaced. When you injure your skin, for example, you must grow new skin cells to replace the damaged ones. The cells near the damaged area begin duplicating, making many copies of themselves, enough to fill in the damaged area. Regular body cells are called ***somatic*** cells. (Remember, "soma" is Greek for "body.") We need this term in order to make the distinction between body cells and the reproductive gamete cells. Only gametes go through meiosis. Regular somatic cells never use meiosis, only mitosis. In mitosis, you end up with two cells that both have 46 chromosomes. In meiosis, you end up with four cells that each have 23 chromosomes.

Formally and technically, the word mitosis means only one part of this duplicating process: the duplication of the nucleus. Our drawing will clearly show what is included in the technical definition of mitosis. However, even scientists and teachers who know perfectly well that mitosis is actually about the nucleus, will still use this word to describe the whole process of somatic cell division. There isn't really a word for the whole process, so "mitosis" has to do double duty!

The organelle that controls mitosis is the ***centrosome***. A centrosome is a pair of centrioles surrounded by a blob of protein goo. Centrosomes are often called "microtubule organizing centers" because they seem to be the focal point of the cytoskeleton network of microtubules. They are also the organelles that create cilia and flagella. Each cell has a centrosome that hangs out near the nucleus. When the cell wants to divide, one of the first things that happens is that the centrosome duplicates itself. The original centriole is always called the "mother" and the duplicate is called the "daughter." (The centriole that was the daughter in the original pair turns into a mother because it produces a daughter. The original mother stays a mother and also produces a daughter.) Then these two new centrosomes then go to opposite sides of the nucleus. The DNA inside the nucleus then duplicates, and all the resulting pairs of chromosomes then line up in the middle of the nucleus. The centrosomes start forming a spindle made of microtubules. The microtubules fasten themselves to the chromosomes (at a place in their middle called the kinetochore). When all is ready, the centrosomes contract those microtubules and pull the chromosome pairs apart. Each side of the cell ends up with a full set of chromosomes and one centrosome.

There are official names for the stages of mitosis as well as the stages of a cell's life when it is not dividing. When a cell is resting, just growing and enjoying cellular life, we say it is in i***nterphase***. ("Inter" means "between.") Some cells spend most of their existence in this phase. Nerve cells (neurons), for example, stop dividing by the end of infancy and then stay in interphase for the rest of your life. Other cells, such as skin cells, are constantly dividing and spend very little time resting.

Interphase can be divided into phases, too. The part of interphase where a cell is really and truly resting and not even thinking about mitosis is called the ***Gap 0 phase***, or just G_0 (G zero). (During G_0 the DNA in the nucleus looks like a pile of spaghetti, and is called ***chromatin***. Later, as part of mitosis, the chromatin will be arranged into ***chromosomes***.)

Gap 1, or ***G_1***, is when the cell has decided it would like to expand and grow and probably go into mitosis. So during Gap 1, the cell makes lots of extra organelles, especially mitochondria that can give the energy necessary to divide. There must eventually be enough organelles for two complete cells.

Once the cell has enough organelles for two cells, it will enter the ***S phase***. The S stands for "synthesis," which means "to make." What is being made is DNA. Every chromosome (though at this point it is still a messy pile of chromatin) will duplicate itself. Now the nucleus will have a total of 92 chromosomes. Also, the centrosome duplicates itself during S phase.

The last part of interphase is called ***Gap 2***, or ***G_2***. During this phase, the cell continues to make organelles and enzymes in preparation for mitosis. At various points in this cycle, there are "check points" where certain protein gadgets are given the task to check and make sure that everything is going well. If one of these little protein robots senses that there are not enough mitochondria, for example, it will send out a signal that means "do not begin mitosis yet." If these little gadgets are not working correctly, a cell might begin to divide when it shouldn't. This could lead to uncontrolled cell growth and the formation of a tumor. Cancer researchers are trying to figure out how to use these mitosis "policemen" to stop inappropriate cell growth.

When a cell is ready to divide, the first thing that happens is that the chromatin (those 92 chromosomes) get organized and go from looking like a pile of spaghetti to tightly organized little bundles called chromosomes. This phase is called ***prophase***. "Pro" means "first," so this is the first stage of mitosis. Also during this stage, the thick nuclear envelope begins to dissolve. The nucleus must dissolve in order to let the pairs of chromosomes go opposite directions.

The next phase is called ***metaphase***. The centrosomes begin to weave the spindle and the chromosomes line up in the middle. This is the easiest phase to find when looking at real cells.

Then comes ***anaphase***, where the centrosomes tug on the chromosomes and split the pairs. Generally, anaphase is identified when the spindle has split and shrunk to where it looks more like a V (or A) on each side. It's hard to say when exactly prophase ends and anaphase begins. All you need to know is that if you are asked to find cells that are in anaphase, look for two V-shaped spindles. (It might be helpful to think of them as looking like the letter A instead of V, because the word "anaphase" begins with the letter A and you can use this as a mnemonic.) "Ana" means "opposite" so anaphase is when the chromosomes are in the process of going to opposite sides of the cell.

The last part of mitosis is ***telophase***. "Telo" means "far," so this is when the chromosomes are now on the far sides of the cell, far away from the middle. (A telescope is something you use to look at faraway objects.)

Notice that mitosis ends before the cell splits in half. Again, the technical meaning of the word mitosis is the division of the nucleus, not the whole cell. When the entire cell splits in half, this is called ***cytokinesis***. You know that "ctyo" means "cell," and "kine" means "motion." The root word "kine" shows up a lot in biology and has various shades of meaning. Here, the motion or action is the cell pinching in at the middle and then splitting to make two individual cells. While the split is happening, the new nuclei are busy re-growing a nuclear envelope. The DNA must not be exposed for longer than necessary. While the envelope is missing, the DNA is vulnerable. (This is another place where cancer treatments, especially radiation, can intervene and try to stop cell division.) These two new cells will then be in interphase, at the beginning of the cell cycle. They will grow for a while then possibly begin the process of mitosis again.

Let's take a close-up look at DNA replication. (Often, the word "replication" is used instead of "duplication," so we'll switch over to that word, but they mean the same thing.) First, the DNA must be unwound and unzipped. As you might guess, there is a specialized enzyme robot that does that job. Its name is ***helicase***, because it unzips the helix. Remember, the rungs on the DNA ladder are made of base pairs (A and T, C, and G) held together in the center by hydrogen bonding. Hydrogen bonding is the best bonding to use in a location where you want the molecules to stick together most of the time but still be able to be separated when necessary. (Remember, water molecules stay together by hydrogen bonding.) The helicase doesn't have to pull any molecules apart or undo any covalent bonds. Hydrogen bonds are relatively weak and easy to separate.

After the helicase unzips the DNA, another specialized robot enzyme called ***DNA polymerase*** is used to make a new matching strand of DNA. DNA polymerase is similar to RNA polymerase, but it does not use uracil. The polymerase runs along the DNA strand, reading the sequence of bases and forming a complimentary strand with the opposites, A with T, C with G. You can see in the picture that the result is two identical strands of DNA. However, there is one small problem that must be overcome.

Remember that DNA and RNA strands have a direction to them. Just like English must be read left to right, DNA and RNA are directional. We don't use left and right or up and down, we use the terms 5' (5 prime) and 3' (3 prime). The numbers 5 and 3 come from the numbering of the carbon atoms in the ribose molecule. This illustration shows how the carbons are numbered. They go clockwise, with number 1 being the first one after the oxygen atom. So number 3 ends up being the carbon that attaches to the phosphate below it, and carbon number 5 ends up being the one that attaches to the phosphate above it. If you imagine the molecule shown here as being just one part of a long string, the 5' end would above and the 3' end below. If you turned the string upside down, the 5' end would then be pointing down instead of up.

Why is this 5' and 3' stuff worth mentioning? Because one of these strands has a replication problem. The DNA polymerase enzyme can only go from a 3' end towards a 5' end. So for one of the strands, the polymerase enzyme slides along with no problems. This strand is called the ***leading strand***. The other strand is "backwards" for the enzyme and is called the ***lagging strand***. The enzyme must work on little sections of the lagging strand, constantly hopping back up to a new piece as it finishes the old one. Your drawing should have this motion indicated by arrows. Each little section of new DNA is called an ***Okazaki fragment***, after the scientist who discovered it. The Okazaki fragments must then be "glued" together by an enzyme called ***ligase***. Yet another specialized enzyme, ***primase***, is used to determine the starting point for each fragment. (It is interesting to think about the fact that these little enzyme task robots were made using the information encoded somewhere on the very DNA that they are duplicating!) Watching a few video animations of this whole process is very helpful.

After the DNA polymerase is done replicating the entire strand of DNA (millions of base pairs long) all the enzymes disengage and you've got two copies of the original. Does DNA polymerase ever make mistakes? Yes, all the time. There are specialized "spell checking" robots that go along checking the work that the polymerase did, and trying to fix mistakes. The process is amazingly accurate given how difficult the task is and how fast it has to occur. However, little mistakes are introduced every time this process happens. These tiny mistakes add up over a lifetime and in the end contribute to the aging process. Mistakes in the instructions for making organelles cause them to work less efficiently. A mistake in a very critical place could be the beginning of a degenerative disease. (Just think-- a mistake could occur in the DNA that codes for the spell checker enzymes themselves!)

Mitosis is the process where body cells (somatic cells) duplicate themselves in order to make new cells. Don't confuse mitosis with meiosis (the process that produces gametes) or with binary fission (how prokaryotes split in half).

| A CENTROSOME is made of two CENTRIOLES surrounded by a blob of protein goo. | The centrosomes form a spindle of microtubules. The pairs of chromosomes are lined up in the middle of the spindle. The microtubules will then contract and pull the pairs apart. |

THE CELL CYCLE

DNA REPLICATION

INTERPHASE is when cell is not actively in mitosis.

Gap 0 = cell is resting
Gap 1 = organelles duplicated
S phase = DNA replicated
Gap2 = enzymes are made

PROPHASE

Chromatin condenses into chromosomes, and nuclear envelope begins to dissolve.

- - - - - - - - - - - - - - - - - -

METAPHASE

Chromosomes line up in the middle.

(Think META---MIDDLE)

- - - - - - - - - - - - - - - - - -

ANAPHASE

Chromosomes pulled apart

(ANA = opposite)
(Or visualize letter A)

- - - - - - - - - - - - - - - - - -

TELOPHASE

Chromosomes are far away and nuclear envelope begins to form again.

(TELO = far; "telescope")

CYTOKINESIS
cell   movement

The cell splits (cleaves) in half.

24: EPIGENETIC MECHANISMS

A zygote is a cell that can become any type of cell. The DNA in the nucleus contains every bit of information that every type of cell will need for the entire lifetime of the organism. A fairly large portion of the DNA contains instructions for embryonic development. Once the body has fully formed, this information is no longer necessary and will be locked away permanently. Also, bodies change and grow over time. When children begin to turn into adults, their bodies will begin to make proteins they never made before.

The nucleus has a way to control what sections of DNA can be opened at what times. The opening and closing of DNA is called *epigenetics*. "Epi" means "outside of" (or "on top of") so epigenetics is on the outside of the genes. A gene is a piece of DNA that codes for one thing. The traditional understanding of a gene was a continuous strip of DNA, running along like a section from a bead necklace. New research has suggested that it is more complicated than this. A gene, even though it codes for one thing, (such as a particular protein), can be encoded as smaller sequences located in different places, not one continuous piece all in one place. The polymerase machinery has to copy the small sections then splice them together to make the mRNA. Or, a gene might be encoded in a continuous strip but using only every other base pair, or every third base pair. Genes have turned out to be a lot more complicated than anyone had imagined. However, it is okay to keep it simple for now and think of a gene as one strip, like a section snipped from a necklace.

The most permanent way to close DNA is *methylation*. The *methyl* group, CH_3, can be used like clip. The methyl clip fastens to cytosines that are almost opposite each other. We have to say "almost opposite" because, of course, cytosines can't be exactly opposite because they always match up with guanine across from them. However, if you have a C-G next to a G-C, the cytosines are close enough that a methyl clip can be placed on the cytosines. This clip prevents those polymerase machines from using the DNA to create RNA. If the polymerase "sled" tries to ride along the DNA, it gets stuck on these clips and has to stop. If the polymerase can't ride along that section of DNA, no mRNA will be made. If no mRNA can be made, no proteins will be made, and the gene has effectively been silenced.

These methyl clips are put on by special enzymes designed to do that job. They have to do this every time the DNA is replicated (during mitosis, for example). DNA replication is complicated enough, but all those little methyl tags have to be added, too!

Another way to prevent gene transcription is *histone modification*. A *histone* is a protein gadget that acts like a spool. DNA is very long and thin and there is a lot of it in the nucleus. It might get hopelessly tangled if it were not for the ingenious way it is wound onto "spools" and then wound again into long chains. Eight "balls" called histones stick together to make one spool. Often, the entire spool is mistakenly called a histone. The spool is properly called a *nucleosome*, but this word seems to be harder to remember than the word histone. Therefore, many people say "histone" when they mean "nucleosome." (We can say "histone spool" and we'll know we are talking about the whole spool, not just one histone.)

The histone spool (nucleosome) has DNA wound around it—145 to 147 base pairs in length, to be exact. There is also a protein gadget that acts like a clip to keep the DNA tightly wound. The clip (or "binding histone") is called H1.

When the DNA is tightly wound, the polymerases can't get in to read the DNA. The histones must allow the DNA to loosen and relax and come unwound a bit. There are little switches (not shown in these pictures, but shown in your drawing) that control the tightening and loosening. The switches look like strings, and can be activated by three different molecules:

1) methyl (CH_3)   2) acetyl ($COCH_3$)   3) phosphate (PO_4)

Acetyl tags make the spools stay open and let the DNA be transcribed. When a gene is actively being used, we say it is being *expressed*. ("Gene expression" is a term you should know.) If the histones are "deacetylated" (acetyls are taken off) then we say that the gene has been "turned off." Methyls and phosphates work in a similar way.

Cells have a "Plan B" for stopping genes from being expressed. Even if the DNA is open and mRNA has already been made from the DNA, it is still not too late to prevent proteins from being made. The cell can manufacture a piece of *microRNA*, or *miRNA*, that will prevent that mRNA from being used. A piece of miRNA is very small, usually only 20 or 22 nucleotides long. It will match one place on the mRNA. The miRNA locks on to that place and sticks there permanently. If the mRNA tries to slide through a ribosome, the miRNA will stop it from doing so. It physically prevents the mRNA from sliding through. The mRNA becomes useless and is eventually chopped up and recycled. Thus, the gene is effectively silenced.

The zygote is a human cell but it is not any particular cell. To become a specific type of cell, such as a skin cell or a muscle cell, all the non-skin or non-muscle DNA must be permanently zippered shut. There are three main ways that DNA can be silenced.

---

1) DNA methylation

Methyl:

This is the most permanent form of locking away information.
Enzymes put methyl tags (CH_3) on cytosines in the areas that are to be locked.

---

2) Histone modification

The histone spools on which DNA is wound can control whether a gene is expressed or not.
("Expressed" means that the information is being used and proteins are being made.)

---

3) Micro RNAs (miRNA)

Micro RNAs are non-coding RNAs whose sole purpose is to mess up RNA. When a miRNA attaches, that portion of the RNA becomes unusable. Thus, gene expression is blocked.

25: EMBRYOLOGY: WEEK 1

The zygote cell is ***totipotent***. ("Toti" means "totally" and "potent" means "powerful or capable.") This cell is capable of becoming any type of human cell, even supporting cells such as the placenta or the amniotic sac. All of its DNA is open; nothing is methylated or closed. It takes an entire day for the zygote to go through mitosis. The two nuclei of the gametes must fuse together and function as one complete cell. This first division is called ***cleavage***. ("Cleave" means "split.") The two resulting cells are joining together by ***gap junctions***. A junction is a place where things are joined. Gap junctions are protein gadgets that might remind you a bit of a sewing bobbin. There are two plates with a tube in the middle. The hollow tube allows the cytosol from the cells to flow back and forth. This allows the cells to communicate. Chemicals made by one cell can flow into another, so the cells "know" what their neighbors are doing. We will see gap junctions again when we look at cardiac (heart) cells. The gap junctions allow the cells in the heart to all beat together in unison.

On day 2 the cells go through mitosis again, making 4 cells. The zona pellucida is still there, so the cells can't get larger. These four cells don't take up any more space than the one zygote did. The cells divide again, making a clump of 8 cells. This turns out to be a critical stage in development and some pre-embryos don't make it past this stage. We know this because of research done during "in vitro" fertilization where eggs and sperm are mixed in a test tube and researchers can watch this whole process happen right before their eyes. They've noticed that some pre-embryos don't develop beyond this point. They'd like to know why this happens so that they can make good choices about which pre-embryos to select for putting into the mother's uterus. But as of the writing of this text, they still have not figured it out.

When the cells divide again, there are enough of them now to make a little clump that looks a bit like a mulberry or raspberry. Since the Greek word for mulberry is "morus" the stage was named the ***morula*** stage. By the time the morula has 32 cells, the ones at the very center might be having a hard time getting enough oxygen. The cells are receiving oxygen (and getting rid of carbon dioxide) by simple diffusion. The cells on the outer edge are getting plenty of oxygen. The ones in the center are in danger of not getting enough. (No cell in our bodies is more than a very short distance from a capillary. That's how many capillaries we have!) Since there are no capillaries yet, this ball of cells must use another strategy. What happens is that it becomes hollow.

The next stage is called the ***blastula*** or ***blastocyst***. ("Blast" means "bud," those things that plants make that turn into flowers and leaves. In this case, the bud will turn into a human being.) The ball becomes hollow and filled with fluid (mostly water). The fluid can flow around and carry oxygen to all the cells.

On approximately day 6 (some charts say day 5) the blastocyst will start to grow a little lump called the ***inner cell mass***. The cells that stay in the outer ring will become a supporting structure called the chorion. These inner cells will differentiate into the placenta, several sacs, and the baby. These inner cells are ***pluripotent cells***. They have undergone a very small amount of differentiation (methylation) but almost all of their DNA is still open. They have many ("pluri") options still open. These are the cells that stem cell researchers like to harvest. ("Stem cells" are cells that have not completely differentiated into a type of body cell. More on stem cells in a future lesson.) Of course, doing this on a human pre-embryo is very controversial. Fortunately, however, the pre-embryos of other mammals (and even some invertebrates such as the sea urchin) can be used for embryology research. In fact, most of what we know about embryology comes from research on non-human embryos.

On day 6 (some charts might say 5 or 7), the embryo finally "hatches" out of the zona pellucida. The cells of the blastula secrete enzymes that soften the zona pellucida, then the blastula has a growth spurt and gets larger very quickly, popping the z.p. The blastula is now free to start growing larger.

It is interesting to note that this first week is pretty much the same for all mammals. This is amazing when you think that a mouse has a gestation period of only 3 weeks. That leaves only 2 weeks to go from the blastula to a baby mouse that has most of the same inner organs that we do! Also, it is interesting to note that some animals (including bears, kangaroos and roe deer) are able to pause gestation at this point and hold the embryo for up to several months until the right season of the year comes around.

Finally, we need to know where this is happening. This whole process takes place in the tube that connects the ovary to the uterus. Ovulation is when an ovum leaves the ovary and gets picked up by the "fingers" of the ***fallopian tube***. The ovum is fertilized by sperm soon after it enters the tube. The sperm must swim all the way up the tube so that only the strongest sperm will survive. This whole first week of development happens as the zygote travels down the fallopian tube towards the uterus. Tiny hairs called ***cilia*** beat in the direction of the uterus, pushing the pre-embryo along. If the cilia are not able to do this fast enough and the pre-embryo gets stuck in the tube, this is called an ***ectopic pregnancy***. ("Ecto" means "outside," in this case meaning outside of the uterus.) This condition is very painful and is life-threatening for the mother. A surgeon must remove the pre-embryo from the tube. Most of the time, the pre-embryo travels down the tube just fine, and enters the uterus on day 5 or 6. It floats around for a while until it bumps into the lining of the uterus and sticks there. When it sticks to the lining of the uterus, it will implant there and begin a new stage of development. If the pre-embryo does not stick anywhere, it will be flushed out of the body when the woman begins to menstruate. (Menstruation is when the temporarily thickened lining dissolves and drains out.) A woman (or any mammal) is not technically pregnant until the embryo implants into the wall of the uterus.

The zygote is a TOTIPOTENT cell. ("Toti" means "totally" and "potent" means "powerful or capable.") In what sense is this cell totally powerful? It can turn into ANY type of human cell, even supporting cells such as the placenta and amniotic sac. All the DNA in this cell is open and accessible. None of it is methylated or closed in any way. As the embryo develops, the cells will become less "potent" and will have much of their DNA closed.

DAY 1   DAY 2

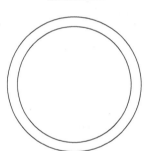 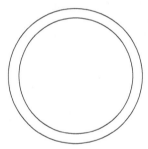

Embryonic cells stick to each other with GAP JUNCTIONS.

The zygote takes an entire day to make the first division.
This split is called _____.

DAY 2.5   DAY 3   DAY 4   DAY 5

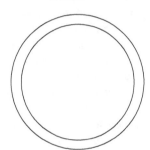

Cells are getting smaller while overall size is staying the same.

This is a critical stage for unknown reasons. Some embryos don't make it past this stage.

DAY 6   DAY 6 or 7

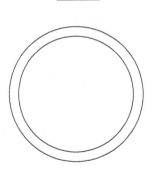

These cells are _____ and are often the ones harvested for use in embryonic stem cell research.

Blastocyst secretes enzymes that soften the zona pellucida, then it enlarges suddenly and breaks free.

Map of where this is happening:

AMAZING FACT: The first week is the same for ALL placental mammals, regardless of how long the gestation period is. (mice: 3 wks, elephants: 2 yrs)
SECOND AMAZING FACT: Some mammals can pause pregnancy at this stage and hold the embryo for several months, waiting for the right season.

26: EMBRYOLOGY: WEEK 2

At the start of week 2, the pre-embryo implants into the **endometrium** (lining of the uterus). Once it implants, it will officially graduate to being called an embryo. The cells of the embryo need an outside source of energy by this point. They have used up most of the energy that was stored by the ovum. (Remember, the ovum had a rich supply of fats and sugar in the cytosol. Also, remember how large the ovum was. The ovum would not have had enough energy to produce hundreds of cells the same size as itself, but since these embryonic cells kept getting smaller and smaller, there was enough energy to go around.)

The embryo needs to tap into the mother's supply of nutrients. Food and oxygen are brought into the uterus by tiny capillaries in the endometrium. (The endometrium also contains glands that produce a temporary supply of nutrients for the embryo.) The mother's capillaries come very close to the embryo, but the mother's blood will never actually touch the embryo. This is good because once the embryo develops a blood supply of its own, it might not be the same blood type as the mother's. Different types of blood (A, B, AB, O and Rh factor) can have bad reactions when mixed. We will learn about blood types in a future lesson.

The side of the blastocyst that touches the endometrium starts growing cells that look like they have little fingers, called **villi**. These villi quickly begin growing into the endometrium. We call this process **implantation**. (If the blastocyst was a plant, the villi would be the roots.) Scientists consider implantation to be the official start of pregnancy. And to celebrate, we can stop saying "pre-embryo" and begin using the term "embryo."

The outside "shell" of the embryo is now called the **chorion**. This word comes from Greek and means "the outer membrane that covers a baby." Since villi are structures we find in other cells of the body, we should really call these villi the **chorionic villi**. The place where the chorionic villi come very close to the mother's capillaries will grow to become the **placenta**. It is important to note that the placenta is a combined structure: one half is from the mother and one half is from the baby.

The chorionic villi began secreting digestive enzymes that eat away at the endometrial cells around them. They dissolve a little hole for the embryo to snuggle into, and the endometrium eventually closes in around it. This provides extra security for the tiny embryo. The mother can jog and dance and bounce up and down all she wants to and the embryo is safe.

The chorionic villi have yet another important function. These cells secrete a hormone called **HCG** (human chorionic gonadotropin), which acts as a chemical messenger. HCG goes to the ovaries and tells the ovarian cells to secrete a hormone that will prevent the endometrium from being flushed out as it usually is every month. (This hormone is called **progesterone**. "Pro" means "for" and "gester" means "gestation." Progesterone promotes pregnancy by keeping the endometrium alive and well.) Most pregnancy tests are testing for the presence of HCG in the mother's blood or urine. If the level of HCG is not quite high enough at the time the mother does a pregnancy test, the result might be a "false negative." Pregnancy tests rarely give a "false positive" because if HCG is detected, it most certainly means that there are little chorionic villi producing it. (It's interesting to think that since all cells in the body have identical DNA, then every cell in the body has the information needed to produced HCG. However, as we learned, most information is permanently locked away by methylation. The information needed to make HCG is only opened by these cells at this particular time. DNA has timing mechanisms built into it!)

Now the inner cell mass begins to change. It organizes into two layers called the **epiblast** and the **hypoblast**. (This is called the **bilaminar disc**.) A cavity begins to grow next to the epiblast and becomes the **amniotic cavity**, which will eventually give rise to the amniotic sac. On the other side of the hypoblast, another cavity forms, which we call the **yolk sac**. Obviously, since newborn babies don't have yolk sacs, this structure will be temporary.

A big change comes when the 2-layer disc turns into a 3-layer disc. This process is not shown in the drawing, but is discussed in the video. The cells in the middle of the epiblast begin dividing quickly and then moving (yes, moving) into the space between the two layers, creating a third middle layer.

A space also begins to open up just inside the outer layer, creating a cavity all around what used to be the blastocyst. We call this the **chorionic cavity**. This is also a temporary structure that will not be seen by the time the baby is born.

The three layers are often called the **germ layers**. ("Germ," in this sense, means "seed." These are the little "seeds" that will grow into a complete human body.) These layers are like a little stack of pancakes. The top pancake is called the **ectoderm** and it will grow into the outer layer of skin (epidermis), hair, nails, and the entire nervous system including the brain. The middle layer is called the **mesoderm** and it will become muscles, bones, connective tissue, blood, kidneys, bladder, gonads, heart, lymph system, spleen, and the bottom layer of skin (dermis). The bottom layer is the **endoderm** and it will become the inner lining of the digestive tract, the lining of the lungs, the liver and pancreas, and many glands including the thyroid and thymus. (We'll meet the thymus when we study white blood cells in lesson 44.) The formation of the 3-layer germ disc is often referred to as **gastrulation**.

In the middle of the mesoderm there is a little dot that represents something called the **notochord**. We will learn more about the notochord in the next lesson.

Cancers are often classified as sarcomas or carcinomas. **Sarcomas** occur in body parts that can be traced back to the mesoderm. Most childhood cancers fall into the sarcoma category. **Carcinomas** are cancers in tissues that came from the ectoderm or the endoderm. Adults are more likely to get carcinomas. Tumors and other cancers are caused by cells that don't know when to stop dividing. Cancer cells keep dividing and dividing and dividing, and body parts begin to malfunction as a result.

Week 2 begins with IMPLANTATION. The embryo attaches itself to the wall of the mother's uterus.

DAY 7 - 8

By the end of week 1, the embryonic cells have used most the energy that the original egg cell had stored up. To get more energy, the embryo will have to tap into the mother's blood supply and use food she has digested.

DAY 8-9

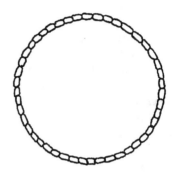

The chorionic villi begin to secrete _____, a hormone messenger molecule that will tell the mother's body that an embryo is present.

DAYS 10 - 14

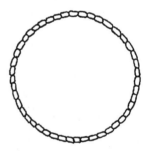

By the end of week 2, there are 3 GERM LAYERS: ("germ" means "seed")

1) _____ will become _____

2) _____ will become _____

_____

3) _____ will become _____

SIDE NOTE: After all the body parts are fully developed, the cells are programmed to stop dividing so fast. "Rogue" cells that continue to divide can cause childhood cancers.

Cancers that arise from the *mesoderm* are called _____.
    Children and teenagers are more likely to get this kind.
Cancers that arise from *ectoderm and endoderm* are called _____.
    This type of cancer is more common in adults.

27: EMBRYOLOGY: WEEK 3

In this drawing, we will focus only on the 3-layer "germ" disc. The disc is still inside the chorion (formerly blastocyst) but it will be too complicated to draw the chorion in all of these drawings. This is typical of embryology drawings, though. They often show just the 3 layers developing. So your drawings will match many of those you might see in a text or on the Internet.

First, let's take a took at the disc from the top. The ectoderm (in blue) is defined as the top layer and the endoderm is then the bottom layer. The bottom layer will eventually curl around to be inside. If the germ disc was already a person, it would be lying on its stomach. The blue layer would be the back (*dorsal* side), and the yellow endoderm would be the stomach (*ventral*) side. Dorsal and ventral are good words to know when studying anatomy. (The word *dorsal* comes from the Latin "dorsum" meaning "back," and the word *ventral* comes from the Latin word "venter" meaning "belly.") Not only do we need top and bottom, we also need to determine left and right, and head and feet.

A little streak begins to form at one end of the disc. This is called the *primitive streak* and it is the very beginning of determining the general body plan. (Sometimes they say this already begins while the disc is still in the 2-stage layer.) At the top of the streak is a little place called the *primitive pit* and above that is a place called the *primitive node*. The side with the primitive streak will be the side that eventually turns into the legs and feet. Above the primitive steak is the end that will become the head. And since we now have head and feet, back and stomach, we can imagine the baby lying on its stomach with its head toward the top of the page, so the left and right sides of the body correspond to the left and right sides of the paper. We will now imagine that we have cut a cross section through the disc and we will look at the inside.

(1) This picture shows the three layers with the *notochord* dot in the middle of the mesoderm. The notochord area will be the "foreman" of the body building process and will tell the other cells where to go and what to become. The notochord secretes a protein called *SHH* that acts as a chemical messenger telling the other cells what to do. *SHH* stands for *Sonic HedgeHog*. Yes, the cartoon character. When this protein was first discovered, they collected some of it and then put it into a fruit fly embryo to see what would happen. The fly ended up being covered with long spikes, like a hedgehog. The cartoon character called Sonic the Hedgehog happened to be popular at that time, so the scientists couldn't resist using the name. Then they found other similar proteins, and classified all of them as "hedgehog" proteins. There's Desert HH, and Indian HH, named after species of real hedgehogs.

(2) The SHH tells cells in the primitive streak to form a groove (the *primitive groove*). This is often called the *neural groove* because it will become the *neural tube*.

(3) The neural groove deepens and begins rolling into a tube. The process of forming the neural tube is called *neurulation*. (So we've got blastulation, then gastrulation, then neurulation.)

(4) The *neural tube* separates from the top layer so that it is embedded in the mesoderm. The cells that were in the top of the fold, (the "crest" of the fold) now migrate down toward the neural tube and are called the *neural crest cells*. They will develop into various type of cells, but mainly they will become nerves and nerve bundles (*ganglia*) in places such as between the spinal vertebrae, in the digestive system, around the heart, and in the head and face. Some of these cells will turn into connective tissues and bones in and around these nerve centers, including the tiny bones of the inner ear. Oddly enough, a few neural crest cells will decide not to become nerves at all and will become *melanocytes*, pigment-producing cells in the bottom layer of the skin. A *neural plate* now becomes visible on the top of the ectoderm. This is the area where the spine forms.

If the neural tube does not close up all the way, but remains open at some point, (like in picture 3), various birth defects will occur. The most well-known of these defects is *spina bifida*. Part of the soft spinal cord never gets covered with protective bone, so the baby is born with what looks like a soft lump on its back. Corrective surgery can be done to try to correct this problem, but the child can still have lasting health issues due to this defect. We know that a lack of *folic acid* (in the B vitamin family) can cause this birth defect. Pregnant women are encouraged to take vitamin supplements that have a high level of folic acid.

(5) Bumps called *somites* begin to appear and the neural plate continues to grow. The somites have areas that will differentiate into the dermis, muscle, and bone of the spine and rib cage. We also see the beginnings of the heart tubes. The heart begins as two tubes that eventually join together. The ends of these tubes will become the large blood vessels on the top and bottom of the heart.

(6) The ends of the mesoderm begin to split, with the top part destined to become an outer "bag" around the organs (made of connective tissue and called the peritoneum), and also appendages (arms and legs). The bottom part becomes all the connective tissues "bags" that will surround the organs in the body cavity.

(7) The disc begins to curl (*embryonic folding*), with the yellow endoderm layer forming the *gut tube* (which will turn into the inner layer of the digestive system). The inside of the yolk sac begins to grow little clumps of cells called blood islands. Each island will turn into a piece of blood vessel, complete with blood cells already inside them. Then the tiny pieces will join up to form a network. The network will grow and then spread into the embryo, joining with other blood vessels that are starting to develop in and around the heart and kidneys. The yolk sac will eventually shrink and disappear.

NOTE: In this drawing we will focus on just the 3-layer germ disc. The disc is still inside the chorion, but we won't see the chorion in this drawing. We will see it again in drawing 28.

TOP VIEW OF 3-LAYER DISC:

(1) Notochord and primitive streak
    Notochord secretes SHH protein.

(2) Neural groove forms

(3) Neural tube begins

(4) Neural tube will become brain and
    spinal cord. Neural crests will be
    peripheral nervous system and more.

(5) Mesoderm begins to differentiate
    and heart tubes appear

Here we see the 3-layer disc with amnion and yolk sac.

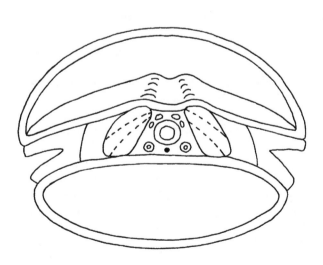

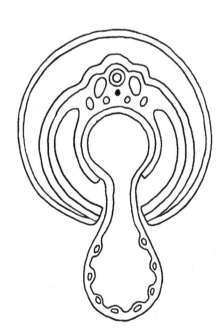

(6) "Somites" appear. (Future spine and associated muscles)
    Mesoderm continues to differentiate and the ends split.

(7) The embryo curls, forming the gut tube and the
    amniotic sac. Blood islands appear in the yolk sac
    and will become complete little sections of vessel,
    with red blood cells already inside.

28: EMBRYOLOGY: WEEK 4

We have two different types of cross sections in this drawing. A *saggital cross section* is a lengthwise view, as if we are cutting the body down the midline from the head to feet, separating it into a left and right side. A *transverse cross section* is like the infamous "magician saws a woman in half" trick. (Or like chopping a carrot on a chopping board.) It is what we usually think of when we talk about cutting something in half. A transverse section of the body could be at any point--head, neck, upper chest, abdomen, or even a leg-- so we need to specify where it is. In our saggital view you'll see a dashed line. This is where our cut is for the transverse section. We will only see what is along that line. Therefore, in our transverse view we will not see the head or the brain or the lower (posterior) region.

In the 21 DAY saggital cross section we again see the embryo in the lining of the uterus. As soon as there is enough of an umbilical cord to hold it securely, the embryo will tear free of the lining and simply be in the uterus itself. The maternal capillaries create a "pool" of blood around the chorion, so that the chorionic villi are surrounded by the maternal blood supply. The baby's blood and the mother's blood will never actually touch, as indicated in this picture. Notice that the embryo is beginning to curl in the head-to-foot direction. This is a very simple picture of the embryo and does not show the somites, the neural tube, the notochord, etc. The main point of this diagram is to show how the embryo is still embedded in the endometrium and its placement inside the chorion and the amnion.

We do see the three germ layers here, and we see that the endoderm has turned into the gut tube, the yolk sac, and a new feature called the *allantois* (al-an-TOE-is). (This word is Greek for "sausage," which the allantois was thought to resemble.) The allantois (number 7) acts as a garbage bag, collecting waste products produced by the cells. When the embryo has its excretory systems up and running (kidneys, bladder, liver) and when the placenta is fully formed, the allantois will no longer be necessary. (However, a tiny remnant of it will remain as a piece of fibrous tissue sitting on top of the mature bladder.)

The 21 DAY transverse cross section is an imaginary cut along the dashed line. The main focus of this diagram is the rearrangement of cells in the three layers, especially in the mesoderm. We see that the somites have differentiated into muscle, dermis and bone tissues. Part of the somites turned into what will become vertebrae. We can't properly call these new vertebrae "bones" because they have not turned into bone yet. They are still made of soft cartilage. As the months go by, cartilage will begin absorbing minerals and turning into hard bone. The middle part of the somites has turned into muscle tissue. These will be the muscles of your back, neck and stomach areas. Lastly, the outer parts of the somites have turned into dermis, the bottom layer of skin.

The ectoderm (blue) has now fully separated into epidermis (the top layer of skin) and the nervous system. The epidermis is now covering the whole embryo, and is joining with the dermis. The neural crest cells (those blue blobs above the neural tube) have turned into nerves that connect the muscles to the spinal cord. The neural tube is in the process of turning into a spinal cord. The notochord is still present, too.

The mesoderm is busy creating kidneys, blood vessels and the heart. Here the heart still looks like two tubes. The tubes will join together in the next few days and begin growing into a heart. Though we don't show it in this picture, a lot of what is going on in the mesoderm areas we colored light red is formation of blood vessels that will connect all the organs. The ends of the mesoderm that we saw splitting in the last lesson are now turning into the *visceral peritoneum* and the *parietal peritoneum*. ("Viscus" was the Greek word for any internal organ. The plural, "viscera" was to the Greeks like saying "guts.") The visceral peritoneum will become all the "bags" that surround each organ. The parietal peritoneum will become the outer bag, which can also be thought of as the inner lining of your skin. (Think of a formal jacket with a silky lining. The fabric of the jacket is your skin and the lining is the parietal peritoneum.) These membranes keep all the organs in place so they don't slosh around, and they also provide a surface to which blood vessels and nerves can be attached. The fact that there are multiple membranes (bags within bags) allows for movement and flexibility of this support system.

The yolk sac has a thin layer of mesoderm around it. The mesoderm produces the blood islands that then produce many tiny sections of blood vessel. The tiny sections join together to become a complicated network of capillaries.

The 26-28 DAY illustration shows the blood vessel development that is going on during week 4. If all the details of embryology were shown in one drawing, it would be too confusing, so here we focus on just the developing vascular system. In this picture, the heart tubes have fused together and are starting to grow into a heart. We don't have any chambers yet, just a bulging tube. However, even at this early stage, the cells begin to contract together in rhythm, the way they will for the rest of this embryo's life. There are some very large vessels above the heart, the *dorsal aorta* and the *cardinal vein*. These will change and grow a lot over the next few weeks. We don't have any lungs yet, so the arrangement of these vessels is temporary. As the vascular system develops, some vessels will grow and enlarge while others will shrink and disappear. Very often, diagrams of the vascular system make the vessels going away from the heart red ("arteries") and vessels going toward the heart blue ("veins"). In the chorionic villi and in the yolk sac, the vessels change from going to/from the heart. Often this can indicate a chemical change, such as picking up oxygen or getting rid of carbon dioxide. In the yolk sac we also have nutrients going into the blood. The yolk sac provides a small amount of nutrition while the placenta is still forming. (The yolk sac will never grow larger than the size of a pea and will eventually be discarded, often getting trapped between the amnion and the chorion.)

Notice that the actual sizes of the embryos on this page are 1-2 millimeters. That's small!

21 DAYS SAGGITAL CROSS SECTION

21 DAYS TRANSVERSE CROSS SECTION

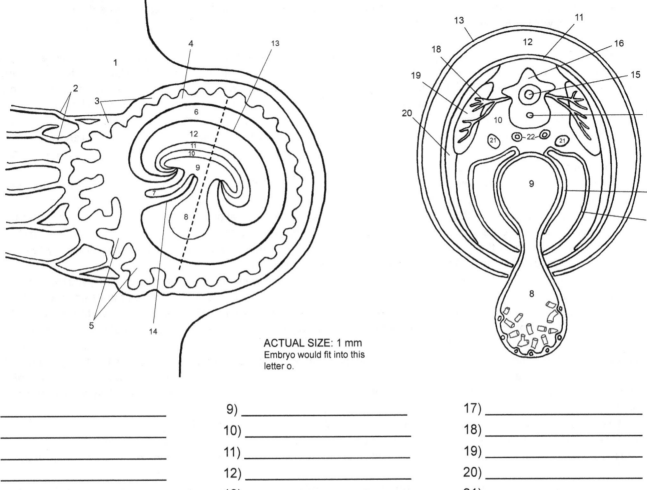

ACTUAL SIZE: 1 mm
Embryo would fit into this
letter o.

1) _____	9) _____	17) _____
2) _____	10) _____	18) _____
3) _____	11) _____	19) _____
4) _____	12) _____	20) _____
5) _____	13) _____	21) _____
6) _____	14) _____	22) _____
7) _____	15) _____	23) _____
8) _____	16) _____	24) _____

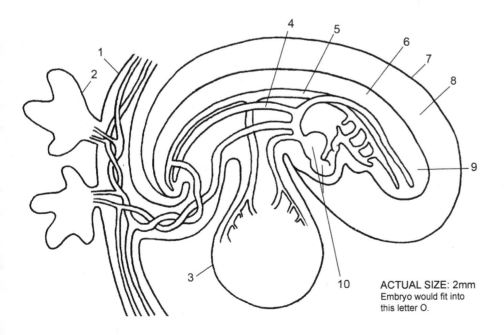

ACTUAL SIZE: 2mm
Embryo would fit into
this letter O.

26-28 DAYS
SAGGITAL CROSS SECTION
SHOWING VASCULARIZATION

1) _____
2) _____
3) _____
4) _____
5) _____
6) _____
7) _____
8) _____
9) _____
10) _____

29: EMBRYOLOGY: WEEK 5

We have two very different types of drawings here. The first one is very symbolic, with blue, red and yellow representing our three layers and what they are turning into. The second drawing will be more realistic, showing the three-dimensional shape of the embryo, similar to what you would see in a photograph.

We have two blue layers in this first drawing: the epidermis and the neural tube. The neural tube can't really be called by this name anymore, however. It is turning into the central nervous system and four distinct areas can be labeled: the spinal cord, the *forebrain*, *midbrain* and *hindbrain*. The forebrain will turn into the *cerebrum*, which is the gray wrinkly part we envision when we think of a brain. The cerebrum is the part that does "thinking." The midbrain will turn into many inner parts, such as the thalamus, the hypothalamus, and the hippocampus. These middle parts are important for regulating things such as appetite, body temperature, and blood pressure, and also for storing memories. The hindbrain will become the lower parts such as the cerebellum and the brain stem. These control basic function such as balance, reflexes, breathing, and the waking and sleeping cycles. The lower part of the neural tube turns into the spinal cord.

The endoderm has differentiated into areas that will become the digestive system and the lungs. Right where the throat branches off into both the esophagus leading to the stomach and the trachea leading to the lungs, there will be a flap called the *epiglottis* that can close over the tube that leads to the lungs while food is being swallowed. The endoderm also contributes to organs and glands that are associate with the digestive system such as the liver and pancreas. In the throat area the endoderm will contribute to the structure of the thyroid and other glands. (NOTE: The endoderm alone does not form all these systems. It becomes the inside layer of all of these. The mesoderm will form the outer part.)

The opening at the end of the gut tube is called the *cloacal* (*klo-AK-al*) *opening*. If you've studied birds, you might recognize the word *cloaca*. A cloaca is an opening that expels both urine and feces. Bird droppings are a mixture of urine and feces. The cloacal opening in a human embryo is a temporary structure and will eventually change into the anal opening. Right now the embryo is not expelling either urine or feces, and the wastes are being collected partially by the allantois. Some molecular wastes are also beginning to go out through the developing placenta. At this stage, both the mouth opening and the cloacal opening are not really open, but are covered with a membrane. In a few weeks the membranes will rupture and then openings will actually be open. The allantois will shrink and retreat to the inside of the embryo, helping to form the top part of the urinary bladder. (A remnant of the allantois will always remain on top of the bladder., but it will not have any function.) The yolk sac is also still visible and functioning at this stage.

We can also see the area where the heart is developing. Although not nearly fully formed yet, the cells are already beginning to contract in rhythm. Remember, these cells are joined by gap junctions so that they can communicate very well with each other. This diagram does not show the many blood vessels that have already formed. We saw them in the last drawing lesson.

The second drawing shows us a more realistic view of the embryo. We can see the bumpy somites, and we can see lumps that are the growing heart and liver. A tiny eye spot and ear spot are just barely visible. By the end of week 5 there are little bumps, or "buds," that are the beginnings of arms and legs. The end of the embryo is sometimes called the "tail bud," though it is not a tail. It will shrink as the legs grow. The amnion now looks more like a sac. Notice that the yolk sac is outside of the amniotic sac.

(NOTE: If you look at 5-6 week embryo pictures on the Internet, be aware that many of them will not show you the yolk sac, even though it is definitely still present at this stage.)

END OF WEEK 4 (28 DAYS)
SAGGITAL SECTION

1) _____
2) _____
3) _____
4) _____
5) _____
6) _____
7) _____
8) _____
9) _____
10) _____
11) _____
12) _____
13) _____
14) _____
15) _____
16) _____
17) _____
18) _____

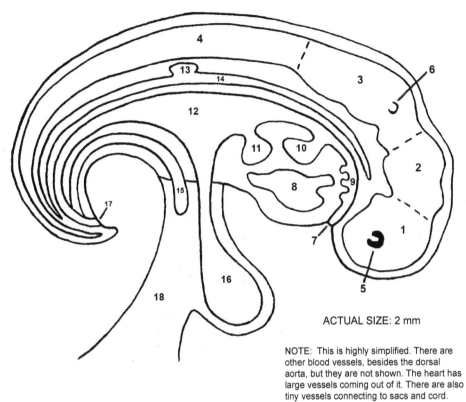

ACTUAL SIZE: 2 mm

NOTE: This is highly simplified. There are other blood vessels, besides the dorsal aorta, but they are not shown. The heart has large vessels coming out of it. There are also tiny vessels connecting to sacs and cord.

END OF WEEK 5 (33-35 DAYS)   EXTERNAL VIEW

1) _____
2) _____
3) _____
4) _____
5) _____
6) _____
7) _____
8) _____
9) _____
10) _____
11) _____
12) _____

TWINS!

Visual	Description	How many amniotic sacs?	How many placentas?	Result
	2 eggs, 2 sperm			
	1 egg, 2 sperm			
	1 egg, 1 sperm Zygote splits during first cleavage			
	1 egg, 1 sperm Morula splits			
	1 egg, 1 sperm Inner cell mass splits			
	1 egg, 1 sperm Bi-layer disc splits			
	1 egg, 1 sperm Bi-layer disc almost splits but not completely			
	2 eggs, 2 sperm Zygotes merge together. (Resulting individual will have 2 genomes.)			
	1 egg, 1 sperm One of the pluripotent cells mutates, then all of its "daughter" cells will carry that mutation.			

30: TISSUE TYPES and EPITHELIAL TISSUE (part 1)

Now that we've finished our study of cells, we are ready to see how cells cooperate to form tissues. Tissues will then combine together to form organs. Organs are connected in various ways to make our major body systems such as the respiratory system or the digestive system. Finally, the systems are all interconnected so that they all function together as a unit— a whole body.

All cells in the bod, except egg and sperm cells, can be classified into one of four categories. Just four! With all the many different types of cells, you'd think there would be more categories. Admittedly, it's a stretch for a few of the classifications, especially blood cells. However, this classification system works well enough that it probably won't change any time soon.

The four main tissue types are: *epithelial, connective, muscle, and nervous*.

Epithelial tissue is used as a covering, both outside and inside. The top layer of skin, the *epidermis*, is epithelial tissue, and so is the lining of the lungs and the digestive tract. Epithelial tissue is designed to be replaced often so it can take the wear and tear of touching and rubbing against things. It will also absorb things from the environment, both nutrients and toxins.

Connective tissue is used to bind and support body parts. By definition, connective tissue occurs in a *matrix*. This means the cells are in some kind of solid, liquid, or gel. Connective tissue includes ligaments, tendons and cartilage, but it also includes bone, blood and fat cells. You might be able to guess that bone will be in a solid matrix and blood will be in a liquid matrix.

Muscle tissue is designed for movement. Protein cables called actin and myosin will work together to cause cell fibers to slide past each other. Muscles cells are very strange, forming long "megacells" with hundreds of nuclei. There are three types of muscles tissue: *voluntary* (the ones you can move), *involuntary* (the ones you have no control over, like in your intestines) and *cardiac* (in the heart).

Nervous tissue is made of cells called *neurons*. This type of tissue is designed for communication and uses electricity to pass signals from cell to cell. Nervous tissue is found in the brain, the spinal cord and the peripheral nerves (nerves in the body).

Epithelial tissue is our exterior covering, both inside and out. ("Epi" means "on," and "thele" means "teat or nipple." So originally, back in the early 1700s, the word epithelial meant just the skin around the nipple area, which looks different from regular skin. Eventually, scientists realized that the two types of skin were basically the same and the word was adopted for all skin areas.) Epithelial cells come in three basic shapes: *squamous* (flat and wide), *cuboidal* (like a cube), and *columnar* (like a column). Each type of cell is very good in a certain application, as we will see in the newt lesson. A specialized type of columnar cells is the *goblet cell* that produces mucus.

All epithelial tissue is built on *basement membrane*, which is a layer of protein cables of different types. The connections found in the basement membrane are very mechanical— great examples of protein gadgets. The epithelial cells themselves are held together by adhesion junctions that use *desmosomes*. Desmosomes have *attachment plaques* on the inside of the plasma membrane so that the protein cables won't pop out of the membrane. The plaques help to spread out the pulling force over a larger surface area, thus reducing the pressure at any one point. (Remember, the consistency of the plasma membrane is similar to olive oil— soft and fluid.) The cables that go across between the plaques are actually half-cables coming out from each side. The half-cables meet in the middle, with a connection that might be thought of as biological Velcro®. (Cells can dissolve and rebuild desmosomes very quickly. We'll see this again in later lessons.) On the inside, the plaques are attached to the cytoskeleton that runs all over the inside. Thus, again, we see pulling and stretching forces being spread out over a large area to prevent tearing at any one point. NOTE: There might also be gap junctions present so that the cells can communicate.)

On the bottom, there are *hemidesmosomes* (half-desmosomes) that attach to the basement membrane. The basement membrane is not made of cells, but of protein cables. Mostly, these cables are made of a type of protein called *collagen*. We will take a look at the molecular structure of collagen in lesson 32. There are different kinds of collagens and they are known by Roman numerals. Some sources say that there are 16 types of collagen, but others will say up to 28 have been discovered. Here we see collagens I, III, IV and VII. The top layer of basement membrane is called the *basal lamina* and it has two layers: a top layer consisting of protein hooks and a bottom layer made of fibers of collagen. (These hooks are made of proteins called *laminins*.) Under the basal lamina there is a layer called the *reticular lamina*. (Remember, "rete" is Latin for "net," so when you see a word starting with "reti-" expect to see some kind of network.) The reticular lamina is made of more collagen fibers, and looping fibers woven throughout it, connecting up to the laminin hooks. The basal lamina and the reticular lamina together make the basement membrane.

Beneath the basement membrane, you will often find another layer of connective tissue. In the skin, it will be the dermis layer. Remember that skin is made of two layers, the epidermis and the dermis, with the basement membrane in between. Only the epidermis is made of epithelial cells. The basement membrane in skin is often where a blister will form. Those protein cables get broken as your skin rubs back and forth against something (such as a rake handle). The epidermis separates from the dermis and a space opens up between the two. The immune system senses damage and immediately fills the space with fluid that has many immune cells in it. As the blister heals, new epithelial cells and new basement membrane will grow and will replace the old, damaged ones.

There are many different types of cells in the body, but all of them can be classified into one of the following categories:

1) _____

_____

2) _____

_____

3) _____

4) _____

EPITHELIAL TISSUE (part 1)

All epithelial tissue is built on _____.

Cells are held together by _____, using _____.

31: EPITHELIAL TISSUE (part 2)

Epithelial tissues are classified using basically two characteristics: the shape of the cells and how deep they are stacked. Each type of tissue is just what is needed in certain areas of the body. Notice that all types of epithelial tissue are anchored to basement membrane.

The first category we will consider is the "simple" category. Simple means that there is only one layer of cells. **Simple squamous** tissue is a single layer of squamous cells. Squamous cells are very flat, so a single layer of these cells is perfect for places where you want to transfer gases and nutrients from one side of the cells to the other. Two places where you would find this type of tissue are the lining of the lungs (where you want to transfer oxygen and carbon dioxide) and the walls of capillaries (where you want to transfer oxygen, carbon dioxide, nutrients and wastes). Epithelium in blood vessels has a special name: **endothelium**. The endothelial cells <u>don't</u> form perfect water-tight bonds between themselves. These cracks between the cells will allow small things to leak through. This leakiness is helpful in many situations, as we will see in future lessons. However, the capillaries of the brain are different; they do not leak. This lack of leakiness is called the "blood-brain barrier." These tighter junctions in the brain capillaries is to make it very difficult (hopefully impossible) for bacteria, viruses and harmful large molecules to get into the brain.

Simple cuboidal tissue has one layer of cuboidal cells. Since these cells are thicker than squamous cells, you don't expect them to do much transferring of gases or nutrients. Cuboidal cells are usually specialized for secretion and absorption. Secretion means they make some kind of product, and absorption means they take something in. This type of tissue is found in glands and in the lining of the tiny tubules in the kidneys. Some simple cuboidal cells have microvilli. ("Villi" means "little fingers.") The purpose of the microvilli is to increase surface area. This is especially helpful to cells that are involved in absorption of some kind.

Simple columnar tissue often has goblet cells, microvilli, and/or cilia. Goblet cells produce mucus. Microvilli, as we've already stated, serve to increase the surface are of the cell. Cilia are tiny hair-like structures that can move. The cilia "beat" in rhythm producing a sweeping effect. This is most clearly seen inside the Fallopian tubes where the ovum must be swept along, down the tubes and toward the uterus. Simple columnar is also found along the inside of the digestive tract.

Stratified tissues have more than one layer. **Stratified squamous,** the type of tissue found in the **epidermis** of the skin and in the lining of the mouth, has many layers of cells, with the youngest cells at the bottom and the older cells at the top. The bottom layers, just above the basement membrane, keep making more cells all the time and these cells gradually move upwards.

There are two types of stratified squamous tissue: 1) **keratinized**, and 2) **non-keratinized**. **Keratin** is the name of one of the proteins found in the cytoskeleton. Remember, there are three sizes of cytoskeleton fibers: the large microtubules, the intermediate filaments, and the tiny microfilaments. Keratin is an intermediate filament. It is keratin fibers to which the desmosomes attach. In keratinized tissue, the cells begin to fill up with keratin as they rise to the surface. By the time they reach the surface, the cells have lost most or all of their organelles, including the nucleus, and they are basically an empty shell stuffed full of keratin. Since keratin is a waxy substance, the keratin-filled cells give the surface of the skin a fairly waterproof texture. In non-keratinized epithelium, the cells stay alive all the way to the top and do not fill up with keratin. The top cells still flake off easily, though, and are constantly being replaced, just like in keratinized epithelium. An example of non-keratinized epithelium is the inside of the mouth.

Stratified cuboidal tissue is found primarily in glands such as salivary glands and mammary (milk-producing) glands because this type of tissue is very good at secretion. It is usually only two cells thick, unlike stratified squamous.

Stratified columnar tissue is not as common as the other types of epithelial tissue. It is harder to find examples of this type, but it can be found in small amounts in the eye, the throat, the uterus, the urethra (tube leading out of urinary bladder) and the salivary glands. As with simple columnar tissue, stratified columnar is good for secretion, and it is also good for protection, as it is several layers thick.

Pseudostratified columnar tissue is a major feature of the lining of the **trachea** (pipe leading down into lungs). "Pseudo" means "false," so pseudostratified looks like it is stratified, but it isn't. In true stratified tissue, cells are stacked on top of cells so that the cells in the top layer are not touching the basement membrane. Pseudostratified isn't stratified because each and every cell in pseudostratified is touching the basement membrane, even if it does not look like it. (The fact that their nuclei are at various levels adds to the illusion that they are stratified.) We've got to trust the professional biologists who have examined these tissues under an electron microscope and can assure us that all the cells are touching the basement membrane. Most of the cells in pseudostratified are either goblet cells or are ciliated cells. The mucus and the cilia work together to be a kind of housekeeping service, sweeping dust and dirt up and away from the lungs. When the mucus gets to the top, you either swallow it or cough it up. (Those goblet cells really starting cranking out the mucus if you get some irritating particles down your airway. Soon after you start coughing, you find that suddenly there's a large volume of mucus to cough up!)

Transitional epithelial tissue is found in the urinary bladder. It is designed to stretch and then snap back to normal again. As the bladder fills up, tiny nerves sense the stretching that is going on and send signals to your brain that tell you that you need to empty your bladder soon.

Designed for gas and nutrient exchange because it is very thin. EX: lining of lungs, inside of blood vessels and capillaries

Specialized for secretion and absorption. Some have microvilli. EX: glands such as thyroid, pancreas. Also found in kidney tubules.

Is specialized for secretion and absorption, and also for pushing things along. Can have cila or microvilli.   EX: lining of stomach and intestines, Fallopian tubes

2 Types:

1) KERATINIZED: top layers hard and dead. EX: skin

2) NON-KERATINIZED: top layers soft and alive EX: inside of mouth

Designed for secretion,
EX: salivary glands, mammary glands
NOTE: usually just 2-3 layers

Secretes and protects.
Not very common
EX: eye, throat, uterus, urethra, salivary glands

Nuclei are at different levels.
EX: lining of trachea

Designed to stretch.
EX: Only found in one place: urinary bladder.

32: CONECTIVE TISSUE OVERVIEW, and COLLAGEN

The definition of a connective tissue isn't that it must connect something; some types of connective tissue do connect things. The definition of connective tissue comes from the microscopic view of these tissues. Connective tissues are made of three things: 1) specialized cells, 2) ground substance, and 3) protein fibers. In general, the **specialized cells** create the protein fibers and much of the ground substance. In other words, the specialized cells create the environment in which they live. The **ground substance** is the "background stuff" that everything is immersed in, and can be a solid, liquid or gel. **Protein fibers** can be **collagen, elastin, or reticular fibers**. Reticular fibers are basically very thin threads of collagen.

There are three types of connective tissue, with three subcategories under each. (The number three shows up a lot, as you will see!) **Fibrous** connective tissue is divided into **loose** (areolar), **dense**, and **adipose** (fat). **Cartilaginous** connective tissues are **hyaline, elastic or fibrocartilage**. The last category is a catch-all for everything else, and I have called it "other." These others include **bone, blood and lymph**. Don't worry that you don't know what all these tissues are. The important thing right now is to get this chart written so that you can come back to it later on, as we study each kind of connective tissue.

Let's start with a close-up look at **collagen**. The word collagen comes from the Greek word "kolla" which means "glue," and "gen," which means "to make." So collagen is...a glue-maker? Actually, this name makes sense if you know that for centuries, scrapings from animal hides were used to make glue, because skin has a lot of collagen in it. Collagen from animal hides is used today to make gelatin ("Jello") which is very gummy and sticky, like glue.

The smallest unit of collagen is a polypeptide (chain of amino acids) in an alpha helix shape. Three of these alpha helices are bound together to form the complete triple-helix collagen protein. (Every third amino acid in the chain is **glycine**, the smallest amino acid. Its small size helps the strands wind together very tightly.) The triple-helix proteins are assembled inside the endoplasmic reticulum of cells called fibroblasts. In order for the stapler enzyme inside the ER to be able to do its job and assemble these proteins, it needs a molecule of vitamin C (ascorbic acid) to attach to its active site. An extra molecule that is necessary for an enzyme to be able to do its job is called a **coenzyme** or **cofactor.** If there is a shortage of vitamin C, the body will not be able to produce good quality collagen, and a condition called **scurvy** will result. Many sailors died of scurvy in past centuries, before nutrition was recognized as a possible cause of disease.

The triple-helix collagen proteins exit the ER in vesicles and go over to a Golgi body where they are tagged with sugar tags that mean "put me outside the cell." The collagen proteins exit the Golgi and are taken (inside vesicles) to the plasma membrane where exocytosis occurs and they are put outside. Once outside the cell, collagen proteins begin to link up with each other and make larger strands. The first larger stand they make is called a **microfibril.** Then some microfibrils join together to make a **fibril**, and finally some fibrils are bound together to make a **fiber.** (Sometimes the fibers are bound into even bigger bundles.) This "bundles of bundles of bundles" organization is used in other tissue types, too, such as muscle. It is the ideal structure for achieving great strength.

Fibroblasts are cells that make collagen. Since they also secrete most of the things that form the ground substance, the fibroblasts really create their entire environment. The ground substance is made of 90 percent water 10 percent something called **proteoglycans**. The "proteo" part means "protein" and the "glycan" part means "sugar" (like "gluco"). A proteoglycan molecule looks a bit like a bottle brush. This picture shows many proteoglycans attached to a long "rope" molecule. The hairs of these brushes are long chains of sugars that are very hydrophilic. Thus, they attract lots of water molecules to the area. That is where the 90 percent water comes from. The proteoglycan molecules draw the water in and hold it there. The high water content of the dermal layer of skin (under the basement membrane) is partly what makes the skin so soft and resilient. (One of the first symptoms of serious dehydration is that the skin doesn't bounce back after you press it.) One of these brush molecules is called **chondroitin sulfate**, a very popular   nutritional supplement that you can find in any pharmacy or grocery store.

The little brushy things are the proteoglycans. The orange stripe in the middle is hyaluronic acid.

The long rope molecule that all the bottle brush molecules are attached to is called **hyaluronic acid.** (Hi-ah-lure-ON-ic) The ends of the hyaluronic acid molecules are fastened to the collagen fibers. In many parts of the body, the space around the cells is filled with these ropes and bottle brushes. This space, all around the cells, is often called the **extracellular matrix**. "Extra" means "outside" in this case. Other things can float around in the extracellular matrix, too. White blood cells are often found here.

Hyaluronic acid is sometimes taken as a nutritional supplement by people who are wishing to avoid problems with connective tissue. The theory is that by supplying plenty of hyaluronic acid, the connective tissues will take these molecules in and use them to build extracellular matrix, and thus stay strong and healthy. Cosmetic companies have found a way to make hyaluronic acid into a facial cream. Claims are made that using these creams will slow down or reverse wrinkling. Does it work? One researcher in Japan believes it does. He moved to Yuzurihara, a small town north of Tokyo, in order to study its population of exceptionally healthy and long-living people. The elderly in this town look youthful, rarely get sick, and often live to be over 100. The researcher concluded that besides having an excellent diet and lifestyle, these people also happen to have available an ideal food source for longevity: a species of potato that is very high in hyaluronic acid. A lifetime of eating a diet high in hyaluronic acid (and also lots of green vegetables) has made diseases of aging very rare in this population. Many elderly people in this village reach their 90s without ever having been in a hospital.

CONNECTIVE tissue is made of 3 things:
1) _____, 2) _____ and 3) _____

The _____ and the _____ make the _____.

The protein fibers can be made of 1) _____, 2) _____ or 3) _____.

There are three types of connective tissues, and several categories under each:

_____   _____   _____

1) _____   1) _____   1) _____

2) _____   2) _____   2) _____

3) _____   3) _____   3) _____

COLLAGEN is a protein cable made of three separate polypeptide chains (alpha helices).
Every third amino acid is glycine, the smallest amino, so that the triple helix can be wound very tightly.

One type of specialized cell is the **FIBROBLAST**, which makes collagen proteins and exports them (using exocytosis) outside the cell, where they then join together and make collagen fibers.

Fibroblasts also made the ground substance which is a mixture of water (90%) and glycoproteins (10%).

Fibroblasts live 2 to 3 months.
They multiply rapidly after an injury.
Scar tissue is a result of very active fibroblasts.

33: LOOSE CONNECTIVE TISSUE
(part 1 of FIBROUS)

Loose connective tissue is very abundant in the body. Skin is the most obvious example of where loose connective tissue is found, but you can also find it surrounding blood vessels, nerves and all organs. It is the connecting link between epithelial tissue and muscle tissue, and it lies just beneath the surface in your lungs and intestines. The word *areolar* ("air-ree-OH-lar," or "a-REE-oh-lar") is often used to describe loose connective tissue because this word means "airy" and, as you will see, there seems to be lots of open space in this tissue. This space is actually filled with *proteoglycans* (those bottle brush molecules) that hold lots of water. Some of the immune cells are able to swim around in this watery space, a bit like protozoans do in a pond.

Fibroblasts are the specialized cells in all types of fibrous connective tissue, including loose connective tissue. They secrete both the protein fibers and the ground substance. There are three types of protein fibers in loose connective tissue: *1) collagen fibers, 2) elastic fibers, and 3) reticular fibers*. Collagen fibers are very strong. Elastic fibers, as their name implies, are stretchy. Collagen and elastin together make this tissue both strong and flexible. Reticular fibers are very thin and form a delicate network for capillaries, nerves, and immune cells. (Remember, "rete" means "net.") Without this reticular network, all these parts would either slosh around or fall down to the bottom because of gravity. The reticular fibers keep the cells in place. Reticular fibers are actually very thin collagen fibers.

The ground substance in this type of tissue is the same as we saw in the last drawing. The fibroblasts make all the *proteoglycans* ("bottle brush" molecules) that are attached to long "ropes" made of *hyaluronic acid*. Proteoglycans are very small and can't be seen with a regular microscope like fibroblasts and collagen fibers can. The proteoglycans are like sponges and soak up lots of water. This means that the "empty space" you see in this drawing is actually full of water (not air, despite the fact that areolar means "airy").

Adipose cells (*adipocytes*) can be found in and around loose connective tissue. Adipose means "fat" so we can also call them fat cells. Fat cells are cells that have HUGE storage vacuoles (vesicles) filled with triglycerides. (A vacuole and vesicle are basically the same thing: a sphere made of phospholipid membrane. When a vesicle is very large and is filled water, air, or fat, it is often called a vacuole.) Remember those 3-legged lipid molecules from lesson 3? That's what is inside these vacuoles. The vacuole gets so large that it just about fills the entire cell. The other organelles are still there, but they are all pushed to one side and squished into a small space. The primary job of an adipocyte is to store energy. Because of this stored energy, the body can go for a fairly long time without taking in food. Equally important, adipose cells provide protective padding around delicate parts of the body.

The cells that float or move around in loose connective tissue are cells of the immune system—white blood cells. We will study these in greater detail when we get to the lessons on blood cells, but this is a good place to introduce a few of them.

Macrophages are "big eaters." ("Macro" means "big," and "phage" means "eat.") They eat pretty much anything that you don't want in your body: bacteria, viruses, dead or injured cells, and dirt. Macrophages know not to eat normal, healthy cells, but will gobble up just about anything else. They are your body's recycling unit. They can move around like an amoeba, oozing about by stretching out blobby "arms." They eat things by using the process of endocytosis. They patrol in this type of tissue, looking for invaders or garbage that needs to be dealt with.

Neutrophils are also eaters, though their name does not have "phage" in it. We'll find out what their name means in a future lesson. Neutrophils are one of the "first responders" at the scene of an injury. They are there immediately, ready to gobble up any foreign invaders that try to take advantage of the injury to enter the body. Neutrophils have a very strange multi-lobed nucleus that allows them to get through the thin cracks between the endothelial cells that form capillaries.

Lymphocytes come in two varieties: T cells and B cells. These two types of cells work together to identify and tag foreign invaders. Some types of T and B cells are memory cells that can remember how to fight pathogens you've encountered before. (A pathogen is something that makes you sick.) T and B cells look identical under a microscope.

Mast cells are the most notable immune cells in loose connective tissue. They are responsible for starting the *inflammatory* process. Though excess inflammation can be a problem, a normal amount of inflammation is necessary to get the healing process started. Swelling helps to keep the injured area isolated and to reduce blood loss. The cells that kick off the inflammatory process are the mast cells. They contain thousands of tiny vesicles filled with a chemical called *histamine*. When some of the mast cell's surface receptors receive signals that there is trouble in the area, their vesicles all go through rapid exocytosis, meaning they burst open at the cell's surface and spill out their contents—histamine and cytokines. The histamine causes the capillaries to dilate (get bigger) and become leaky. Lots of water and proteins leak out of the capillaries and produce the swelling. The histamine also stimulates nerve cells and makes them send messages of pain (and sometimes itching) to the brain. The *cytokines* are chemical messages that go out to other immune cells, telling them to come to this area and help out. We'll meet these other immune cells in future lessons.

These vesicles filled with histamine are often called *granules*. There are several kinds of immune cells that look like they are filled with little blobs, or "grains," and they are classified as "granular" cells. The mast cell is one of these granular cells. We will meet other granular cells in lesson 40. Right now you just need to know that when we talk about granules we mean these little vesicles filled with chemicals. You are likely to meet the word granules in books and websites about immune cells.

_____ cells secrete _____ and _____

_____ fibers are tough and strong. The _____ fibers are stretchy.

_____ fibers serve as an anchoring network for capillaries, fat cells and immune cells.

_____ are the "big eaters," swallowing _____

_____ are also eaters and gobble up pathogens. _____ are white cells (T and B cells)

that recognize and tag foreign invaders. _____ cells start the _____ process.

The ground substance is made of "ropes" of _____ to which are attached "bottle brushes" called

_____ which act like sponges and soak up _____.

HOW MAST CELLS START INFLAMMATION:

Mast cells are covered with receptors of all kinds that can detect just about any kind of irritation (pathogens, injury, allergens).

Inside, there are thousands of vesicles (often called _____) that are filled with _____ and _____. If the receptors are activated, the vesicles merge with the membrane (exocytosis) and dump their contents outside the cell.

HOW INFLAMMATION HAPPENS:

1) Mast cells are triggered by _____ or _____ or _____, which causes them to _____ their _____.

2) The _____ causes capillaries to _____ and become _____. This causes swelling (_____).

3) _____ also sends signals to nerve endings, causing _____ and/or _____.

4) The granules also contain _____ which are chemical messengers that go and recruit more white blood cells to come to this area.

34: DENSE and ADIPOSE FIBROUS CONNECTIVE TISSUE and CARTILAGINOUS CONNECTIVE TISSUE

Before you start this drawing, go back and look at the top of drawing 32. Look at the three categories of connective tissues. They are fibrous, cartilaginous and other. Notice that there are three categories under each of these categories. In drawing 32 we looked at just the first category of fibrous tissue: loose, or areolar, tissue. In this drawing we will finish the other two types of fibrous tissue, then do all three types of cartilaginous tissues. All these tissues will seem very similar to you and might be hard to keep categorized properly in your mind. It is helpful to remember that FIBROUS tissues have specialized cells called FIBROBLASTS, and CARTILAGINOUS tissues have specialized cells called CHONDROCYTES. Look for the word root "fibro" in the first pair, and then notice that the words in the second pair both start with the letter "C."

Fibrous connective tissues have **fibroblasts**. We met the fibroblast cell in the last lesson in the loose connective tissue, so you already know that these cells produce the protein fibers and ground substance that surround them. They make both collagen and elastin, but in proportions that are appropriate for their location in the body. Tissues that need to be very strong and just a little flexible have a lot of collagen and less elastin. Tissues that need to be very stretchy have more elastin than collagen.

Dense fibrous connective tissue can be either **regular** or **irregular.** The word regular is being used here to mean something like "straight and orderly." The collagen fibers in **regular dense** tissue are mostly parallel, going in just one direction. This makes these tissues very strong in one direction—the direction in which the fibers run. They can sustain a pulling force in that one direction, so they are perfect for tendons and ligaments. **Tendons** connect bone to muscle and **ligaments** connect bone to bone. There are very few elastin fibers in tendons and ligaments; they are mostly collagen. Another place you find dense regular tissue is in a protective layer that surrounds the brain. (This layer is called the **dura mater**, which is Latin for "tough mother.")

Dense irregular tissue has fibers that are less parallel and go in several directions. It is not super strong in any one direction, but it is still fairly strong in all directions. Again, it is mostly collagen with only a little elastin. You find this kind of tissue as a protective layer around bones and muscles. (We will see when we study muscles that they are all in bags. The ends of the bags become tendons.) Dense irregular is also found in the dermis layer of skin. As we will see when we draw skin, the top layer of the dermis is loose (areolar) connective tissue and the bottom layer of the dermis is dense irregular tissue. The dense irregular layer is what gives skin its strength. Small tears in the collagen fibers of this layer result in permanent stretch marks visible on the surface of the skin.

The third type of fibrous connective tissue is **adipose tissue**. This is actually a variation on loose connective tissue (remember, we saw adipose cells there!) but when there are many adipose cells in a large clump, it is called adipose tissue. There are two types of specialized cells in adipose tissue: fibroblasts that make collagen fibers and **adipocytes** that store fat. Adipocytes far out number the fibroblasts. An adipocyte is a real cell and has a nucleus and other organelles, though they are squished off to one side. The most prominent organelle is the **storage vacuole** (a large vesicle) that is filled with triglycerides (those little hangers with three fatty acids hanging down). The adipocytes take extra fatty acids that the body doesn't need to burn immediately, and puts them into storage. Adipose tissue can be found under the skin and around the internal organs where it provides insulation and protection. It also provides padding on places that need it (like where you sit down). Females have more adipose tissue than males; the extra adipose padding is part of what defines a face or a body as feminine.

CARTILAGINOUS

The specialized cells in cartilaginous tissue are called **chondrocytes** (KON-dro-sites). Each chondrocyte is surrounded by a little "pool" called a **lacuna**. (To help remember this name, try associating it with the word "lagoon," which is a ring of water around an island.) Cartilaginous tissue, unlike fibrous, has no capillaries and no nerves. This type of tissue occurs in places where there is a lot of rubbing, so you really would not want nerves here. The three types of cartilaginous are hyaline, elastic, and fibrocartilage.

Hyaline cartilage looks shiny and sort of a clear white color. (The word root "hyalus" came from Greek and means "glass.") If you've ever looked at the ends of chicken bones, you'll have seen this cartilage. Hyaline is found on the ends of long bones, the ends of ribs, and also in the nose and the trachea. That hard tube in your neck (the trachea) is made out of hyaline cartilage. Also, during embryonic development, the bones begin as hyaline cartilage. During the fourth month of pregnancy, the baby's cartilage bones begin hardening. Minerals such as calcium, magnesium and phosphorus are brought into the cartilage, transforming it into bone.

Elastic tissue has a lot of elastin protein and not much collagen. Obviously, this makes it very flexible. Elastic tissue is what makes the underlying structure of our ears. It is also found in the epiglottis, that flap that covers our trachea (windpipe) as we swallow, so that food does not get into our lungs. Obviously, you have nerves in your ears, but they are found in the fibrous connective tissue that surrounds the cartilage.

Fibrocartilage is very tough. It is an excellent shock absorber. You'll find fibrocartilage between the vertebrae bones in the spine and also in the crack where the pelvic bones come together in the front. In these two places you need a tissue that can withstand a lot of compression (being pushed together). The fact that the pelvic bone has a gap filled with cartilage is very important in females. During pregnancy, hormones cause cartilage to soften, allowing for stretching and bending of places like this crack in the pelvis. Otherwise, the bones might break.

34

FIBROUS CONNECTIVE TISSUE (part 2: Dense and Adipose)

Fibrous connective tissues have specialized cells called _____.

CARTILAGINOUS CONNECTIVE TISSUE

Cartilaginous tissues have specialized cells called _____.

Cartilaginous tissues have no _____ and no _____.

35: BONE CELLS and the OSTEON

Though bone may not seem like a connective tissue, we will see that it does indeed qualify according to our three requirements. Bone has specialized cells called *osteoblasts* and *osteocytes*. ("Osteo" means "bone.") These specialized cells secrete *collagen fibers,* just like fibroblasts and chondrocytes do. The ground substance in bone is a *solid*. (In the loose and fibrous connective tissue the ground substance was more of a gel.) The solid texture in bone comes from the minerals that are deposited between all the collagen fibers. The two main minerals that you find here in bone are *calcium* and *phosphorus*.

Osteoblast cells are very similar to fibroblasts because they secrete collagen. Their secretion is more "organized," though, because they can make layers of fibers that are all going in the same direction. They can alternate the directions of the layers forming something like plywood, where you find alternating layers of wood grain going in perpendicular directions. This alternation of directional layers is what gives plywood its strength. The alternating layers of collagen fibers in bone also provide a great deal of strength. The collagen fibers in bone could also be likened to the steel re-bar rods that are put into concrete. The re-bar allows concrete to survive small amounts of bending without cracking the cement. Collagen in bone also provides some flexibility in the midst of a mineral cement.

Osteoblasts work as a team. They are fastened together by both gap junctions and tight junctions. The gap junctions allow for communication and the tight junctions give a leak-proof barrier. As the osteoblasts move along they leave a web of collagen fibers behind them. The minerals don't fill in immediately, so for a short time there is an area called *osteoid* that is not solid yet. The osteoblasts also secrete another specialized protein that attracts minerals and causes them to solidify into that collagen web.

Once in a while, one of the osteoblasts will get left behind and become stuck in the solid bone and will change its shape so that it has very thin finger-like projections. After this happens, the cell is no longer called an osteoblast but is called an *osteocyte*. Osteocytes are surrounded by a *lacuna*, just like chondrocytes are. (This is another similarity between cartilage and bone.) The lacuna goes all the way around the osteocyte, even around its long projections. The areas of the lacuna that surround these projections are called the *canaliculi,* or "little canals." The finger-like projections of osteocytes reach out until they touch the projections of other cells. Since the osteocytes are stuck in the solid bone, the contact they have with other cells through these projections is the only contact they have with the "outside world." The osteocytes can communicate with each other through these projections, and they can also share oxygen and nutrients. Osteocytes near blood vessels pick up the nutrients and pass them along to osteocytes that are further away from the vessels.

Osteocytes arrange themselves into concentric rings (rings inside of rings). The rings are called *lamellae*. A set of rings is called an *osteon*. The osteon is considered to be the basic unit of solid bone. In the center of an osteon is a space reserved for blood vessels. This space is called the *central canal*, or the *Haversian canal* (named after the discoverer, Mr. Havers). A tiny vein and artery share the space with a lymph vessel and a tiny nerve. Since there are nerves running through osteons, bones can experience pain. The lymph vessels help to get rid of wastes. (The small lymph vessels eventually connect to the larger lymph network which dumps the lymph fluid back into the blood system.) The tiny vein and artery bring oxygen and nutrients to the bone cells, but it is only the cells closest to the vessels that actually absorb the nutrients. Those inner osteocytes then have to pass the nutrients along to the cells in the outer rings. This system works okay on a small scale, but there is a limit to how far nutrients can be transported. The maximum size of an osteon seems to be about 200 microns, about the same size as an egg cell (the largest human cell).

Osteons are constantly being re-built. The minerals stored in compact bone can be taken back out again and put into the blood stream if necessary. The calcium level in the blood must be kept at a constant level, and if calcium is not being supplied through the food that is coming into the stomach, the body must get the calcium from somewhere else. There are special cells called osteoclasts (which we will meet in a future lesson) that dissolve the minerals out of the collagen and put them back into circulation. After the osteoclasts have destroyed an area of bone, osteoblasts must move in and replace it. This cycle of destroying and rebuilding bone tissue is called *bone remodeling* and it happens continuously throughout your life. Newer osteons form complete circles. Older ones get covered over by the new ones so they no longer look perfectly round. Remember, also, that ostens are more like tubes than circles. The lamellae are cylindrical, like straws.

BONE CELLS and the OSTEON

35

Bone is classified as a connective tissue, so it has:

1) Specialized cells (called _____ and _____)

2) Protein fibers (called _____)

3) Ground substance, which is a _____ made of _____ such as _____ and _____ .

OSTEOBLASTS secrete collagen, then fill the empty spaces with minerals. Cells are cuboidal when active, flat when not.

Osteoblasts can become **OSTEOCYTES**.

When osteoblasts are done, they have created an **OSTEON**.

The inner cells pick up nutrients from blood vessels then pass them along to outer cells (through canaliculi).

Maximum diameter is about 200 microns (.2 mm)

36: BONES MAKE BLOOD CELLS

The ends of the long bones contain **red marrow**. Red marrow is the site for blood cell production. In babies, red marrow fills all the bones, so all bones are involved in blood production. As we grow up, the marrow in the middle of our long bones changes to become yellow marrow, made of mostly adipose cells, but the ends of the bones remain full of red marrow throughout our life. Other bones that have blood-producing marrow are the ribs, vertebrae and pelvic bones.

The red marrow is found within an intricate network of bone, called **trabecular bone**. ("Trabecula" is Latin for "little beam.") Trabecular bone looks very much like a sponge that has become hard, so it is often called **spongy bone**. Red marrow fills all the empty space in spongy bone. Like all bone, spongy (trabecular) bone is manufactured by osteoblasts, some of whom become trapped and turn into osteocytes.

Red marrow contains many different kinds of cells. There are some adipose cells, but not nearly as many as in yellow marrow. The most important kind of cell in red marrow is the multipotent stem cell called the **hematopoietic** stem cell. ("Hema" is Greek for "blood," and "poiein" is a Greek word meaning "makes.") From the hematopoietic stem cell come all the different types of blood cells. Where did the hematopoietic stem cells come from? They've been there from earliest days, when the embryo was only a few weeks old. Some of the embryonic cells differentiated into these stem cells and have stayed as stem cells ever since. The stem cells never get used up because they themselves don't become other cells. They make clones that then turn into other cells.

The first decision a hematopoietic stem cells makes is whether to become a **myeloid** stem cell or a **lymphoid** stem cell.

The **myeloid stem cell** can differentiate into more types of stem cells. (Remember that every time a stem cell differentiates, it takes a step "down." There is no going back—there's no un-differentiating.) One option the myeloid cell has is to become a **mast cell**. We've already seen this cell in action. It is found in loose connective tissue, close to the capillaries. The mast cell has lots of vesicles filled with histamine. When injury occurs, the mast cell releases its histamine. The histamine acts on the cells of the capillaries (the endothelial cells) and causes the capillary to expand and to leak. Water and proteins leak out and create swelling (edema). This is part of the normal inflammatory response and is necessary for healing.

Another option open to the myeloid stem cell is to become a **megakaryoctye**. ("Mega" means "large," and "karyo" means "nucleus" so these cells have a large nucleus.) Ironically, these large-nucleus cells will form tiny cells with no nucleus. These tiny cells, called **platelets**, are actually classified as cell fragments, not actual cells. Platelets are essential to blood clotting and without them we would bleed to death from even a tiny cut. The technical name for platelets is **thrombocytes.**

The myeloid stem cell also has the option of turning into an **erythrocyte**. ("Erythro" is Greek for "red" so in English we would just say "red cell.") Like platelets, erythrocytes have no nucleus, but unlike platelets, erythrocytes are actual cells, not fragments. There are very strange cells, though, because the nucleus and other organelles disappear in order to make the red cell more efficient at its job (carrying oxygen from the lungs and delivering it to cells). We will learn more about red cells in drawing 39.

Another option for a myeloid stem cell is to turn into a type of stem cells that will produce three different cells: **basophils, eosinophils, and neutrophils**. These three cells look a bit similar but have distinctly different jobs. The word "phil" on the end of these names is one you have seen many times already. You will remember that it means "love." So what do these cells love? Stains. They tend to soak up either acidic or basic (alkaline) stains. Basophils soak up basic (alkaline) stains and as a result turn blue. Eosinophils soak up a stain called eosin which is acidic, turning them bright red. Neutrophils don't soak up either very well, leaving them light pink. Basophils are very similar to mast cells. In fact, they are pretty much the same cell, except that mast cells are found in tissues and basophils are found in the blood. Basophils are also responsible for allergic responses. Both basophils and eosinophils are key players in fighting parasites. Yes, it is a yucky thought, but sometimes bodies can be invaded by tiny parasitic worms. Eosinophils are the number one parasite fighters of the immune system. Neutrophils have a much different role — they are a type of phagocyte ("eating cell"). They float through the blood, looking for invaders to gobble up. They are part of your immediate response to pathogens. Many of them go to sites of infection, then end up as "pus." (Pus is mostly dead neutrophils.)

The myeloid cell can also become a **monocyte**. Monocytes then differentiate into either **macrophages** ("big eaters") or **dendritic cells** ("dendros" is Greek for "branch"). We saw macrophages in loose connective tissue. Dendritic cells have a similar job, eating up pathogens, and then presenting pieces of these invaders to the T cells. We'll learn more about this soon.

The **lymphoid stem cells** can turn into **dendritic cells**, or they can become another type of stem cell which will produce cells we call **lymphocytes**. **B cells** are so named because they mature right there in the <u>B</u>one. Some B cells will become "memory" cells because they retain information about how to attack viruses and bacteria you have come into contact with in the past. (This is what gives us our immunity to viruses after we have had them once.) **T cells** leave the bone and mature in the <u>T</u>hymus, a gland in the upper part of your chest, in front of your heart and lungs but under your ribcage. There are several kinds of T cells, including killer Ts, helper Ts and suppressor Ts (also known as regulatory Ts). **NK cells** are **Natural Killer** cells. They are also part of your immediate response to invaders. We will learn more about lymphatic cells in future drawings.

One last vocabulary word you should know is **granular**. Some white cells look like they are filled with little particles, or granules. (The granules are little vesicles.) These include basophils, eosinophils, neutrophils and mast cells. They are sometimes called granulocytes. The other cells, those without granules, are called agranular, meaning "without granules."

Blood cells are made in the ends of the long bones, and also in the red marrow of ribs, vertebrae and pelvic bones.

HEMATOPOIETIC STEM CELLS make all kinds of blood cells.
(He-MAT-o-po-ee-ET-ic)
"Hema" is Greek for "blood," and "poiein" is Greek for "makes."

("HEMATOPOIESIS" is the process of making blood cells.)

37: BLOOD (as a tissue)

Blood can be classified as a connective tissue because it has specialized cells, ground substance and protein fibers. The specialized cells are numerous (as we saw in drawing 36) and include red cells, white cells, and platelets. The ground substance in blood is liquid, and we call it *plasma*. For the protein fibers, we actually have to stretch the definition a bit. Yes, there are fibers, called *fibrinogen*, but these proteins are made in the liver, not by the blood cells. There are also non-fibrous proteins in blood, too.

In order to see what is in blood, a machine called a *centrifuge* is used. A centrifuge is a bit like the carnival ride where you sit in swings that spin around and up, but the centrifuge spins the test tubes about about 2000 revolutions per minute, whereas the carnival ride does about 5 rpm. The elements of blood settle out according to their density. Plasma is at the top, red cells go to the bottom, and white cells and platelets end up as a narrow band in the middle. About 55% of blood is plasma, 45% is red cells and less than 1% is white cells and platelets. Sometimes people are very surprised to learn what a small percentage of blood is made of white cells, because they are so important in fighting infection. But the system works!

In the last drawing we did a survey of blood's specialized cells. In this drawing we will look at the other elements of blood. The ground substance that all the cells are floating in is called *plasma*. Plasma is 91% water. About 7% of plasma is various proteins, and about 2% is "solutes." Solutes are tiny molecules such as nutrients (such as glucose), wastes (such as urea and ammonium), gases (such as oxygen and carbon dioxide), vitamins (water soluble ones), and some minerals.

The proteins in blood fall into basically four categories: 1) *fibrinogen*, 2) *clotting factors*, 3) *albumins*, and 4) *globulins*.

1) Fibrinogen, as its name suggests, creates *fibrin*. ("Gen" is Greek for "create.") Fibrin is made of protein "strings" which are used to form blood clots. "Blood clots" sounds like something you don't want to have. You may know people who take medications to prevent blood clots. However, blood clots are a natural body defense. If blood did not clot, a tiny cut would keep on bleeding until all our blood was gone. Clotting is necessary to keep our blood inside. When the body tries to keep the blood inside, this is called *hemostasis* ("hemo" is blood, and "stasis" means to keep steady). In hemostasis, blood stays in. We will learn more about clotting in the next drawing. We will see the *clotting factors* at work, too, using a chain reaction known as a *cascade*.

2) *Clotting factors* are tiny proteins that activate fibrinogen and platelets. The clotting factors themselves need to be activated by chemical messages from injured cells. When an injury message goes out, the clotting factors then initiate the clotting process. Because the clotting factors are not active until they receive a signal, they can safely float in the blood alongside fibrinogen and platelets without activating them. The clotting factor we will meet here is *thrombin*, which activates fibrinogen.

3) The *albumin* proteins are like little taxi cars, transporting things through the blood. About half of all blood proteins fall into this category. Why would you need taxis in the blood? Hydrophobic substances are "afraid" of water, and since plasma is 91% water, we have a problem. Hydrophobic substances include fats, hormones, some ions, and some proteins (such as bilirubin, which we will meet later). These substances need a way to go through the blood without touching water. Albumins have little "pockets" where hydrophobic substance can tuck in and hide while they ride through the blood.

4) The *globulins* are of several different kinds: *alpha, beta and gamma*. (Greek letters A, B and G.) Globulins do look somewhat like a "glob" because their shape is not linear, like fibers, but more round, like a blob. The albumins also look fairly blobby. The alpha and beta globulins are very similar. Like the albumins, they act like little taxis for substances that don't like being in the watery blood. They transport things like fats, cholesterol, hormones, and mineral ions. The beta globulins have another role, and that is to help dissolve clots once they are not needed anymore. HDL (high density lipoprotein) is an alpha globulin that transports fats back to the liver. LDL is a beta globulin that takes fats away from the liver. You want more HDL than LDL (more fats being returned to the liver) so HDL is often called the "good" fat and LDL is seen as "bad." LDL isn't really bad, but must be kept in balance.

Gamma globulins are also known as *immunoglobulins* or *antibodies*. "Anti" means "against" so they must be against some kind of bodies. The bodies they are against are any bodies (foreign particles) that do not belong to your body. Antibodies are made by B cells and act a bit like tags that say "NOT SELF" and can be stuck onto invaders. You can have antibodies to just about anything: viruses, bacteria, parasites, pollens, and even foods. It is then up to other kinds of white cells (such as T cells and macrophages) to actually destroy the invaders.

More about fibrinogen:

Fibrinogen, blood's little fibers, are short proteins that have the capability, under the right conditions, of joining together to make large, strong fibers called *fibrin*. The shape of fibrinogen is a bit like a barbell, having a narrow middle and large ends. Fibrinogen fibers can only be activated by clotting factors, notably *thrombin*. When there is an injury to a vessel, a signal will go out that activates a whole series of clotting factors, causing a chain reaction which, as a final result, changes the fibrinogen molecule. The protein called thrombin will activate sites on the middle of the fibrinogen molecules, allowing them to stick together. Immediately, strands of fibrin begin to form. Fibrin will become a strong mesh that will catch and hold platelets and blood cells, forming a clot. On the surface, where we can see the clot, we often call it a scab, especially when it dries out.

BLOOD (as a tissue)

Blood is classified as a connective tissue because it has:

1) _____ : _____

2) _____ : _____

3) _____ : _____

A centrifuge can separate blood into its 3 parts: plasma, "Buffy coat," and red cells.

PLASMA:

SIDE NOTE: Blood has a pH of 7.4.

1 _____ (%)

2 _____ (%)

    1 Fibrinogen:

    2 Clotting factors:

    3 Albumins:

    4 Globulins:

        (1) α alpha--

        (2) β beta--

        (3) γ gamma--

3 _____ (%)

FIBRINOGEN -- the "fibers" in blood

Fibrinogen is a protein
molecule that looks like this:

We can simplify the shape like this:

Clotting factors in the blood can act on fibrinogen to form _____,
but only after a special message is sent.

Message goes out:
"Help! Blood is
leaking out!"

Both float in blood, but don't interact
because fibrinogen has a "safety cap"
on its active site.

THROMBIN is told to take
the safety caps off the central
active site of fibrinogen.

The fibrinogens bond with
each other to form strings
of FIBRIN.

38: BLOOD: THE CLOTTING CASCADE

Hemo**sta**sis is the mechanism by which our blood **sta**ys in our bodies. ("Hemo" is Greek for "blood.") When a blood vessel is injured, our cells react immediately, forming a clot that plugs up the hole. The clot stays in place until the surrounding cells have repaired the damage. Then the clot dissolves so blood can start flowing again. Hemostasis has two steps: a "Platelet Plug" then a "Fibrin Fabric."

NOTE: Some sources will give three steps, adding "vasoconstriction" before the platelet plug. This means the muscles around the vessel contract, shrinking the diameter of the vessel. Less blood flows through, so less blood is lost.

Step 1: Platelet Plug (Primary Hemostasis)

When endothelial cells (the epithelial cells that line the inside of blood vessels) are injured, the connective tissues underneath are suddenly exposed. Normally, collagen never comes into contact with blood. The only time that collagen comes into contact with blood is when something goes wrong.

Floating in the blood are platelets and clotting factors. (We saw in a previous drawing that platelets are pieces of a larger cell called a megakaryocyte. The clotting factors were made by the **liver.**) The platelets have little receptors on their surface that function like hooks that can grab onto collagen. Normally, platelets do not come into contact with collagen, so these receptors are not active. When something does go wrong, and collagen is exposed, platelets begin sticking to the collagen strands. Enough platelets stick that a clump, or "plug," begins to form. One of the clotting factors, called **von Willebrand Factor,** is very helpful at this point. It acts like a glue between the platelets and the collagen. The plug is much stronger with vWF there to help out.

Another important event in this stage is **platelet activation**. When the receptors on the outside the platelets sense that they are grabbing onto something, several changes occur. The most obvious change is their shape. The platelets grows arm-like things that we might call "branches." In Greek, "dendros" means "branch," so we call their new shape their **dendritic form**. These arms will help them stick together. While their shape is changing, they are also releasing chemicals that they have been storing in vesicles: calcium ions, and clotting factors including thrombin. **Calcium** is needed to activate thrombin. Note that we get some thrombin here, but will get a lot more at the end of step 2.) Additionally, inactive receptors on their surface become activated so they can hold on to collagen and to other platelets even more strongly.

Step 2: Fibrin "Fabric" (Secondary Hemostasis)

In the second step, the tissues will form a proper clot, which will be strengthened by fibrin stands. The net of fibrin strands will begin to catch other kinds of cells, besides platelets. It will catch red cells, and some white cells. The presence of red cells is what gives a clot (or scab) its reddish color. The arrival of white cells will be helpful because they will kill germs that have gotten into the cut. If the population of white cells (especially neutrophils) becomes so large that we can see it, we call it "pus."

As we know, to make fibrin out of single fibrinogen units, a protein factor named **thrombin** is necessary. Thrombin activates quite a few other proteins and it even activates molecules called **antithrombin** and **Protein C**, which will begin to undo what thrombin has done. During the healing process, the clot will need to be dissolved. Thrombin is "thinking ahead" to that time, and has already begun producing the factors necessary for that process.

The complicated process that leads to the formation of thrombin is called the **coagulation cascade**. Outside of biology, "cascade" means "waterfall." Waterfalls begin small at the top, and then spill down, getting bigger at each level. In biological cascades, there are just a few molecules at the beginning of the process, then, at each step, more and more molecules are affected. The first molecule undergoes a change that affects a second molecule. That second molecule changes the third, the third changes a fourth, and so on. It can be helpful to think of a "domino rally" where you set up a line of dominoes and then knock over the first domino. Each domino falls against its neighbor and, one by one, the whole row goes down. In the coagulation cascade, the dominoes are proteins called **coagulation factors** and are known not by names but by Roman numerals. Instead of getting knocked down, they are activated." Activation often involves a change to the shape of the molecule. When activated, each factor goes and activates the next factor down the line. The final result of the coagulation cascade is the activation of thrombin which turns fibrinogen into fibrin. The thick strands of fibrin then form the sticky fabric that holds the clot together.

There are two "streams" that join together in the coagulation cascade. The shorter branch starts outside the vessel, with the release of a protein called **tissue factor**, also known as factor III (3), from the surface of body cells near the injured vessel. Tissue factor combines with factor VIIa (7a) and they activate factor X (10). This stream is the "quick start" of the system and occurs outside of the vessel. Once things are rolling, thrombin itself can begin activating factors 7, 10 and 11. It is important at this point to note that several of these factors, namely 2, 8, 10 and 11, require a cofactor (a helper molecule) to function properly. This cofactor is **vitamin K**, a vitamin found in dark green leafy vegetables. The action of vitamin K can be blocked by a chemical called **warfarin** (named after the Wisconsin Alumni Research Foundation, the group that sponsored its discovery in 1940). In high doses, warfarin has been used as a rat poison. The rats bleed to death, but hopefully back in their nest, not on your back porch. In small doses, warfarin can be used as a medicine to prevent life-threatening blood clots. The most common warfarin drug is called **Coumadin**. It is very likely that someone you know takes Coumadin, as it is a widely used medication.

If your body can't make factors 8, 9 or 11, you will have a disease called **hemophilia** in which you can bleed to death even from a small cut. This disease is often called "the royal disease" because Queen Victoria of England carried the gene for it and passed it to several of her children who then married into other royal families around Europe. (Nowadays patients can get injections of the missing clotting factors.)

The longer pathway, the one that starts with factor XII (12) is also kicked off by injury that exposes the collagen that surrounds the blood vessel. This pathway occurs inside the vessel, so it is often called the "intrinsic" pathway.

Why this complicated system? Why can't factor 12 just activate thrombin? The cascade allows for a **geometric increase**. At each level, there is a huge increase. If factor 12 had to activate all the thrombins, the process would go too slowly. The cascade actually works faster than factor 12 alone. Speed is important when your body is racing to stop bleeding.

The body's way of dealing with damage to blood vessels is called _____

The PRE-STEP is: _____

STEP 1: _____ ("primary hemostasis")   INACTIVE PLATELET

                                                                    ACTIVE PLATELET

1) INJURY: When endothelial cells rip, collagen is exposed.
2) ADHESION: Platelets stick to collagen. (VWF comes over and acts like glue.)
3) ACTIVATION: Platelets are activated and release more clotting factors plus calcium.
   Changes occur which allow the platelets to stick better.

This is the "dendritic" form.

*Don't forget-- platelets can also be called "thrombocytes."

STEP 2: _____ ("secondary hemostasis")

Turning fibrinogen into fibrin is a many-step process called a CASCADE. ("Cascade" means "waterfall.") The proteins are called COAGULATION FACTORS and are known by Roman numerals. Viamin K is needed here.

Cascades allow for geometric increase:

NOTE: When steps 8, 9 or 11 don't work, we call it Hemophilia A, B and C.

The fibrin is made of fibrinogens.

NOTE: Warfarin (rat poison) blocks the action of vitamin K. When used as a medicine, warfarin is called Coumadin.

THROMBIN also makes ANTI-THROMBIN and PROTEIN C, which act to dissolve the clot. (Protein C inhibits VII and V.)

What activated fibrinogen?
The clotting factor _____.

39: RED BLOOD CELLS (ERYTHROCYTES)

We've already met erythrocytes in lessons 36 and 37. We know that they come from stem cells that were produced by the hematopoietic stem cell. (Erythrocytes are sort of like the "great-grandchildren" of the hematopoietic stem cell.) As the erythrocytes mature, they begin to lose their organelles. When fully mature, red cells have no nucleus (and therefore no DNA), no ER, no Golgi bodies and no mitochondria. Mitochondria use oxygen for their electron transport chains, so mitochondria would be a detriment to the purpose of red cells, which is to carry oxgyen. Surprisingly, though, they can live like this, with no organelles, for about 3 months. After about 100 days, the red cells begin to express protein tags on the surface of their plasma membrane that mark them as old cells. As the cells pass through the liver and spleen, they go past macrophages that are looking for these tags. The macrophages grab the old red cells and eat them. The erythrocyte's parts are recycled and used to build new cell parts, perhaps even new red cells.

The number of erythrocytes that the bone marrow produces is controlled primarily by the kidneys. Since all the blood filters through the kidneys many times per hour, this is a good location to have an oxygen monitoring system. Special cells in the kidney can sense the oxygen level in the blood, and if it begins to fall, a substance called **erythropoietin** will begin to be produced. The erythropoietin is taken, via the blood, to the bone marrow where it affects the hematopoietic stem cells. It tells these stem cells to begin producing more erythrocytes. More red cells will mean more "vehicles" to carry oxygen. This process takes several days. Athletes can take advantage of this, training at high altitudes, then competing at lower altitudes. Bone marrow can make about 2 million red cells per second, giving you a total of about 20 trillion red cells in your blood at any given time.

An erythrocyte contains about 250 million molecules of **hemoglobin**. Hemoglobin is a protein made of four separate proteins joined together in a square-looking quaternary structure. The "-globin" part of the word refers to these globular proteins. The "hemo-" part of the word refers to "**heme**," a molecule that holds an iron (Fe) atom. Each of the four proteins in hemoglobin holds one heme molecule and each heme molecule holds one iron atom. These iron atoms are what makes your blood red.

The structure of heme is very similar to the structure of chlorophyll. Chlorophyll has a magnesium atom where heme has an iron atom. The magnesium atom in chlorophyll is the "action site" where the energy from a photon of light can be transfered to electrons. For some types of anemia (low iron level in blood) patients are given chlorophyll as a nutritional supplement. The body can pop out the magnesium and transform the chlorophyll in to heme.

The iron atom in heme can loosely hold a molecule of oxygen, O_2. (Oxygen always goes around in pairs.) It is important that the oxygen is only loosely held because it must be able to jump off quickly when it gets to a place that needs oxygen.

When hemoglobin is recycled, the globin part is reduced to amino acids. The heme molecule is removed and the iron atom is taken out. The iron goes into a "taxi" called **transferrin**, which circulates in the blood, making the iron atoms available to any cells that need them. The rest of the heme begins to be broken down by cells in the liver. After the first break, heme becomes **bilirubin**, a yellow molecule that gives bruises their yellow color. Bilirubin is broken down further into molecules that are yellow and brown. These are eventually excreted in urine and feces and are what give them their characteristic colors.

Erythrocytes express hundreds of proteins on their outer surface. Most of these proteins don't affect medical procedures such as blood transfusions, but a few of them are critically important. The most important proteins are called A, B, and Rh. A cell that expresses A proteins on its surface will also have antibodies to protein B in the plasma. A cell that expresses B proteins on its surface will have antibodies to protein A. An antibody is a tag designed to stick to a foreign particle that gets into the body. (The foreign particle is often called the **antigen**.) So the anti-B antibodies will cling to the B proteins on a cell that expresses B. The A antibodies will stick to the A proteins. This can cause quite a mess. The mess is called **agglutination** and on a microscope slide it looks like clumping or coagulation of the blood.

A person whose cells express both A and B proteins on the surfaces is said to have AB type blood. The AB's won't make either kind of antibody. A person whose red cells don't express either A or B is said to have type O blood and will have both A and B antibodies in their plasma. Your blood type is something you inherit from your parents. It is one of the most studied topics in genetics and lots info is available online if you want to know more.

The **Rh** protein was first discovered in the **Rhesus** monkey. Biologically, humans are primates and thus share many biological similarities with apes and monkeys. A red cell either has the protein, or doesn't. If it has the Rh protein, it is said to be "Rh positive." If the protein is missing, you are "Rh negative." A person who is Rh negative won't automatically produce anti-Rh antibodies, though. The antibodies will only be produced if the immune system comes into contact with Rh positive red cells. The two occasions where this might happen are transfusions and pregnancy. Donated blood is carefully typed ahead of time, so this is not a problem you run into much with transfusions. Pregnancy is still an issue, however.

If an Rh negative mother is carrying an Rh positive baby, there is a chance that a few of the baby's red cells might leak into the mother's bloodstream, causing her immune system to start producing Rh antibodies. Normally, the placenta does keep the blood system separate, but if even a few cells leak over, that is enough to sensitize the mother. Also, during the messy birth process, it is fairly easy for the blood to get mixed up. If the mother's Rh antibodies got into the baby's blood, they would attack the baby's red cells, endangering the baby. This can be prevented by giving the mother an injection of artificial antibodies that "cover up" the Rh factors on any of the baby's cells that might leak over. The mother's immune cells won't "see" the Rh proteins since they are covered up, and thus no Rh antibodies will be made.

RED BLOOD CELLS (erythrocytes)

Erythrocytes are produced by the myeloid stem cells in bone marrow.
The **KIDNEYS** control how many are produced. Low oxygen levels in the blood cause the kidneys to make a
substance called **erythropoietin** which acts as a signal to the myeloid cells to differentiate into more red cells.
This process takes a few days. (This is what happens when you adjust to higher or lower altitudes.)

ERYTHROCYTE   HEMOGLOBIN   "HEME"

Iron (Fe) attracts a molecule of oxygen (O_2) and holds it loosely.

Oxygen can leave heme when passing by cells that need oxygen.

Does NOT have: _____

_____

DOES have _____
hemoglobin molecules.

Bone marrow can make _____
red cells per _____!

The body has _____ red cells
at any given time.

Red cells live for _____
Old cells are eaten by _____
in the liver and spleen.

RECYCLING of hemoglobin:

GLOBIN   HEME

Bilirubin

GLOBIN is broken down
into amino acids.

Heme is broken down in several steps.
After the first break, it is called "bilirubin."
It is further broken down into yellow and
brown molecules that eventually go out
in the urine and feces.

Iron (Fe) is taken out
of heme and put into
transferrin "taxis" to
float in the blood and
be available to any
cells than need iron.

BLOOD TYPES

An erythrocyte has hundreds of proteins on its surface. The most critical ones are ____, ____ and _____.

The surface protein called "Rh factor" was named after the _____.

40: LEUCOCYTES (part 1): GRANULOCYTES

First, notice the spelling difference between this title and the title on the template page. Here is a word that can, and is, spelled both ways with equal frequency. It doesn't matter whether you use a "c" or a "k" in the word leucocyte/leukocyte. Some people claim that this is a British/American difference, but apparently both spellings show up on both sides of the Atlantic. So you can choose which way you think looks nicest and go with it. Personally, I am partial to the "c" spelling, (for no logical reason at all) but I might use both at various times in this course. (NOTE: A few words require one or the other, such as the word "leukemia" which must be spelled with a "k.")

The hematopoietic stem cells differentiate into myeloid stem cells, and then one option that the myeloid cell has it to become a granulocyte cell. These cells are called *granulocytes* because when you look at them under the microscope they have little dots, or granules inside them. Technically, these structures are vesicles, and are filled with chemicals waiting to be released.

All the granulocytes have very strange-looking nuclei. They have 2 or more *lobes*, with thin strands connecting them. This arrangement is probably to help them get through tight spaces. Immune cells move around among other cells, and can even leave the blood and go into tissues by squeezing through the tiny cracks between the epithelial cells. Having the nucleus split up into two or more small sections probably makes it easier for the cell to move in narrow spaces.

Basophils were discovered in 1879 by a German physician named Paul Erlich. A year earlier, Erlich had discovered the mast cell, which is very similar to the basophil. The name "basophil" means "base lover," as these cells will take up an alkaline (basic) stain, turning them dark blue. The basophil's granules (vesicles) are filled with several chemicals, of which the most important to know is *histamine*. Histamine kicks off the inflammatory response, by dilating blood vessels so that plasma and proteins leak out into the tissues. (If this happens to much and too fast, your blood pressure can drop suddenly, causing you to faint.) Histamine also makes you itch because it irritates the nearby nerve cells. Basophils also make a chemical called *heparin*, which slows down the clotting process. The *basophil* normally floats around in the blood and does not permanently reside in tissues like the mast cells do. However, if there is trouble in a tissue somewhere, the basophils can be called in to help. They can squeeze out of the capillaries and get into tissue if needed. However, to keep things straight in your mind, think of basophils in blood, and mast cells in tissues.

Antibodies called *IgEs* stick to the outside of basophils. (B cell lymphocytes make these antibodies.) The bottom portion ("base" of the Y) sticks to the cell, and the top "v" part attracts antigens. (The word *antigen* originally started out as meaning something apart from the body, something distinctly "not self." When the body attacks one of its own parts that part is called an auto-antigen. "Auto" means "self.") Often, substances like food molecules, mold spores or pollen grains can stick to these IgEs and cause the basophil to release its histamine. Thus, basophils are involved in allergic reactions. Basophils, along with mast cells, are responsible for the intense "anaphylactic" (life-threating) allergic reaction that is often associated with allergens like bee stings and peanuts.

The *mast cell* was officially discovered and named by Paul Erlich, but had definitely been observed by several scientists in previous years. Erlich thought maybe this cell had something to do with being well-nourished so he named it in honor of the fattening food (mast) that farmers gave their livestock. (Even famous scientists can make mistakes!)

We've already met the mast cell in loose connective tissue, which is where it lives all the time. It does not float around in the blood like basophils do. Mast cells start the inflammatory process by releasing histamine, which dilates blood vessels and makes them leak. Histamine also irritates your nerve endings, causing them to itch. Mast cells are covered with IgE antibodies (made by B cells). When antigens stick to the antibodies, the mast cell begins to release histamine. Mast cells can also be triggered by any kind of trauma, even something as simple as slapping your skin or pressing something hot against it. Mast cells, like all immune cells, release chemicals called cytokines that act as messages to other cells.

Eosinophils got their named from their "love" of an acidic stain called eosin. (Some people prefer to call them *acidophils*.) Erlich was the person who stained and named these cells in the 1870s, but they had already been noticed by microscopists for several decades. Though Erlich was wrong about the function of mast cells, he guessed correctly the function of the eosinophils and the role that the granules played.

Eosinophils are best known for their role in fighting parasites, especially worms. Yes, worms. Worms have been a reality of life for humans for thousands of years, and still are a reality for millions of people today. Eosinophils attack worms and their eggs. When a parasite is discovered, chemical messages go out to recruit eosinophils to that site and to tell the bone marrow to start producing extra eosinophils. Hundreds (or thousands) of eosinophils will surround the parasite, releasing toxic chemicals as close to the worm as possible, hoping to cause damage.

Eosinophils migrate quickly out of the blood and into tissue. Their favorite places to hang out are in the digestive system, the lungs, and the skin. People who have overactive eosinophils often have problems with these body parts. (Eosinophils seem to play a significant role in asthma, but their exact role is still not completely understood.) It's strange that these cells could stir up trouble, because overall their role is to counteract histamine and help to clean up the "mess" that basophils and mast cells make. The counteracting substance is called *histaminase*, which neutralizes histamine.

BASOPHILS (.5% of leukocytes in blood)

1) Normally, they float around in the blood, but they can be recruited into tissues if other cells "call" for them to come.

2) Are extremely similar to mast cells

3) Vesicles are filled with histamine (and other chemicals). Histamine dilates blood vessels and makes them leak.
(If this happens too fast and too strong, you get a sudden drop in blood pressure and you faint.)

4) Basophils have many IgE antibodies (from B cells) attached to them. IgEs trigger release of histamine when antigens bind to them.

5) Basophils also have the ability to "call" other white cells to come and help, including eosinophils, neutrophils and basophils.

MAST CELLS (not in blood)

1) Found only in tissues, not in blood. We met these in loose connective tissue, sitting next to capillaries.

2) Are extremely similar to basophils.

3) They start the inflammatory process when endothelial cells are damaged.

4) Mast cells are covered with IgE antibodies (from B cells). When allergens bind to the IgEs, histamine is released.

EOSINOPHILS (2% of leukocytes in blood)

1) Attack parasitic worms and their eggs. (3 billion people in the world have some kind of worm infection. Some worms are not very harmful and people just live with them.)

2) Vesicles are filled with histamin**ase**, which neutralizes histamine.
(clean up from basophils!)

3) Other vesicles have chemicals that are helpful for fighting, but can cause damage to body cells, too. For example, eosinophils have be shown to be very active during asthma attacks.

NEUTROPHILS (60-65% of leukocytes in blood)

These guys are so amazing that they need their own page... (They make several kinds of chemical weapons!)

41: LEUCOCYTES (part 2): NEUTROPHILS

Neutrophils, as their name suggests, are fairly neutral when it comes to stain that microscopists pour over them. They don't particularly "love" either basic or acidic stains. They prefer to stay as neutral as possible, though they do look a bit pink after the lab technicians are done with them. Like the other granulocytes, neutrophils also have a multi-lobed nucleus that will allow the cell to slip through tight cracks and narrow spaces.

Neutrophils are very abundant, as high as 65 percent of the white cells in your blood. (NOTE: If you do some online searches, you will find that these percentages will vary. The percentages quoted here are the most commonly used ones.) Our bone marrow makes about 100 billion neutrophils every day. The marrow keeps a reserve ready in case of emergency, storing up to five times as many neutrophils as there are in circulation. These cells have a very short life span of only a few days. (If you've ever seen "pus" you've seen billions of dead neutrophils.) Neutrophils can eat up to about 50 bacteria in their short lifetime. When neutrophils die, macrophages come and eat them, recycling all the parts.

Neutrophils float around in the blood almost as if they are a police force on patrol, constantly scanning for trouble spots where they might be needed. Our cartoon neutrophil has a red cross on its forehead, reminding us that it is a "first responder" and is very often the first leucocyte at the scene of an infection. Often, the neutrophils clean up the infection while it is very small and we never know what happened. Sometimes an infection will progress to the point that we become aware of it, but there are many tiny invasions every day that are taken care of by our neutrophils without our knowledge.

Our cartoon neutrophil has hair made of little bumps that represent receptors. All cells have numerous receptors, of course, but we are emphasizing them here because the neutrophil's specialized receptors allow it to sense all kinds of pathogens. Cells don't have eyes, of course, so the only way they can know what is in their environment is through their receptors. A neutrophil needs to be able to sense bad things around it and go after them. Other parts of the immune system help the neutrophil to know what needs to be eaten by tagging pathogens with **opsonins**. This word comes from the Greek "opsonein" which means "getting ready to eat." In ancient Greece, side-dishes were knows as "opsons," and the Greeks thought of them as helping you to get ready to eat the main course. The immune system opsonins are tiny chemical tags that "taste good" to neutrophils. Neutrophils will eat anything covered with opsonins. Antibodies are a type of opsonin, and the IgG is especially useful as an opsonin. The two prongs of the Y stick to the pathogen and the base of the Y matches a receptor on the neutrophil. (Remember, B cells make Ig antibodies.) Other opsonizing molecules are made by the liver, and are part of a system called "complement" which we'll talk about in a later lesson. A protein called **C3b** is the most well-known protein opsonin made by the liver.

Neutrophils normally float around in the blood. When body cells outside the capillaries are under attack by invaders, they can bring neutrophils to the area by the following process. The nearby endothelial cells put out "hooks" that can grab the passing neutrophils. Once the neutrophils are stopped, they begin squeezing through the cracks between the endothelial cells. They can make enzymes that dissolve the junctions between cells, opening the gap even wider. (Then the endothelial cells go to work fixing the gap.) Once in the **interstitial space** (the space between body cells) the neutrophils begin "sniffing out" the invaders. Neutrophils can move on their own, a bit like an amoeba, so they can actually chase down the pathogens. They can "smell" both opsonins that may have been attached to the invaders, and also chemical trails being left by the invaders themselves. Once they get close enough, they engulf the pathogen by the process of **phagocytosis** (endocytosis of large things). This process brings the pathogen into the cell inside a vesicle. Vesicles containing engulfed pathogens are called **phagosomes**.

Once the pathogen-containing phagosome is brought inside the neutrophil, chemical weapons are employed. The neutrophils makes several kinds of chemical weapons, most of which involve oxygen. A general term for these oxygen-based weapons is Reactive Oxygen Species (ROS). The "bullets" on these weapons are often called **free radicals**.

1) Super-oxide: This is an oxygen molecule, O_2, with an extra electron stuck onto it. A special enzyme sticks on this electron. The electron is like a bullet that will go flying off, damaging whatever it hits.

2) H_2O_2 (hydrogen peroxide): You can see that if we took off one oxygen, we'd have water, H_2O, a very stable molecule. So how does that extra oxygen even stay on the molecule? Not very well, and it can go flying off very easily as a bullet called "singlet oxygen." A single oxygen is a very unhappy atom and wants to steal two electrons from anything it bumps into. In this case, stealing electrons from a pathogen will help to destroy the pathogen.

3) HOCl: (hypochlorous oxide, a form of bleach) Hydrochloric acid, HCl is a relatively stable molecule, so, again, we have a single oxygen that can fall off very easily. The enzyme robot that makes this molecule is light green in color. Since this weapon is made particularly in response to bacteria, if your mucus is greenish, it is likely that you are fighting a bacteria rather than a virus.

4) Digestive enzymes: Like all cells, neutrophils have lysosomes that are filled with digestive enzymes that can dissolve just about anything. If a lysosome merges (joins) with a phagosome, its enzymes get dumped all over the pathogen helping to destroy it.

5) Hiding iron: Neutrophils can gather up iron atoms so that the bacteria can't use them. Bacteria need iron for many of their cellular processes, as do most forms of life. (The Lyme disease bacteria B. Burgdorferi is a rare exception—it uses manganese.)

NOTE: A few types of bacteria can survive being in a phagosome by preventing the neutrophil from dumping in those chemicals. (Survival tricks like this are called **virulence factors**. Often, virulence factors can be shared with other bacteria, like secret information being passed around.) Fortunately, most bacteria can't survive being in a phagosome.

ADDITIONAL NOTE: Free radicals can damage body cells, too, not just pathogens. The body must have ways to protect its own cells from free radical damage. Substances that can absorb free radicals are called **antioxidants** ("against oxygen"). Some natural substances can be antioxidants, such as vitamins C and E, and the mineral selenium. Your body makes a powerful antioxidant molecule called **glutathione**. Glutathione is able to absorb dangerous electrons and not be destoyed itself.

NEUTROPHILS (60-65% of leukocytes in blood)

1) Our body makes about 100 billion per day.

2) We have 5 times as many neutrophils in reserve (in marrow mostly) as we do in circulation.

3) They float in blood until needed in tissues. When they get chemical signals that they are needed, they leave the vessels by squeezing through the cracks between the endothelial cells.

4) Lifespan: a few days

5) "Pus" is mostly dead neutrophils.

HOW THEY GET FROM BLOOD INTO TISSUES:

Neutrophils engulf pathogens by phagocytosis. Then they use chemicals to kill them.

Chemical messages are sent out by cells in distress. The endothelial cells put out "hooks" to slow down and catch the neutrophils that are floating past.

The neutrophils then squeeze through the cracks and get into the *interstitial* space. (the "empty" space between cells)

The neutrophils have chemicals that can dissolve the junctions between the endothelial cells, in order to make the crack larger. The endothelial cells then quickly repair the damage.

Neutrophils can sense the bacteria, but are also helped out by "yummy" tags placed on the invaders by other parts of the immune system.

NEUTROPHILS make 3 oxygen-based weapons:

1) **"Super-oxide"** is an oxygen molecule, O_2, with an extra electron stuck on (by a special enzyme). The electron will go flying off like a bullet, striking the invader. Super-oxide is a very common "free radical" found in your body. It kills pathogens, but it can also damage your cells.

2) **Hydrogen peroxide, H_2O_2** (yes, the same stuff that is in your First Aid kit for sterilizing wounds). The "bullet" here is the second O molecule.

3) **HOCl**, a form of "bleach." (hypochlorous acid) When the neutrophil is making a lot of this, your mucus turns green. (The enzyme that makes it is greenish in color.) Green mucus suggests bacterial infection rather than viral.

Other strategies:

4) **Digestive enzymes:** Lysosomes filled with enzymes can merge (join with the phagosome and dump enzymes all over the pathogen.

5) **Iron:** Neutrophils can hide (Fe) from bacteria. Bacteria die without iron.

42: LEUCOCYTES (part 3): MACROPHAGES

Macrophages start out as *monocytes*. We saw monocytes listed in our chart of blood cells back in lesson 36; it came from the myeloid stem cell. Half of our body's supply of monocytes are stored in the spleen and the other half are found in the blood. Monocytes can eat pathogens the way that neutrophils can (and they even have chemical weapons inside) but they are not as efficient at eating compared to neutrophils. In other words, they eat more slowly. The monocyte's main job isn't eating, so it's okay if it's slow. Its main job is to turn into either a *dendritic cell* or a *macrophage*.

So here's the pathway for making a dendritic cell: hematopoietic stem cell, myeloid stem cell, monocyte, dendritic cell. Dendritic cells get their name from their long skinny branches. They stay small, only about 10-15 microns, about the same size as the other white cells. The dendritic cell's job is to eat pathogens, digest them, then put tiny pieces of the pathogen in little clips pinned to its outer surface so that T cells can come and look at them. (Or rather, feel them, since T cells don't have eyes!) This process of eating pathogens then showing their pieces to T cells is called *antigen presentation*. Cells that do this are called *Antigen Presenting Cells*, or *APCs*. Often they will even be called "professional" APCs. (Where the amateurs are is anyone's guess!) Dendritic cells are found in large numbers in the skin and in the digestive tract, two areas that come into contact with things from the outside world, which is a good place to have cells that process antigens. Dendritic cells are also found in lymph nodes, which makes sense because this is where T cells hang out, too, and the antigens are being presented primarily for T cells.

The other option for monocytes is to turn into a macrophage. We met a macrophage back in the drawing on loose connective tissue, so we know that macrophages can be found in body tissues. Macrophages can even adapt themselves to inhabit certain types of tissues. Of course, this requires that scientists make up new names for them so that you can have more terms to learn. When we meet them in the lungs, they are called "dust cells." In the liver, macrophages are called Kupffer cells, in the skin they are Langerhans cells, and in the brain they are called microglia. Some scientists think that a type of bone cell called an osteoclast is also a type of macrophage but this is being debated. Osteoclasts are the opposite of osteoblasts, since they destroy bone and dissolve the minerals out of them, whereas the osteoblasts build bone and add minerals to them. Since osteoclasts spend their lives eating and recycling bone tissue, we can at least say they are certainly "big eaters" like macrophages. More on osteoclasts in the next module!

Macrophages have lots of lysosomes so they can digest almost anything that gets into your body. They will even try to eat things that they can't digest. Macrophages are your body's bottom line when it comes to cleaning up. If a macrophage can't deal with it, you are in big trouble. One thing macrophages can't break down is the mineral asbestos. Asbestos fibers are fireproof, so in past decades there were used extensively for insulation in buildings or inside appliances. Little did anyone know back then, but when microscopic asbestos fibers get into the lungs, they cause permanent and devastating damage. The poor macrophages get skewered by the sharp asbestos fibers. Since they can't digest them, those fibers stay in the tissues, stabbing all cells they come into contact with. People with asbestos fibers in their lungs often end up getting lung cancer eventually. Nowadays, asbestos is never used in places where people might get exposed to it.

Macrophages have basically three jobs.   First, as we already know, they are "eaters." They eat pathogens of all kinds, old erythrocytes (this takes place in the liver), dirt, debris, all cellular messes, and old or sick cells, especially neutrophils. Remember, neutrophils only live a few days and if they've been fighting germs, at the end of those days they are filled with dead, or maybe even still living, pathogens. What happens to the old neutrophils that are full of germs? Macrophages are enough larger than neutrophils that they can engulf them and digest them. Macrophages can grow to be 40 to 50 microns, compared to 10-15 for a neutrophil. For really big pathogens, macrophages can merge together to make a super macrophage that is over 100 microns in diameter.

Macrophages determine whether a neutrophil needs to be eaten by requiring what we are calling the "CD31 handshake." (This term is my own, so you won't see it elsewhere.) The "handshake" uses those CD31 hooks we drew on both the neutrophil and macrophage. A macrophage can "catch" a neutrophil by using this hook. It holds the neutrophil there, as if in a strong handshake, and won't let it go unles the neutrophil can give the correct chemical password. If the neutrophil is young and healthy, it will be able to give the correct chemical response and the macrophage will let it go unharmed. If the neutrophil is old or sick (or full of pathogens) it will have trouble giving the chemical response. If it fails to give the response, the neutrophil is engulfed and digested.

The second job of the macrophage is to present antigens to T cells. Like dendritic cells, macrophages are "professional" APCs. They digest pathogens and then put the tiny pieces on their outer membrane so that T cells can examine them. The little protein gadget clip used to post the pieces on the membrane is called the Major Histocompatibilty Complex, or *MHC*. We will learn more about this molecule in the next few lessons. Since MHCs are also used by body cells to display samples of proteins inside of them, we can help ourselves remember what MHC does by using the mnemonic "My House Cleaning." These clips show what is going on inside the cell.

Lastly, macrophages secrete a lot of cytokine messages in order to communicate with other cells. Messages that are passed between white cells are called *interleukins* ("Inter" means "between," and "leuko" means "white.") Interleukins 4 and 12 are especially important when macrophages are talking to T cells. They will tell the helper T cells what kind of helper to turn into.

Monocytes in the blood can go into tissues and differentiate into either macrophages or dendritic cells.

MONOCYTES (about 5% of leukocytes in blood)

-- 50% of them are in spleen
-- They can do phagocytosis, but are less efficient than neutrophils.

DENDRITIC CELL: Similar to macrophage except that it stays small, and is less involved in secreting cytokines (messenger molecules).

Dendritic cells are "professional" **A**ntigen **P**resenting **C**ells (APCs). They eat pathogens by phagocytosis then put tiny pieces of the antigen on their plasma membranes so they can show them to T cells in the lymph nodes.

Many dendritic cells are found in the skin, the lining of the intestines, and in lymph nodes and spleen

MACROPHAGES

Macrophages are found in all body tissues.
In some tissues they go by other names:

— Skin: Langerhans cells
— Liver: Kupffer cells (Macrophages in the liver clear out bacteria and also eat old red blood cells.)
— Lungs: "dust cells"
— Brain: microglia
— Bone: (osteoclasts?) This is being debated...

MACROPHAGES have basically 3 jobs:

(1) Phagocytosis of:
  1— pathogens
  2— old red blood cells (by macrophages called Kupffer cells in liver)
  3— dirt, debris, all cellular messes
  4— old or sick neutrophils

The "CD31 handshake" is when a macrophage grabs a neutrophil and won't let it go until it gives a chemical password. Sick or infected neutrophils will not be able do do this. They will get eaten.

(2) Presentation of antigens to T cells (and B cells)
(An antigen is anything that is "not self.")

Pieces of digested pathogen get attached to a "clip" called MHC II, and then moved to the surface of the membrane where they can interact with T and B cells.

MHC = Major Histocompatibility Complex (histo = tissue) but think "My House Cleaning"

(3) Release of cytokines (messenger molecules), especially the kind called "interleukins" (IL). Interleukins are numbered, such IL1, IL6, IL10. This is how white cells talk to each other and coordinate their actions.

43: LEUCOCYTES (part 4): LYMPHOCYTES: B CELLS

We've met antibodies in several lessons already. Now it is time to meet the cells that make them: **B cells**. B cells are lymphoid cells, which means they came from lymphoid stem cells. (You can look back at the chart in lesson 36 if you want to review the lymphoid and myeloid stem cells.)

B cells stay in the **B**one marrow to mature. (Oddly enough, technically, the "B" is not really for "Bone," but for "Bursa," an organ found in birds, where B cell maturation was discovered 1956.) Before they mature they are called "naïve," which means they have no experience and don't know what to do yet. (Those double dots over the "i" means you pronounce the "i" separately from the "a." You will see this word spelled both with and without the dots, but even if the dots are missing, you still say "nah-eve" or "ni-eve." We'll probably switch over to the no-dots spelling soon, as it is easier to type.) Maturing just means that they become capable of making one particular kind of antibody. One B cell makes one kind of antibody. Just one. It can make a million of them, but they will all be identical "clones." In fact, a whole bunch of identical antibodies are called mono<u>cl</u>onal antibodies. (Medicine names that end in "-mab" are made of <u>mono</u>clonal <u>a</u>ntibodies. Example: Rituxi<u>mab</u> is an antibody that attaches to a protein called CD20, which is found on the outside of B cells. The antibody sticks to the B cells so that macrophages will eat them. This medicine is used to reduce the total number of B cells in the body.) B cells, as a group, can produce over 10 million different shapes of antibodies, but each B cell makes only one shape. Most antibody shapes will not find anything to match up with and will never be used. Only a few will fit with an antigen. Once a fit is found, that B cell will be told to produce a lot of them.

Antibodies are protein gadgets. They are very small, too small to see without an electron microscope. When we draw them as Y's sticking to things, they are WAY out of scale! Antibodies are made the same way any protein is made, by transcribing sequences from the DNA, then having ribosomes translate them into protein chains. The difference between antibodies and any other protein is that the transcribing process allows for some mutations to occur. Only B cells have this special process where the bases C, G, T, and A are allowed to be scrambled randomly. When mutations occur in regular cells, or in other parts of the B cell, it is very bad for the cell. Cells have machinery akin to spell checkers and editors that try to eliminate mutations. But here, mutations are allowed because each B cell must produce an antibody with a unique shape. The only way this can happen is to use the process of random mutations.

Antibodies are made of 4 pieces of protein. The two longer ones are called **heavy chains** and the two smaller ones are called **light chains**. The base of the antibody Y is called the **constant region** and does not change. There are 5 different types of bases: A, D, E, G and M, and each one is used for a different purpose in the body. (We've already met IgEs and we know that the E base sticks to basophils and mast cells. None of the other bases stick to cells.) The top "V" part of the Y is the **variable region**, so this part varies from B cell to B cell. (This is where the mutation was allowed.) The top of the variable region is called the **antigen binding site**, the part that sticks to the antigen (not-self). The antigen sticks as a result of shape matching, like a piece going into a puzzle, or a key into a lock.

Antibodies have more than one function. First, they can act as an annoyance to a pathogen, by sticking to it. One antibody isn't annoying, but if you are completely covered in them, that's very annoying. (Imagine having 5,000 clothespins clipped to your body and clothes and you can't get them off.) Second, antibodies can cause clumping, called **agglutination**. Clusters of antibodies and antigens get stuck together, making the antigens non-functional. These clusters can be called **immune complexes**. The body will then try to get rid of these clusters. Large clusters will probably have to be eaten and digested by macrophages. Third, antibodies can attract neutrophils and macrophages. When they do this, we call them "**opsonins**," after the Greek word "opsonein," meaning "getting ready to eat." Neutrophils and macrophages love to eat things with opsonins (in this case, Ig's) stuck to them. The Ig's act like little candies.

After the B cells are mature and capable of producing their antibody, they mostly migrate to lymph nodes, but might also end up in the spleen (which is like a large lymph node) or the tonsils, or a few other minor places that you don't have to remember. They sit and wait. For what? For a T cell to come and let them know if their antibody is needed. We'll see this happen in a future lesson.

Lymphoid stem cells differentiate into 3 types of cells: B cells, T cells and NK cells.

B CELLS: Stay in the **B**one marrow until they mature, then they migrate to lymph nodes (also spleen and tonsils). Part of the maturation process is to "learn" to make one particular type of antibody. Antibodies are also called immunoglobulins, or Ig's)

The DNA in the nucleus will direct the production of one kind of antibody. Antibodies are super small, too small to see with a regular microscope. The are the size of enzymes.

Our bodies make10 million different shapes! (most will never be used)

Mature B cells go to lymph nodes to sit and wait until they are needed. These cells are called NAÏVE because they don't know what to do yet.

Q: What do antibodies do?
A: Stick to antigens.
Q: What is an antigen?
A: Anything "not self"

Antibodies function as "flags" to alert other cells to the presence of an intruder.
They also act like signs that say "Eat me!"

44: LEUCOCYTES (part 5): T CELLS

First, more information about how regular body cells function.

To understand how T cells work, we first need to learn a little more about how body cells work. Cells are like little houses with no windows. How will their neighbors know what is going on inside? What if a thief is inside? (for example, a virus)

The body has a roaming police force that constantly scans for trouble. The police cells (killer T's and NK cells) will kill any cell that cannot prove that everything is okay inside. Body cells must "clean house" and cover their outer membranes with samples of the proteins that are floating around inside. If the cellular police detect an intruder or a sickness (such a virus or a cancer cell) they will kill the cell so the problem does not spread to other cells.

There are always proteins floating around in the cytosol of the cell. They are part of the cytoplasm. A tiny shredder machine called a **proteasome** chops these proteins into tiny bits. (This is the same shredder machine that recycles mis-folded proteins have have been scrapped.) The little shredded pieces might be as small as only 5-10 amino acids. Then these protein bits go through a portal and into the ER (endoplasmic reticulum). Meanwhile, a ribosome is making a polypeptide and "spitting" it into the ER. This protein will fold up to become a protein gadget clip, called **MHC**. This little MHC clip is designed to hold a piece of shredded protein. Once this clip is loaded with a sample of protein, it is put into a vesicle and shipped to the surface of the plasma membrane. The clip inserts into the membrane and sits there, displaying the piece of protein it is holding.

MHC stands for **Major Histocompatibilty Complex**. "Major" means big or important, "histo" means "tissue," "compati-bility" means "getting along together," and "complex," in this sense, means a clump made of several smaller parts. There are two kinds of MHCs, and they are usually numbered with Roman numerals 1 and 2: **I and II**. Regular body cells make and use MHC I. We'll see soon that antigen presenting cells use MHC II. It is the interaction of these two MHCs that allows the immune system to know what is going on inside a cell and also to identify intruders.

MHC II is found only on **"professional" antigen-presenting cells (APCs)** which include macrophages, dendritic cells and some B cells. We will see how MHC II works in the next lesson.

An important side note about MHC I is that it is responsible for rejection of transplanted organs. The MHC molecule can have a lot of minor variations that don't affect its functionality. Each person can have as many as 50 different variations of MHC. You might want to think of MHC as being made of colored beads. There would be lots of ways you can arrange the colors and still have a functional MHC molecule. Each person has their own unique selection of colors and patterns. The closer you can get to matching the MHC patterns of the transplanted organ to the MHC patterns of the patient who is receiving it, the better. You really want to fool the immune system, if possible, into thinking that the transplanted organ should be accepted as part of the body. You may have seen registries where people can give a tissue sample that will go into a national database. The donor's MHC pattern will be analyzed and recorded. Then, doctors can use this database to try to find a good match for their patients. (Obviously, these databases can only be used for things like bone marrow transplants, where the donor doesn't have to die.)

NOTE: Just in case you see this term and wonder what it is, "HLA," Human Leukocyte Antigen, is another name for MHC.

To understand how T cells work, we first need to learn a little more about how body cells work. Cells are like little houses with no windows. How will their neighbors know what is going on inside? What if a thief is inside? (a virus)

The body has a roaming police force that constantly scans for trouble. The police will kill any cell that cannot prove that everything is okay inside. Body cells must "clean house" and cover their outer membranes with samples of the proteins that are floating around inside. If the cellular police detect an intruder or a sickness (virus or cancer) they will kill the cell.

WHAT'S GOING ON IN THE PICTURE ABOVE:

1) There are always proteins floating around in the cytosol.

2) A tiny organelle called a proteasome acts like a shredder and chops these proteins up into tiny pieces.
The shredded pieces are 5-10 amino acids long.

3) The tiny protein bits go through a portal and into the ER.

4) Meanwhile a ribosome is making a polypeptide, spitting it into the ER.

5) The polypeptide folds up and forms an MHC clip shape.
A protein is attracted to the clip part and sticks to it.

6) The ER puts the "loaded" MHC clip into a vesicle.

7) The vesicle goes to the plasma membrane and joins with it (exocytosis). That's how the MHC gets to the surface.

The little "clips" are called MHC molecules:
Major **H**istocompatibility **C**omplex. (My House Cleaning!)
-- MHC I is found on all body cells
-- MHC II is found only on APCs (antigen-presenting cells)

45: LEUCOCYTES (part 6): T CELLS, continued

Now we need to learn about the MHC clip on the APCs (Antigen Presenting Cells). The APCs are macrophages, dendritic cells and sometimes B cells, but here we will focus on the macrophages since they have a special relationship with T cells.

First, let's see how macrophages present pathogen samples to T cells. The macrophage eats a pathogen and begins digesting with by merging the phagosome with a lysosome. With neutrophils, we saw chemical weapons getting poured into the phagosome, but here we don't want to damage the pathogen's molecules. We just want to chop the pathogen into tiny bits. The merged vesicle is called a *phagolysosome*. (Textbooks make a big deal about knowing this term.) Meanwhile, an MHC II clip is being manufactured by the ER, similar to what we saw in the last drawing. What is different about MHC II manufacturing is that, unlike MHC I, the MHC II must have a safety clip attached to it so that it does not pick up any stray proteins. When the vesicle holding the MHC II merges with the phagolysosome, then the safety clip comes off and a piece of pathogen sticks to the binding site on the MHC II. The process then proceeds the same way as before, with the vesicle merging with the plasma membrane so that the MHC is then stuck to the outside of the cell.

T cells have receptors that look a lot like antibodies. *TCRs* (T Cell Receptors) can stick to a piece of pathogen if the piece happens to exactly match the shape of the receptor. The probability of a T cell finding a match for its receptor are actually very low. Most T cells will never find a match and, thus, never be used. However, a few will find a match. These few will then begin rapid mitosis so there are a lot more of them with a correct match.

A T cell is able to sense either MHC I <u>or</u> MHC II (not both). Some T cells have a specialized protein called *CD4*, which matches MHC II. ("CD" stands for "Cluster of Differentiation" which is just a fancy science way of saying "clump we've identified." They kept finding little clumps of proteins on the outsides of certain types of cells, and started numbering them. This was the 4th.) T cells with this CD4 protein are able to "feel" MHC II, and "know" that there is a macrophage behind the protein sample. Remember, cells don't have eyes or brains. They can't see the macrophage and they don't know that the bit of pathogen protein they are feeling isn't part of an actual pathogen. They have to sense the MHC II clip. CD4 cells are often called "T helper" cells because their role will be to help other T cells and also B cells to become activated. T helpers are not able to directly kill pathogens themselves.

Other T cells have a protein called *CD8*. (Apparently proteins 5, 6 and 7 were discovered in the interim.) CD8 matches with MHC I, so a T cell with this CD8 gadget will "know" it is touching a body cell. This type of T cell has often been called a "killer" cell because it has a weapon that shoots perforin "bullets" that puncture a hole in the cell membrane. Additionally, this T cell can then shoot toxins into the cell through this hole, so recently scientists have begun calling them "cytotoxic" T cells instead of just "killers."

The main thing to remember from this lesson is that CD4 T cells are called "helpers" and they recognize MHC II on antigen-presenting cells such as macrophages. CD8 T cells are called "killers" or "cytotoxic" and they recognize MHC I on body cells. CD8 cells are capable of killing infected body cells directly, using "bullets" and toxins that we will meet in lesson 47.

NOTE: Sometimes they put a little + sign next to the 4 and 8, like this: CD4+ CD8+ This plus sign just means that they would test "positive" for this protein if you tested them in a lab.

NOTE: As long as we have some extra space, this might be a good place to add a general note about all this complicated stuff we are learning. Why learn all this? Isn't this a bit too much? Actually, there's a good chance you might actually use this information at some point in the future, even if you won't be going on to study medical science. For example, you or a family member might have to make a decision about how to deal with an autoimmune illness and you'll need to know how certain medications work on the immune system. You'll have to decide what kind of doctor to go to and whether you want to try natural alternatives to pharmaceuticals that have lots of side effects. Or perhaps you'll find that knowing some of the immune basics will prevent you from being scammed by someone selling a supplement that they claim fixes the immune system. If you can see that their science is bad, don't waste your money. If nothing else, you'll have an appreciation for how incredibly complex the body is. It's far more complex than what is being presented here. Some researchers spend their entire career investigating one small part of the immune system and still don't know even half of what there is to know. Every time a new discovery is made, we find out how much more we don't know!

Now we need to learn about the MHC clip on the APCs (antigen presenting cells such as macrophages).

1) A pathogen gets eaten (phagocytosis). The vesicle containing the pathogen is called a phagosome.

2) The phagosome merges with a lysosome which contains digestive enzymes.

3) This merger is called a phagolysosome. The lysosome digests the pathogen and dissolves it into small pieces. The pieces are 15-20 amino acids in length.

4) Meanwhile an MHC II protein is coming out of the ER. (It was made by a ribosome, of course.) Notice the little safety device on the MHC II, making sure the clips stays open until it meets a pathogen protein.

5) The phagolysosome merges with the vesicle containing the MHC II. The safety device falls off and a piece of pathogen protein sticks to the MHC II.

6) The vesicle merges with the membrane.

7) The MHC II is on the outside of the cell.

T CELL RECEPTORS:

T cells have receptors on their membranes that will lock on to either MHC I or MHC II (not both). This means there are two different types of T cells.

46: LEUCOCYTES (part 7): T CELLS, continued again

Like all blood cells, T cells are "born" in the bone marrow. They come from the hematopoietic stem stem after it has differentiated into a lympoid stem cell. After T cells mature (meaning they've developed their T cell receptors and become capable of recognizing MHC) they migrate to the *thymus*. The thymus is a lesser-known organ and many people are unaware of its existence. This is partly due to the fact that in adults it is not very active any more. Most of its activity is during childhood.

The thymus is located under the ribcage, just above the heart. It has a blobby shape, but usually looks like it has two lobes, left and right. During childhood it is very large, but as we grow older, it shrinks and starts to become more "fatty" and less active. Inside the thymus are several types of specialized cells. One type is designed to catch T cells as they float by in the blood. This is how the T cells "migrate" to the thymus. They are carried along in the blood until they are "grabbed" by these cells in the thymus. Other specialized cells check to see if each T cell can recognize an antigen. If a T cell is defective and not capable of recognizing antigens, it will be destroyed and recycled. Other cells check the T cells to see if they will react to body cells that are presenting normal body proteins. If the T cells react to normal body proteins, they are also destroyed. (NOTE: Sometimes errors occur and T cells that react to body protein escape. These T cells are called *auto-reactive* because they react to "self" ("auto" means "self"). This causes *autoimmune* reactions where T cells go around attacking body cells.)   What happens in the thymus is often called "training" of T cells, but in fact it is more like "screening." About 98 percent of all T cells fail the screening process and are destroyed.

After leaving the thymus, T cells go to lymph nodes and sit there waiting until they are needed. Since they have not yet been activated, and have no "experience," they are called *naïve T cells*. The CD8 T cell has something that the CD4 does not. It has what we are calling a "perforin gun." The CD8 has the ability to shoot tiny proteins like bullets, puncturing a hole in the plasma membrane. After the hole is punctured, it can also shoot toxin molecules through this hole and into the cell. We will learn more about this in the next lesson.

Now, a note about pathogens. Pathogens can live inside or outside of body cells. Some can do both, but more often they choose one or the other. You can see the list of *extracellular* and *intracellular* pathogens on the drawing page. The body needs to know where a pathogen is lurking so that it can launch the appropriate response.

The decision of where to look for the pathogen is made by the macrophages who ate and presented the pathogen. (The science of exactly how this happens is still being researched.) Somehow, the macrophage "knows" where the pathogen is, and is able to pass this information along to the T cells. When a CD4 T helper cell comes along that has a receptor that matches the pathogen sample (being held in the MHC II clip) it will "lock on" to it and then wait for a signal from the macrophage. The macrophage will then secrete a *cytokine* (messenger molecule) that somehow tells the T helper where the pathogen is located, either outside or inside the body cells.

If the cytokine message tells the T helper that the pathogen is hiding inside the body cells, the T helper will then go and alert T killer cells. (We call this kind of T helper "1" and can write: Th1.) If the cytokine message tells the T helper that the pathogen is located outside the body cells, the T helper will then go and alert B cells. (We call this kind of T helper "2" and can write: Th2.) As we already know, the B cells make antibodies that will attach to the pathogens, causing clumping and also attracting phage cells to eat them.   The response of the B cells is called the *humoral response* because it occurs in a "humor" which is an old-fashioned word for body fluid.

An interesting side note about the Th1 versus Th2 response is in the disease known as leprosy, caused by a bacteria called *Mycobacterium leprae*. This bacteria likes to live inside of cells, so the best immune response would be to use Th1 helper cells that go and alert killer T's to kill infected cells. When this happens, the disease is very mild and usually the person's immune system is able to fight it off adequately. However, some people's immune systems make a mistake and use the Th2 pathway instead. This means that B cells will be alerted to make antibodies. Antibodies can't get into cells, so antibodies are pointless when the pathogen is hiding inside cells. When a body makes this mistake and uses the Th2 response to leprosy, the result is the devastating form of the disease that we are more familiar with.

T cells are "born" in the bone marrow, like all leukocytes, then they migrate to the **T**hymus to mature.
In the thymus the T cells will...
1) differentiate into either CD4 or CD8 cells
2) be tested to see if they will not attack body cells, but will attack pathogens (98% fail and are discarded!)

(T cells don't "know" to go to the thymus. They have special receptors that match those found on the epithelial cells in the thymus. The T cells float in the blood stream until they come into contact with the thymus cells, then they stick there.)

When CD4 and CD8 cells "graduate" from the thymus training school, they go sit in lymph nodes and tonsils until they are needed. (Most will never be needed.)

Since they have no work experience, they are called NAIVE T cells.

Now we must talk about the bad guys: PATHOGENS. They can be viruses, bacteria, protozoans, yeasts, or multicellular animals like worms. Some of these critters stay outside your cells and some like to get inside and hide. Don't forget that cells don't have eyes-- they can't see the pathogens in or out of cells! Yet they must decide where to look for the pathogens.

The decision is made by the macrophages who ate the pathogens. They tell the T helper cells whether to activate the B cells (to fight pathogens hiding outside the cells) or to activate CD8 T cells (for pathogens inside cells).

Examples of pathogens that generally stay outside of cells (extracellular):

-- anthrax (affects livestock)
-- most E. coli
-- cholera (in intestines)
-- meningococcus
-- clostridium (in intestines)
-- Borrelia (Lyme)
-- some strep (group A)
-- staph (S. aureus)

Examples of pathogens that like to hide inside of cells (intracelliular):

-- all viruses
-- Listeria (found on food)
-- some strep (group B)
-- candida yeast
-- Yersinia (the plague)
-- Salmonella (food poisoning)
-- mycobacterium (TB)
-- malaria (in erythrocytes)

47: LEUCOCYTES (part 8): How T cells work with B cells

We start with a review of how macrophages work. They eat pathogens, digest them, and put the pieces onto MHC II clips on the outside of the cell so that T cells can come and inspect them. The type of T cell that interacts wtih macrophages is usually the naive helper (CD4) T cell. (Note: I'm switching over now from "naïve" to "naive," not bothering with those two dots.) The cell is naive because it does not have a job yet. If the naive helper's receptor matches the piece of pathogen, it will then begin to change. What it turns into will depend on a signal from the macrophage.

The T cell will need to know whether the pathogen is an intracellular or extracellular pathogen. If the pathogen is intracellular and is hiding inside of body cells, the T helper will need to alert killer T cells who can kill infected body cells. This is called the *cell mediated response*. If the pathogen is extracellular and is located outside of the body cells, the helper T will go to the B cells and tell them to flood the body fluids with antibodies. This is called the *humoral response*. (Textbooks emphasis these terms, so it is good to know them.) The term "humoral" comes from the word "humor," a Medieval word for body fluids. In the Middle Ages they believed that there were four essential body fluids: blood, phlegm, black and yellow bile. These controlled not only your health but also your personality. They had awful remedies for getting rid of "excess" fluids. Even as late as the 1700s, doctors killed George Washington by letting him bleed, in order to get rid of "extra" blood. We don't use the word "humor" anymore for body fluids, but it does show up in several anatomy terms, such as "humoral."

In the cell mediated pathway, the naive Th0 (T helper zero) differentiates into a Th1 (T helper 1) and finds a cytotoxic (Tc) killer T that is waiting for instructions. A killer T recognizes bad proteins on the MHC I on body cells. It knows this is probably an infected cell that needs to be killed, but it waits for confirmation from a helper T. If the helper T has also recognized this bad protein, it will send a cytokine message to the killer T that says to go ahead and kill the cell. The Th1 cell sends cytokine messages back to the macrophage encouraging it to speed up. It also sends cytokine messages to itself telling itself to make many clones. Most of these clones will be active killers, but a few will become memory cells.

Killer T cells do their killing by releasing chemicals very near to the surface of the cell membrane. One of these chemicals is called *perforin* and is released as individual sub-units that will self-assemble into a tube. It's like an instant hole that puts itself together automatically. Once there is a hole in the cell membrane, chemicals (often called toxins) known as *granzymes* go through the hole and into the cell. The granzymes start a cascade process where one molecule affects another, which affects another, which affects another until the final molecule is released. This final molecule is a DNA shredder, and it destroys all the DNA in the nucleus. Once the DNA is gone, the cell can no longer make proteins and it slowly dies. Slow and contained cell death is ideal because then the surrounding healthy cells are not disturbed. (Then who comes along and cleans up the mess? Macrophages, of course.)

The virus that causes HIV AIDS attacks cells that have CD4 on their surface. This is mainly T helper cells, although macrophages and dendritic cells also express small amounts of CD4. As the population of T helpers begins to shrink, there are fewer and fewer helpers to interact with the T killers. The T killers find infected cells and are waiting to kill them, but without the helpers to give the "Go!" signal, they never get around to killing the bad cells. Thus, pathogens start taking over.

If the macrophage tells the naive Th0 cell that the pathogens are extracellular, the T cell will then turn into a Th2 and begin interacting with B cells. This strategy makes sense because B cells make antibodies that are released into body fluids and this is where extracellular pathogens are. B cells will have already discovered the pathogens and will have started making IgM antibodies against them. IgMs are made of a group of 5 antibodies attached at their stems. This large, somewhat clunky-looking shape is good for the initial stage of finding and binding antigens, but in the long run, the IgG form will be better. An IgG looks like a single Y. To ramp up the fight against the pathogen, the B cell should switch from making IgMs to making IgGs. However, it will not make this switch unless it gets a signal from a Th2 cell. The Th2 sends cytokine messages to the B cell, allowing it to start "class switching." Once the B cell switches over to IgGs, it will no longer be able to make IgMs. It is a permanent switch.

The B cell also begins to make many clones of itself. Most of the clones will be *plasma cells*, actively making antibodies. A few will become memory cells so that if this pathogen is ever encountered again, the immune system will be better prepared to deal with it. Usually, we are never aware that our body encounters the pathogen a second time. The pathogen is dealt with so quickly that we never actually "come down with" the infection. B memory cells are very long-lived, often enduring for our entire lifetime.

There are some other options for naive T cells, not just Th1 and Th2. Occassionally they will turn into **T regulatory cells**, which used to be called T suppressors. Tregs tell the killers to stop attacking once the infection is under control. The Th0 might also turn into a Th17. **Th17** cells are the opposite of Tregs—they encourage inflammation and immune activity. Th17s used to be thought of as a malfunction of the immune system, but now they are seen as a cell that plays a helpful role in some situations. **Th3** cells encourage B cells to make IgA antibodies. IgAs are like two Y's joined at their bases. They can't stick to body cells, so they can't cause inflammation. Their purpose seems mainly to be washing out antigens from the respiratory tract and the digestive tract. IgA antibodies bind and hold antigens, then get washed out with mucus secretions.

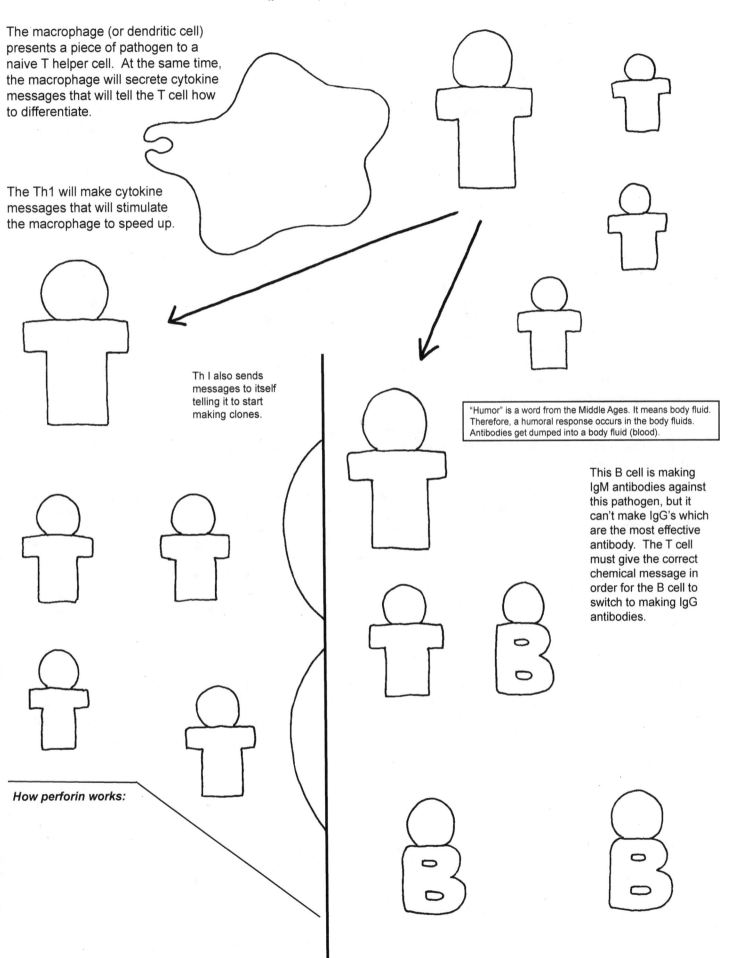

The macrophage (or dendritic cell) presents a piece of pathogen to a naive T helper cell. At the same time, the macrophage will secrete cytokine messages that will tell the T cell how to differentiate.

The Th1 will make cytokine messages that will stimulate the macrophage to speed up.

Th I also sends messages to itself telling it to start making clones.

"Humor" is a word from the Middle Ages. It means body fluid. Therefore, a humoral response occurs in the body fluids. Antibodies get dumped into a body fluid (blood).

This B cell is making IgM antibodies against this pathogen, but it can't make IgG's which are the most effective antibody. The T cell must give the correct chemical message in order for the B cell to switch to making IgG antibodies.

How perforin works:

48: IMMUNE SYSTEM OVERVIEW

The body has several layers of defenses against pathogens. The first layer consists of physical barriers that attempt to keep germs out. The most obvious barrier is the skin. The epidermis does a pretty good job of keeping bacteria and parasites from entering the body. It is usually only when we get a cut or scrape that invaders get in. The lining of the respiratory tract (nose, trachea, lungs) keeps germs out by secreting a mucus layer. In the trachea the cells also have cilia that sweep the mucus up and away from the lungs. The digestive tract has a mucosal layer, as well. The stomach secretes a strong acid that can kill most pathogens. In the eyes, tears flush out germs, and the tears themselves have some anti-microbial properties. The nose, with its constant mucus production, does quite a good job of washing out not only pathogens but dust and dirt particles, too.

The second layer of defense is called the innate immune system. The root word "nat" means "born" so this is the system you are born with. The innate system is all set up and ready to go and does not need any training. (Since the T cells are trained in the thymus, they would not be part of this system.) The cells that are usually considered to be part of the innate (non-trained) system are the basophil, the eosinophil, the mast cell, the neutrophil, the NK (natural killer) cells, and usually the macrophages and dendritic cells, as well. Some texts also include the liver as part of this system, since it produces proteins that help the other cells do their jobs.

The NK (natural killer) cell is a lymphocyte, like the T and B cells. In fact, some scientists considered it to be basically another type of T cell. Its physical appearance and function are a lot like the killer T cells. The NK cell's main job is to feel the outside of body cells, checking for MHC I. It can detect bad proteins attached to MHC I and then directly kill the cell. NK cells do not need to check with T helpers first, the way killer T cells do. Perhaps that is why they are called "natural killers." NK cells can do something that Tc cells can't do—they can detect the <u>absence</u> of MHC I. If a cell is sick it might not be able to manufacture normal tags such as MHC. Also, some viruses can cause cells to stop expressing MHC I. The NK cell "feels" the surface of the body cell and if it does not feel any MHC I, it will kill the cell. (The HIV virus is very sneaky, because it causes the cell to put out just enough MHCs so that the NK cell does not get suspicious.) The NK kills cells the same way that the killer T does, using a ***perforin*** "gun." The perforin makes a hole in the cell, then the ***granzymes*** (the "toxins") go in and start a cascade process that ends with the DNA in the nucleus being shredded. This is called ***programmed cell death***, or ***apoptosis***.

The macrophages and dendritic cells are special members of the innate team because they are also sort of members of the adaptive team, as well. We've already seen how these cells work, presenting antigens to the T cells. The liver is also often listed as a member of the innate immune system. We'll discuss it in just a minute.

The adaptive system has three names, which can be very confusing. These three names are used equally, and you have to be familiar with all of them and know that they are the same thing: ***adaptive, specific, acquired***. "Adaptive" means that this system adapts and changes over time. "Specific" means that its function aims at specific pathogens, not all pathogens in general. "Acquired" means that you gradually acquire it over time, as your body comes into contact with various pathogens in your environment. We've already met the members of the adaptive immune system: the T cells and B cells. We've learned how they are able to react to a very specific pathogen, through the recognition of the T cell receptors and the B cell's antibodies. Remember, antibodies are very specific and can only attach to one particular molecular shape. We've got millions of T cells and B cells, each one with a slightly different shape on its receptor or antibody. Hopefully, when a pathogen comes along, one of the shapes will match. What allows you to acquire and keep this system over time are the memory cells that are produced.

The liver produces many different kinds of proteins that help the immune system to be more efficient. These proteins, as a group, are called ***complement***. (It is strange that they use this word in the singular form, instead of saying complement<u>s</u>. However, strange as it is, that's the way it is.) The complement proteins are numbered using the letter C: C1, C2, C3, etc. These proteins will function in a chain reaction manner, somewhat similar to the chain reactions we see in the coagulation cascade and in apoptosis. Just like coagulation proteins, these complement proteins float in the blood in their inactive form and are not activated unless they are needed.

Complement proteins accomplish four things:
1) They can activate mast cells.
2) They make ***Membrane Attack Complexes*** (similar to the ring that perforin makes).
3) They can act like opsonins, making macrophages and neutrophils want to eat them (and whatever they are stuck to).
4) They can stick pathogens together in a clump so that they are easier for the phagocytic cells to eat. (***agglutination***)

C3 is the only complement protein we'll take a close look at. C3 has two sections, **a** and **b**. A scissor enzyme can come along and separate C3a from C3b. C3a can go over and stick to mast cells, beginning the inflammation process. C3b goes and attaches to the plasma membrane and attracts C5, C6, C7, C8 and C9. These molecules self-assemble and form a ring, much the same way that the perforin molecules do. The ring formed by these complement proteins is called the ***membrane attack complex***. The main difference is that the membrane attack complex is not followed up by toxins. There aren't any granzymes here, just a big hole. If lots of membrane attack complexes are launched, hopefully the pathogen's membrane will end up with enough holes that it will collapse.

The body has several layers of defenses. Scientists have identified three basic layers and given them names.
They are: 1) physical barriers, 2) the innate (non-specific) system, and 3) the adaptive (specific) system.

1) **PHYSICAL BARRIERS** that can block pathogens from entering the body.

2) **THE INNATE SYSTEM** (also known as the NON-SPECIFIC system) ("Nat-" is Latin for "born," so innate means you are born with it.)

Natural Killer cells are lymphocytes (related to T cells). They mature in bone, thymus, tonsils, spleen and lymph nodes. NK cells have multiple types of receptors and can sense both bad antigens and missing MHC I on body cells.

3) **ADAPTIVE SYSTEM**
(also known as **SPECIFIC** or **ACQUIRED**)

The liver produces tiny proteins called COMPLEMENT (written in the singular, though this sounds strange!). These proteins function as a CASCADE so the response can be fast, hopefully faster than the rate at which the pathogens can multiply! Just like coagulation proteins, the complement proteins float in the blood waiting until they are needed. Once the first protein is activated, then the cascade starts and all the others are activated. Most of the complement proteins are named with the letter C (C1, C2, C3... C9).

Complement proteins accomplish 4 things:

1) Activates _____

2) _____

3) Acts as _____ ("eat me" tags)

4) _____ (sticks pathogens together into clumps)

C3 is a key protein:

The "b" part sticks to a membrane and attracts C5, C6, C7, C8 and C9.

The trigger can be when C1 binds to Ig's that are bound to an antigen

49: THE NEURON

Nervous tissue is designed for basically one function: transmitting electrical signals. The cells that actually do the work of transmitting are called **neurons**. Neurons need lots of help, so they have a whole crew of supporting cells that protect them, nourish them, and repair them. In this lesson we will look at the anatomy of just the neuron and one supporting cell.

The neuron we've drawn is a **motor neuron**. ("Motor" means "movement.") This means that it will likely be attached at one end to a **muscle fiber**. Neurons shown in textbooks are almost always motor neurons. This type of neuron is only found in the **peripheral** nervous system, which means everything outside of the brain and spine. **Sensory neurons** are also found in the peripheral nervous system. Some sensory neurons take in information directly, such as pain and pressure sensors. Pain sensors have free nerve endings that are easily irritated and result in signals being sent to the brain, where they are interpreted as pain. Pressure sensors have some padding around them so that they don't register signals as pain. Some sensory neurons connect to specialized cells that can sense things like light, smell and taste. The sensory cells then transmit a signal to the sensory neuron. Sensory neurons usually have their cell body in the middle (this is called a **unipolar** arrangement).

Neurons have basically three parts: a cell body called the **soma,** some **dendrites**, and an **axon**. The soma contains the nucleus, ER, Golgis, mitochondria, lysosome, ribosomes, etc. The dendrites are often long and skinny and look a bit like tree branches, which is how they got their name. The job of the dendrites is to sense in-coming electrical signals. The axon is long and thin and can be anywhere from a few millimeters long to almost a meter long. The place where you find very long axons is in your leg, running from the base of your spine down to your toes. Axons carry electrical signals all the way to the ends of the cell. The end of the cell branches out, but not is not quite as branch-like as the dendrite end. At the end of each branch is a little knob-like thing. This knob doesn't have an official name and goes by quite a few names: terminal knob, synaptic knob, terminal button, axon terminal, synaptic terminal. Take your pick; all of these seem to be used equally. Here, we'll use the term **terminal knob**. The terminal knobs are full of mitochondria and also vesicles waiting to release their chemicals. Terminal knobs are usually connected to either a muscle fiber or the dendrites of another neuron. We will talk more about the terminal knobs in a future lesson.

The axons of neurons found in the body (not in the brain or spine) have their axons surrounded by **Schwann cells**. These protective cells are incredibly thin and very long, and are rolled around the axon like a piece of paper can be rolled around a pencil. One end of the Schwann cell is thicker because it contains the nucleus and organelles. The thick end stays on the outside. The thin part of the cell is so thin that it is basically nothing more than the plasma membranes with a lot of fat (mostly cholesterol) molecules in the middle. This line-up of protective Schwann cells is known as the **myelin sheath**. You hear the term myelin sheath more often than you hear about Schwann cells. It can be easy to forget that the sheath is actually made of cells. The Schwann cells have two jobs. First, they insulate the axon the way electrical wires are insulated by their plastic or rubber coatings. The neuron's job is to conduct electricity, like a metal wire, so it needs to be insulated. Second, the Schwann cells can help an axon to regrow if it gets severed. The Schwann cells will stay in place and acts as tubes to guide the severed ends of the axon back together, if possible. This healing process is slow and can take weeks or months.

If an axon begins to lose part or all of its myelin covering, the neuron will not be able to transmit electrical signals. One result can be that muscles are no longer able to move properly. The disease known as Multiple Sclerosis involves a breakdown of myelin.

The mitochondria in a neuron are critically important to their functioning. They start out in the soma (cell body) and travel down the axon to the terminal knobs. Then, after a while, they are transported back up the axon and return to the cell body where they are recycled and made into new mitochondria. This transport of mitochondria up and down the axon is done by those tiny motor proteins that look like they are walking. They carry the mitochondria along the microtubule "roads." Microtubules (part of the cytoskeleton) go all the way down through the axon. Also, vesicles filled with chemicals are manufactured in the ER and Golgi bodies and then carried down to the knobs by the motor proteins. Researchers think the vesicles can also be taken back up and recycled, but this is a very new discovery and there is still a lot they don't know.

When mitochondria are not properly transported up and down the axon, the results can be devastating. There is a special molecule whose job it is to get the mitochondria from the "down track" to the "up track." If this molecule is not produced, the neuron will malfunction. This is one of the things that goes wrong in neurological diseases such as ALS (Lou Gehrig's disease). Even if the problem seems very small, like the mitochondria not being able to get turned around in the knob, big problems can be the result.

This is a **MOTOR NEURON** (often connected to muscle fibers).
Only found in the peripheral nervous system (not in brain or spine).

SENSORY NEURONS connect to our five senses and transmit information to the brain.

INSIDE AXON

Mitochondria and vesicles are moved along by motor proteins on microtubule "roads" getting them to areas that are experiencing a lot of action potential and therefore need lots of ATPs.

CROSS SECTION of **Schwann cell** shows how the cell wraps around the axon many times, forming an insulating sheath that keeps the sodium ions inside. The plasma membrane of a Schwann cell contains a very high proportion of lipids including a lot of cholesterol. These inner layers are called the MYELIN SHEATH. "**Myelin**" is often defined as an "insulating lipid substance" but it is important to remember that it is also the plasma membrane of the Schwann cell.

The axon's terminal knobs connect either to a muscle fiber or to the dendrites of another neuron.

50: NERVOUS TISSUE in the PNS

The big divide in classification of nervous tissue is the PNS versus the CNS. These abbreviations are standardly used in all texts. **PNS** stands for **Peripheral Nervous System** and **CNS** stands for **Central Nervous System**. The central nervous system is the brain and the spinal cord. The peripheral nervous system is everything else (arms, legs, chest, face, etc.). Each type of nervous tissue has special features that help it to maximize its efficiency.

Neurons in the PNS have slightly different shapes according to what they do. They all have the same three parts (soma, dendrites and axon) but these parts can be arranged differently. The **motor neuron** has the dendrites coming off the soma, and the axon going out one side. Many motor neurons have their cell bodies inside the spinal cord, and their axon terminals attached to a muscle fiber. Their job is to relay signals from the brain that are telling the muscle fiber to contract. This is the type of neuron we drew in the last lesson. Neurons that have multiple "processes" (i.e. "things") sticking off the soma are called **multipolar** neurons. Another vocabulary word associated with this type of neuron (as if we needed another one) is the word **efferent**. Efferent means taking information from the brain and relaying it out to the body. The opposite of this is **afferent** (as if we needed another similar word). Afferent neurons take messages from the body to the brain.

A **sensory neuron**'s job is to pick up information from the outside world and carry it to the brain. Sensory neurons have dendrites that are adapted to one of the senses. For example, the dendrites of some sensory neurons are found in skin and receive signals that are interpreted by the brain has pain, pressure, cold or heat. Sensory dendrites in the nose and tongue and are activated when certain chemicals touch the receptor cells. Sensory dendrites in the eye are connected to other cells that are activated by light. In the ear, there are several types of sensory dendrites, as the ear is involved in not only hearing sounds, but balance, too. The cell body of a sensory neuron is located in the middle of the cell, which gives it a rather odd appearance. Neurons that look like this are called either **bipolar** or **unipolar**, depending on whether there are two things sticking off the cell body (bipolar) or only one (unipolar). (NOTE: Bipolar neurons have no connection with the brain syndrome called Bipolar Disorder. "Bipolar" is a general term that means something with two sides.) Recently, scientists have begun to call unipolar cells **pseudounipolar**, because on close inspection it is more complicated than seeing just "one thing" sticking off the soma. (If you are not going to be studying anatomy in the future, don't worry too much about learning these terms.)

Axons can be very long. The longest axon in the body runs from the base of the spine all the way down to the foot. Each axon only controls a very small muscle fiber, so you must have millions of axons in order to control your millions of muscle fibers. Instead of having these millions of axons running every which way all over your body, the axons are bundled together in a very organized way.

The term **nerve fiber** is used when we talk about an axon and its coverings. Technically, the "axon" is just the long skinny part of a neuron and does not include any wrappings. Of course, we already know that most axons in the PNS are surrounded by Schwann cells that protects and insulates. Around the Schwann cells there is another protective layer called the **endoneurium**. (This is another word you don't need to memorize unless you will be studying anatomy in the future.) The endoneurium is made of connective tissue. It's like the paper wrapper around a sandwich and serves to hold everything in place. (When we studied connective tissue we mentioned that it is found all over the body; this is a place that is rarely mentioned.) So a nerve fiber is an axon with its myelin sheath (Schwann cells) and its connective tissue covering.

Nerve fibers are then bundled together into **fascicles** *(FASS-ick-uls)*. The fascicle also has a connective tissue covering, very similar to the one surrounding the nerve fiber. The covering around the fascicle is called the **perineurium** ("peri" means "around"). Again, not a word you need to worry about unless you will be going on to study anatomy in greater depth. The fascicles are then bundled together to make **nerves**. Nerves also contain some tiny blood vessels and connective tissue to hold everything together, and they have a covering called the **epineurium** ("epi" means "on top"). Nerves can contain nerve fibers (axons) from both efferent and afferent neurons.

Nerves are usually found running alongside blood vessels. A "neurovascular bundle" is a nerve, an artery and a vein held together by connective tissue. When a nerve is cut, some of the special fluid found inside these connective tissue coverings can leak out. This fluid can be sensed by MRI imaging, making it possible for doctors to determine the location and extent of damage to nerves.

A **reflex arc** is an arrangement of motor and sensory neurons that allows fast reaction time by processing the signal in the spine instead of going all the way to the brain. A sensory neuron picks up a signal, relays it to the spinal cord where an interneuron connects it to a motor neuron that then sends a signal out to a muscle. For example, if your finger touches a hot stove, the hot and pain sensors are triggered and the signal is sent to the spine, through the interneuron and over to the motor neuron which is connected to a muscle that immediately moves in order to get your finger away from danger. Another signal is sent to the brain so that you will realize what happened, but by the time your brain understands what just happened, your finger will have already pulled away from the heat.

You can see in the cross section of the spine that the cell bodies of the sensory neurons are all located in the same area, creating a lump that we call a **ganglion**. Since this ganglion is attached to the dorsal (back) part of the spinal cord, we call this lump the **dorsal ganglion**. There isn't a matching ganglion in the front because the cell bodies of the motor neurons are located inside the spinal tissue. Can you see the lump that they form inside the spinal cord?

Neurons in the PNS (Peripheral Nervous System) are specialized for the jobs they do. The size and length of the axon and the location of the soma can vary.

Motor neurons usually run from spinal cord out to a muscle fiber. (The fancy word for this type of neuron is **EFFERENT**.)
Motor neurons have many "processes" sticking off the soma, so they are called **MULTIPOLAR**.

Sensory neurons send signals from senses to brain. (The fancy word for this type of neuron is **AFFERENT**.)
Sensory neurons can be either **BIPOLAR** (two "processes" sticking off soma) or **UNIPOLAR** (or PSEUDOUNIPOLAR) with one "process" off soma.

A NERVE is a bundle of _____, which is a bundle of _____.
A NERVE FIBER is made of an _____ and its _____ and also the _____.

Nerves usually run alongside blood vessels.

Afferent and efferent nerves can be connected by an INTERNEURON, so that they form a REFLEX ARC.

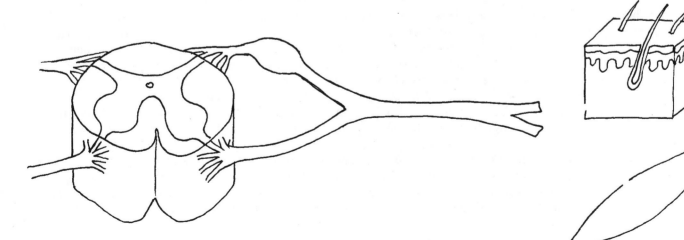

51: NERVOUS TISSUE in the CNS

The **CNS (Central Nervous System)** consists of the brain and the spinal cord. This drawing shows a slice of brain tissue. You would find similar cells in the spinal cord but they would be arranged differently. Neurons are the primary cells of the CNS, but they are quite outnumbered by the supporting cells all around them. The supporting cells are called **neuroglia**. *(nur-o-GLEE-ah)* ("Glia" means "glue.") Some of these supporting cells really do act like glue, keeping the neurons tightly in place.

In our drawing, the neuron is colored green. In some places you can see (green) dendrites coming very close to some (green) axon terminals. These connections between dendrites and axon terminals will be discussed in the next lesson. This is where the electric signal is passed from neuron to neuron, though the signal must be translated into chemicals in order to jump the gap.

These neurons look like they have Schwann cells along their axons, but in fact these cells are called **oligodendrocytes**. ("Oligo" means "few.") In the demo we colored them blue. Oligodendrocytes have a few "arms" that reach out and grab sections of axon, covering and insulating them. The oligodendrocytes also help to stabilize the neurons and prevent them from moving around. Despite their name ("few branches") some oligodendrocytes have been observed to be holding on to as many as 50 neurons. At birth, only some of your neurons have oligodendrocytes around them. The process of **myelination** (putting myelin sheaths around all the axons) isn't complete until you reach 25-30 years of age.

The cells that are best at stabilizing the neurons are the **astrocytes**. ("Astro" means "star.") In our drawing, they are yellow. The astrocytes use their "arms" to hold not only the somas and axons in place, but also, very importantly, those connections between dendrites and axon terminals. The astrocytes surround these connections and keep them from slipping apart. If these connections were to get torn apart, there would be disastrous consequences for the body—it would be like tearing the electrical wiring out of a building. The elevators and lights would not work. In the body, it's not elevators and lights—it's things like muscles and memories. The astrocytes are also responsible for feeding the neurons and taking away their wastes. They do this by connecting some of their "arms" to a blood vessel so that they can absorb oxygen, water, and nutrients. These substances travel though the astrocyte and then go out through other arms that are touching neurons.

The cells that we've colored orange are called **microglia**, and are the only neuroglial cells that have "glia" in their name. This is a bit ironic because these glial cells don't act like glue at all. Microglia are the macrophages of the brain, and are constantly moving around, looking for pathogens to gobble up. Microglia are the only immune cells found in the brain; you won't find neutrophils or T cells or basophils or any of the other immune cells found in the rest of the body. Microglia have extra receptors that normal macrophages don't have, allowing them to be more efficient at recognizing and eating pathogens. When microglia aren't eating, their arms look very skinny and branch-like. Once they get activated and start into eating mode, the branches disappear and they look much more like regular macrophages (kind of blobby).

Like macrophages, the microglia are also responsible for all clean up jobs. They digest damaged cells and get rid of any molecular debris they find in the interstitial spaces (between cells). They also help to prune the electrical wiring by eating old connections that have not been used and are therefore deemed unnecessary. Getting rid of unused connections makes the neuronal network more efficient. Some researchers have observed microglia doing what looks like tapping on neurons and their connections, as if to inspect them. The more we learn about microglia, the more intelligent they seem to be!

Ependymal cells *(eh-PEND-i-mahl)* form the lining of fluid-filled spaces called **ventricles**. We will learn more about ventricles when we take a look at the brain as a whole. Ventricles are "empty spaces" in the brain that are filled with a fluid which is produced, in part, by ependymal cells. The ependymal cells have cilia on the side that faces the ventricle. The cilia are used to move the fluid around. The fluid flows in and out of four ventricles, and then goes out to the spaces around the outside of the brain and spinal cord. Eventually, the fluid is reabsorbed back into the body and the water and minerals are recycled.

The brain is full of blood vessels. Large vessels come in, then split off into smaller and smaller branches, eventually forming tiny, microscopic capillaries. The capillaries are where the gas and nutrient exchange occurs. Astrocytes attach themselves to the outside of the capillaries to absorb nutrients. Another type of cell also attaches itself to the capillaries. **Pericytes** wrap around the capillaries in order to help control blood flow and also to help keep large things from leaking out. They are like helper cells for the endothelial cells. The endothelial cells and the pericytes together form what is known as the **Blood Brain Barrier (BBB)**. The endothelial cells in the brain are stitched together very tightly, much tighter than in the rest of the body. In body tissues, the capillaries are often exposed to histamine from mast cells that causes them to dilate and get leaky. This should never happen in the brain. (Good thing there are no mast cells in the brain!) Tiny molecules such as glucose and amino acids are absorbed from the blood by the endothelial cells, then transfered to the astrocytes, who then transfer them to the neurons. The essential things that neurons need are all small enough to be obtained in this way. The BBB keeps harmful substances out of the brain. In the rare case of a pathogen getting into the brain and causing a condition that needs medical treatment, doctors must give the patient a substance that will cause the brain capillaries to be a little bit leaky so that medicines can get into the brain tissues. This is only done in an emergency, as you really don't want to disturb the BBB. Because the BBB is so efficient, it is difficult to find medicines to treat long-term brain problems.

The CNS (Central Nervous System) consists of the brain and the spinal cord. This is a drawing of the cells of the brain. Most of these cells would be in the spine, also, but the arrangment might be different.

Neurons (transmit electrical impulses)

Oligodendrocytes (act like Schwann cells)

Astrocytes (protect and nourish neurons)

Microglia (macrophages of the brain)

Ependymal cells (secrete fluid into ventricles)

Endothelial cells (form capillary walls)

Erythrocyte (carries oxygen)

Pericytes (wrap around vessels, regulate blood flow)

52: THE ACTION POTENTIAL and THE SYNAPSE

Neurons are the focal point of both the PNS and the CNS since they are the ones that carry the electrical impulses that go back and forth between the brain and the body. There are several basic mechanisms in the neuron's axon that work together to accomplish the task of transmitting an impulse: the sodium-potassium pump, ion channels, and neurotransmitters in the synapse.

All along the axon's plasma membrane there are many sodium-potassium pumps. These pumps use one ATP every time they pump. One complete pumping action pumps 3 Na^+ ions outside the membrane and brings 2 K^+ ions inside. The net result is one more positive ion outside than inside. This means that eventually there will be more positive charges outside than inside. Another way of looking at it is that the inside is now more negative than the outside. Therefore, you often see the inside of the axon labeled with a negative sign and the outside with a positive. This is confusing to students when they see that both ions being pumped having positive charges. The negative charge on the inside is in comparison to the outside, which has even more positive ions. What the ion pumps have done is set up a gradient. Remember, atoms want to be equally distributed everywhere. Now that we have most (but probably not all) the Na^+ ions on one side and the K^+ ions on the other, what will happen if we open ion channels that will let them travel across the membrane? Yes, the Na^+ ions will rush in through **sodium gates**, and the K^+ ions will rush out via **potassium gates**. This rushing in and out is called the **action potential**. This active rushing of ions IS the electrical signal we keep talking about neurons carrying.

After the Na^+ ions have all rushed across the membrane, we now have the reverse situation—we have the inside of the axon being positive and the outside being negative. Once the K^+ gates open and the K^+ ions rush across, we no longer have a positive inside, but we are not quite back to where we started. To be able to do another action potential, the axon must restore the original amount of negativity inside. So the sodium-potassium pumps go to work and start pumping Na^{+i} ions out and K^+ ions in. Once everything is back to where we started, this is called the **resting potential**. The resting potential is when the axon is all set and ready to go for another action potential. The word "potential" is appropriate here because in this state, the axon has the potential to carry a signal, but has not done so yet. (For those of you who like details, in this resting state, a very precise voltmeter will register the inside of the axon at about -70 millivolts, mV. After the action potential has fired, the outside will then register at about +40 mV.)

The axon we are showing here doesn't have any Schwann cells around it. Something very interesting happens under the Schwann cells. The action potential can kind of "jump" through the insulated Schwann areas, without having to do all the rushing in and out of ions. The rushing of ions occurs only at the nodes of Ranvier, where there are gaps between the Schwann cells. Since it seems that the action potential can "jump" from node to node, the Latin word for jumping, "saltare," was used to form a word for this neuron action: **saltatory**. Since we also use the word "conduct" for electricity, the complete term for this jumping of the action potential from node to node is **saltatory conduction**. (For a visual demonstration of this, see the video listed on the lesson page.)

When the action potential has gone all the way down the axon it ends in the terminal knobs. This electrical impulse, the action potential, cannot jump the gap that exists between the terminal knobs and the dendrites of the next neuron. (This area where knobs meet the next neuron is called the **synapse**. The actual empty space between is called the **synaptic cleft**.) Instead, the action potential causes calcium ions to rush into the knobs and thereby causing the waiting vesicles to fuse with the membrane and spill their contents into the gap. We saw this use of calcium in module 1 when we studied fertilization of the ovum. Calcium ions cause vesicles to do exocytosis. The neurotransmitter chemicals in the vesicles can be of various types. (For more information on neurotransmitters, see the additional info on the lesson page.) Each transmitter will have a matching receptor on the sides of an ion channel embedded in that dendritic plasma membrane. (NOTE: Sometimes axon terminals touch the soma, too, not just dendrites.)

Some neurotransmitters fit into receptors on Na^+ channels. Others stick to receptors on K^+ receptors. If two neurotransmitter molecules stick to a Na^+ channel it will open to allow an influx of Na^+ ions. Since this is what happens in an action potential, the result will be that this next neuron will be encouraged to start a new action potential. (Mulitple Na^+ gates have to open to have a new action potential start. One is probably not enough.) If neurotransmitters stick to a K^+ channel, then K^+ ions will flow in and... nothing will happen because having a lot of K+ on the inside is part of the resting potential, not the action potential. This action of K^+ ions is called an **inhibitory** response because it discourages, or inhibits, the next neuron from beginning a signal. The action of the Na^+ ions is called an **excitatory** response because it encourages, or excites, the next neuron into action. If enough Na^+ ions flow into this next neuron, an action potential will begin in the hillock and move down into the axon. One role of the soma is to collect all the various inputs coming from all these ion channels and sort of "sum them up" and then generate—or not generate—a new signal.

This process is, of course, more complicated than described here. For exaxmple, there are a number of neurotransmitter chemicals that have different actions in the synapses. Also, there are **enzymes** waiting in and around the synapses that capture and either destroy or recycle the neurotransmitters. The neurotransmitters have to be disposed of immediately so they don't keep acting after the signal is finished. They, also, need to be reset every time. And all of this (everything on this page) happens in a fraction of a second!

NOTE: You will need to use the cut-off strip at the bottom of the info page for lesson 11.

Electrical signals start in the hillock, travel down the axon, and end up in the axon terminals.

BEFORE ("RESTING POTENTIAL")

The resting potential is maintained by the **sodium-potassium pump** in the membrane.

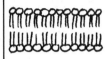

Electrical charge inside the axon is negative *in comparison to* the outside.

DURING ("ACTION POTENTIAL")

Na$^+$ gates open first, allowing Na$^+$ ions to come streaming in. Then the K$^+$ gates open, allowing K$^+$ to flow out.

AFTER (BACK TO RESTING POTENTIAL)

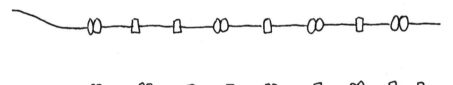

Meanwhile...

Na/K pumps go back to work, restoring the original resting potential.

Vesicles filled with neurotransmitters are waiting in the terminal knobs, and calcium ions are waiting outside.

THE SYNAPSE -- jumping the gap

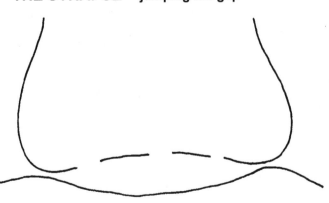

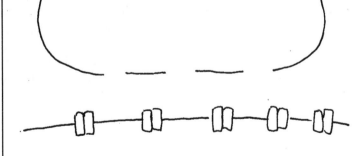

When the action potential reaches the terminal knob, a sudden influx of Ca^{2+} ions causes the vesicles to do exocytosis.

The neurotransmitters cross the synaptic cleft and bind to receptor sites on ion channels. Some neurotransmitters are "excitatory" and will open Na$^+$ channels in order to start a new action potential. Other neurotransmitters are "inhibitory" and will open K$^+$ channels, preventing a new action potential. Enzymes are present, also, for immediate removal of neurotransmitters.

53: MUSCLE FIBERS and the NEUROMUSCULAR JUNCTION

Muscles follow the same organizational format we've already seen several times now: bundles of bundles of bundles. The largest bundle is what we know as a "muscle." A muscle is surrounded by a connective tissue bag called a *fascia (FASH-ah)* that gets thick at one end and turns into a *tendon*. The tendon is what attaches to a bone. The muscle itself is made of bundles of *fascicles, (FASS-i-kuhls)* which are, in turn, made of bundles of *muscle fibers*. A single muscle fiber is very small and you'd need a microscope to see it. Muscle fibers are made of very, very tiny filaments called *myofibrils*. Myofibrils are made of protein "ropes" called *actin* and *myosin*. We've actually met actin already, as it is the smallest of the cytoskeleton filaments. To see actin and myosin you need an electron microscope. The next lesson has more information about actin and myosin.

Motor neurons (the kind we drew in lesson 49) attach to muscle fibers. The axon terminals of one neuron can connect to just a few muscle fibers or as many as 30-50, depending on how big the muscle is and where it is located. (The cell bodies of these motor neurons are usually found in the spinal cord, as you might remember from drawing 50.) A motor neuron and the muscle fibers to which it is attached are called a *motor unit*. They are a unit because the neuron will make all of them work together at the same time.

There aren't any individual muscle cells. During embryonic development, all individual muscle cells fused together to create very long "super cells" with lots of nuclei. So a long muscle fiber actually IS a muscle cell. Sort of. Muscle fibers have all the usual cell parts, especially lots of mitochondria for producing ATPs.

Muscle fibers have some features that other body cells do not:
1) myofibrils filled with actin and myosin
2) a plasma membrane that dips down forming tubes (known as T tubules)
3) an adapted smooth endoplasmic reticulum that looks very different from the smooth ER found in other body cells

Since muscle fibers are so different from other cells, scientists felt compelled to make up different names for their cells parts. The Greek word roots they choose for muscle stuff are "myo" meaning "muscle," and "sarco" meaning "flesh" (similar to "meat"). The plasma membrane was renamed as the *sarcolemma*, the cytoplasm became the *sarcoplasm*, and the smooth ER was named the *sarcoplasmic reticulum*. The places where the sarcolemma (plasma membrane) dips down are called *T tubules*. ("T" stands for "transverse," with "trans" meaning "across.") The T tubules are going to carry the action potential down into the interior of the fiber. As we will see, muscle cells are similar to neurons in that they are equipped to carry an action potential across their membrane. This means that the sarcolemma must be equipped with sodium-potassium pumps, to maintain the resting potential where the outside is more positive than the inside (because of all the sodium ion that have been pumped out).

NOTE: The sarcoplasmic reticulum is not shown in the middle drawing. It would make the drawing too complicated. The SR is shown in the bottom drawing.

The place where an axon terminal connects to a muscle fiber is called the *neuromuscular junction*. Just as with neurons, we have a *synaptic cleft*, a little gap, where neurotransmitter chemicals must cross and start a new action potential on the other side. The neurotransmitter chemical we find inside the vesicles in these axon terminals is called *acetylcholine, ACh*. There are also little enzymes lurking in the gap that will destroy acetylcholine as soon as it has done its job. You would not want ACh building up, as this would make the muscle fibers contract continually, causing a muscle to be unable to relax, even for a second. In fact, this is what happens with some poisons. The poison molecules interfere with the breakdown process and you get too much ACh in the synapse. The muscles go into extreme cramping, even the heart muscle, which causes death. Another poison, *curare*, which is made from a plant that grows in South American (and is used to make poisoned arrow tips), prevents ACh from being able to bind to the receptors on the muscle side of the cleft. This means that the muscles can't contract at all. The muscles become paralyzed and the heart stops beating.

After the vesicles in the axon terminal release ACh into the gap, the ACh molecules stick to binding sites on the sodium channels on the muscle side of the gap. The sodium channels open to allow sodium ions to enter, thus starting a new action potential. However, this action potential will do something different in the muscle fiber. The T tubules (which carry the action potential since they are part of the sarcolemma) lie along thick parts of the sarcoplasmic reticulum that are storing calcium ions. The action potential will cause the SR to release its calcium ions. These ions will flow into the myofibrils and attach themselves to the actin filaments, something will see in more detail in the next lesson.

Muscles are bundles of bundles of bundles. Nerves run through muscles and attach to muscle fibers.

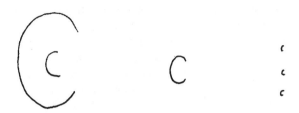

NOTE: Muscles are covered in connective tissue "bags" called *fascia* that taper off into *tendons*.

MUSCLE CELLS are called MUSCLE FIBERS

Muscle cells are called *muscle fibers* because lots of cells join together to make one long fiber.

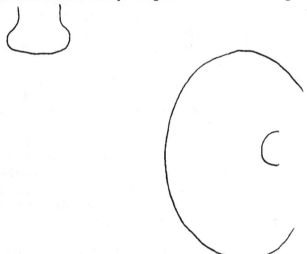

The T tubules carry the action potential down into the fiber so it can reach the myofibrils at the center.

THE NEUROMUSCULAR JUNCTION

This synapse works just like the ones we learned about in the last lesson. A sudden influx of Ca^{2+} ions makes the neurotransmitters flow across the gap and stick to receptors on Na^+ channels on the other side. The Na^+ ions begin an action potential.

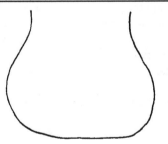

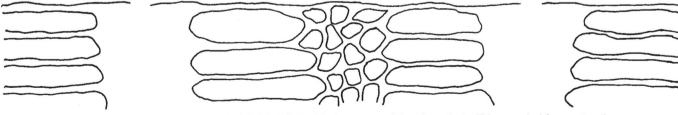

The smooth ER is called the SARCOPLASMIC RETICULUM. It stores calcium ions that will be needed for contraction.

54: ACTIN and MYOSIN

Muscle fibers (cells) are made of myofibrils. Myofibrils are made of two types of protein chains: **actin** and **myosin**. We've actually met actin before when we studied the cytoskeleton in module 1. The smallest filaments of the cytoskeleton are basically made of actin. Myosin filaments are thicker than actin and they have little paddle-like projections sticking off. Sometimes actin and myosin are referred to as thin and thick filaments.

The actin and myosin filaments are organized into short units called **sarcomeres**. Sarcomeres give myofibrils a striped appearance. Each sarcomere is an individual unit that contracts, and when all of the sarcomeres contract at the same time this is what makes the whole muscle contract. All of the myofibrils in a muscle fiber are connected to the same neuron, so they all function together, contracting at the same time. The ends of a sarcomere are called the **Z lines**. They look like thin lines on the sarcomere and they act as scaffolds to which actin and myosin are secured by even tinier protein ropes. The thicker band in the middle is called the **A band**, and it is the area where the myosin fibers are found. This band is darker because of the density of both actin and myosin filaments. Motion (contraction) will happen when the actin and myosin filaments slide past each other and shorten the whole sarcomere.

Actin looks like two ropes twisted together. The ropes are protein gadgets, of course. Sometimes actin is drawn as lines or ropes, but when shown close up, actin is usually drawn to look like two bead necklaces twisted together. On each "bead" there is a binding site where one of the myosin "heads" can bind if the conditions are right. These binding sites are covered by a protein thread called **tropomyosin**. A protein called **troponin** is attached to tropomyosin and helps in this covering process. Troponin is sort like the release button that will allow the tropomyosin threads to roll off the binding sites. When a calcium ion binds to troponin, this triggers the troponin to open up the binding sites by moving tropomyosin out of the way. Where do the calcium ions come from? From the sarcoplasmic reticulum where they are stored. What caused the SR to release the calcium? The release was triggered by the action potential that came down through the T tubules. Where did the action potential come from? It started when the neurotransmitter ACh crossed the synaptic cleft at the neuromuscular junction.

ATPs are used as the energy source for muscle contraction. When an ATP binds to the myosin head, it then splits into ADP and P. When the calcium ions come along, the myosin heads bind to the binding sites on the actin. After they bind, the ADP and the P leave the myosin head. As they leave, this causes a shape change in the myosin head that results in it pushing the actin filament. All the myosin heads do this at the same time, causing the entire actin filament to slide over. This sliding action causes the sarcomere to get shorter (to contract). After this motion happens, a fresh ATP comes over and binds to the myosin head, causing it to let go and go back to its original position. Then the cycle can repeat.

ATPs for muscle contraction come from three places. The first place is NOT the mitochondria, as you might guess. The first source that muscles use is from a molecule called **creatine**. Creatine can hold onto a phosphate, P. An enzyme robot called **creatine kinase** can take the P off creatine and put it onto an ADP, making it into ATP. (A "kinase" is any protein gadget that can take phosphates on and off.) This seems like a very easy and simple way to create ATP— just have an enzyme do the whole thing in one step! So why do we even need mitochondria and their complicated electron transport chains? Both are necessary in the grand scheme of life. ATP from creatine can only last about 5-10 minutes, then it is gone. This is long enough to let you do a short sprint, go up the stairs, carry some heavy boxes, and other short, intense daily tasks. However, if you walk, jog, or swim for over 10 minutes, you will need those ATPs made by the mitochondria. The mitos are really good at cranking out lots of ATPs as long as oxygen is available. That's the thing to remember about cellular respiration (the ETC)-- it needs oxygen. (Creatine does not need oxygen.)

It is possible to exercise harder and longer than your mitochondria can keep up with, depleting the available oxygen in the muscles. Then your muscles will need an alternative energy source. "Plan C" is called **lactic acid fermentation**. Glycolysis splits glucose into 2 pyruvate molecules. The pyruvates will go into the Krebs cycle if oxygen is present. If not, they will pile up in the cytoplasm. Also, after a lot of glycolysis has happened (because of no oxygen), all the NADH "trucks" will be full. (Remember, glycolysis produces not only 2 ATPs but also 2 NADH, also.) With no trucks available, glycolysis will stop.

As a result of the chemical process whereby pyruvates are turned into lactic acid, some NADH trucks are emptied and made available so that glycolysis can continue to take place. The downside to this process is that lactic acid in your muscles does not feel good; it produces that burning sensation. Therefore, lactic acid also serves as a warning signal, letting you know that your muscles are truly running out of oxygen. You feel enough pain that you stop over-doing it and let your muscles replenish their supply of creatine and ATPs in the mitochondria.

Muscle fibers are made of myofibrils. Each myofibril is made of two types of protein filaments: *actin* and *myosin*.
Actin and myosin overlap in such a way that the myofibril appears to have stripes. Dark places are where many fibers
overlap and light places are where few overlap. The repeating patterns are called *sarcomeres*.

SARCOMERES

Myofibrils look stripey because of the
overlapping actin and myosin filaments.

How Ca²⁺ ions allow myosin to bind to actin:

Tropomyosin blocks myosin from binding.

The action potential causes calcium to be released from the SR.
Calcium binds to troponin, which causes tropomyosin to move away.

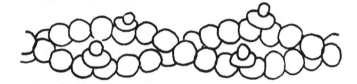

How ATPs are used by actin and myosin:

This is a continuous cycle that
can repeat in a split second.

While the myosin head is not attached, ATP
is "hydrolyzed" (split apart using a water
molecule) into ADP and P.

When a fresh ATP binds, the myosin
head is released and it goes back to its
resting position.

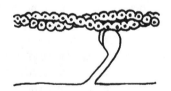

ADP and P are still bound to the myosin
head as the calcium ions roll back tropomyosin and allow the head to bind.

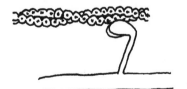

When the ADP and the P leave, the myosin head moves
forward, causing the actin filament to slide the other way.

Where do the ATPs come from?

1) CREATINE is a molecule that holds onto a
P. An enzyme can take the P off, and then put
it onto an ADP, making ATP. No O_2 needed.

2) The Electron Transport Chain (ETC)
This takes place in the mitochondria.
Oxygen must be available so that it can receive
the "tired" electrons at the end of the chain.

3) LACTIC ACID FERMENTATION
is a process that enables glycolysis to
take place over and over again, generating 2 ATPs each time. No O_2 needed.

DRAWING 55: THE LYMPHATIC SYSTEM

The word "lymph" comes from the name of a Roman goddess of water, Lymph, so we can guess that lymph must have some connection with water. In fact, it is mostly water. Lymph is what you get when blood plasma (which is mostly water) leaks out of blood vessels. Sometimes the vessels leak a little (under normal conditions) and sometimes they leak a lot (like when histamine is present). About 90 percent of the fluid makes its way back into the bloodstream again, but about 10 percent does not. The physics of blood pressure (the pressure of osmosis pushing in toward the vessel) causes most of the water to migrate back in. However, the fact that a small amount of fluid can't get back in is anything but a design flaw. Rather, it is a very clever mechanism crucial for the proper functioning of the immune system.

After the plasma leaks out into the interstitial space between cells, its name is changed to **lymph**. The lymph will contain small proteins, waste products from the cells, and also perhaps some viruses or bacteria. Some filtering, cleaning, and recycling needs to be done. This will be accomplished by the lymphatic system. The entry into the lymphatic system is a network of billions of tiny lymph vessels, sometimes called lymph capillaries. The lymph capillaries are found anywhere you find blood capillaries. Unlike blood capillaries, the lymph vessels don't form loops. Blood vessels are basically a closed system, with a continuous circular pattern inside of which the blood flows round and round and round.

Lymph capillaries have a definite starting point. The tip of a lymph vessel is made of a single layer of flat (squamous) epithelial cells that have spaces between them large enough for not only lymph fluid to enter, but all the "junk," too. The lymph vessels also allow entry of immune cells such as macrophages. As the fluid travels down the lymph vessels, it passes through one-way valves (made by epithelial cells). The valves keep the fluid from flowing backwards. The lymph system does not have a pump like the blood circulatory system does (the heart) so keeping the fluid moving has to be accomplished in other ways. The general movement of your muscles throughout the day helps to push the lymph fluid along. Also, though this is a little hard to understand, the pressure difference created by flowing blood at the place where the lymph dumps into the bloodstream also helps to draw the lymph in the proper direction. People with inadequate lymph circulation often get massages to help push the lymph along.

At various points in the body, the lymph vessels must pass through **lymph nodes**. These are filtering centers where the lymph fluid is cleaned up. We will study them in more detail in the next lesson. Groups of nodes are located near joints, in the groin area, in the chest, and in the neck. The **spleen** also functions very much like a lymph node, too, as we will see in the next lesson. Lymph nodes respond to pathogens (viruses and bacteria) and sometimes get a little larger while you are fighting an infection. The nodes can enlarge very quickly, but may take weeks or even months to get back down to their original size.

Microscopic lymph vessels connect to slight larger vessels, which connect to even larger ones, and eventually they all merge together into visible vessels called ducts. The lymphatic ducts will reach their end at the place where they connect to a large vein on the side of the heart. This vein is located underneath the **collar bone**, or **clavicle**, so the vein is sometimes referred to as a **subclavicle vein**. As already mentioned, the action of the lymph fluid dumping into a quickly moving bloodstream is one of the factors that helps to draw the lymph fluid through the lymph system. (As a general rule, lymph is traveling upwards, fighting gravity, which is no mean feat considering that there is no pumping mechanism involved!) After having been through all those nodes, the lymph fluid should be clean and germ-free at this point, so dumping into the blood is not a problem. When it re-enters the blood, its name is changed back to "plasma."

The lymphatic system has other filtering centers, besides the nodes and the spleen. These include the tonsils, the adenoids (up higher than the tonsils), the **appendix** and the **thymus**. We met the thymus in our study of immune cells. It is where T cells go to mature. The thymus is very large during childhood, then shrinks during teen years and is very small during adulthood. The appendix is located on the end of the colon, at a place where there is an extra pouch called the **cecum**. Both the cecum and the appendix appear to be storage areas for good bacteria that are helpful to he intestines. People used to think the appendix was basically useless. Now we know that it is a protected area where good bacteria can survive even through a bout of intestinal disease or infection that wipes out most of the other bacteria. The bacteria in the appendix can then come out and re-populate the intestines.

The lymph system has two sides: left and right.
-- The left side drains both legs, the left arm, the chest and the left side of the head.
--The right side drains only the right arm and the right side of the head.

The lymph system is made of a network of vessels that drain extra fluid from body tissues (the interstitial spaces between cells) into vessels that carry the fluid through lymph nodes and then up (against gravity!) to the top of the rib cage where lymph vessels dump the fluid back into the blood-stream. As the fluid makes its way up through the lymph vessels, it passes through lymph nodes where lymphocytes and macrophages can recognize and/or destroy any pathogens in the fluid. Also, macrophages and dendritic cells from the tissues can intentionally hop into the lymph vessels and drift to the nearest nodes so that they can present their antigens to T cells.

Lymph vessels have one-way valves so that the fluid can't go backwards. The fluid is pushed along through the vessels simply by the motion of our muscles as we go about our daily routines.

56: LYMPH NODES and the SPLEEN

Lymph nodes perform two main functions. They serve as filtering centers where lymph fluid is cleaned, and they also act a bit like army command centers where soldiers from the battlefield can relay information to officers who are in charge of making battle plans and sending out troops. The body has thousands of lymph nodes, ranging in size from a millimeter (the size of this dot .) to 2 cm (the size of a marble).

The shape of a node is very similar to a red kidney bean or a white navy bean. In fact, lymph nodes even share an anatomical feature with a bean: both nodes and beans have an area called the **hilum**. ("Hilum" means "little thing" in Greek.) The hilum on a bean seed is the scar where the seed used to be attached to the wall of the ovary. In a lymph node, the hilum is where the efferent lymph vessels come out and where the blood vessels go in and out. The spleen also has a hilum.

Lymph nodes are wrapped in a "bag" made of connective tissue. Every muscle, vessel, and organ in the body is wrapped in a thin coating of connective tissue. These "bags" can attach to each other to keep the organs tightly in place. The "bag" around a lymph node is called the **capsule**. The inside of a node is divided into compartments. The walls between the compartments are called **trabeculae** (trah-BECK-cue-lae), a word we saw back in lesson 36 when we learned about bone marrow making blood cells. The walls of the compartments in bone are also called trabeculae. Each compartment in a lymph node has a core called the **follicle**. Each follicle has a central area called the **germinal center**. The node is also divided into regions according to distance away from the hilum. The area farthest away from the hilum is the **cortex**. The area just inside of that is the **Paracortex**. ("Para" means "beside.") The area closest to the hilum is the **medulla**. The words follicle, cortex and medulla, are very common words in anatomy and we'll see them again soon.

Afferent lymph vessels bring lymph fluid from the tissues into the node. ("Afferent" comes from the Greek word roots "ad," meaning "towards," and "fer" meaning "to bring.") A node will have about 4 to 10 afferent vessels feeding into it. The afferent vessels enter one of the compartments in the node, and the lymph fluid then flows into the node and around the follicle. In this space around the follicle there are many **reticular fibers**, just like the ones we saw in the loose (areolar) connective tissue. (Remember, "rete" means "net" or "network.") Those reticular fibers act like a scaffolding for immune cells to crawl around on. This area is packed with macrophages and dendritic cells who try eat anything and everything that comes into the node. The macrophages and dendritic cells then present antigens (display pieces of what they ate) to the nearby T cells who live in the outer layer of the follicle. (The T cells matured in the thymus and then came to live in these T cell neighborhoods in the lymph follicles.) The central area of a follicle, the **germinal center**, is where B cells live. The T cells can go over and "talk" to the B cells when they need to. If the T cells have interacted with an APC (antigen presenting cell) they will try to find a B cell that matches that antigen. Remember, as a general rule, B cells can't do anything without permission from the T cells. If a T cell activates a B cell, the B cell will then become a **plasma cell** and will start cranking out antibodies. The antibodies will leave the node through the capillaries or the efferent vessels. Either way, the antibodies will end up in the blood, and the blood will take them to all parts of the body.

A lymph node has blood vessels that come in at the hilum, right near the efferent (exit) lymph vessels. An arteriole (coming from the heart and carrying oxygen) goes into each follicle, then spreads out into tiny capillaries that collectively are called a **capillary bed**. The other side of the capillary bed is drawn in blue, since by that time most of the oxygen has been used up. When these capillaries get a bit larger they are called venules. The venules join together to form a larger vein that exits at the hilum. Some of the blood vessels in a lymph node are a very special kind: High Endothelial Vessels, or HEVs. These capillaries are made of very thick ("high") endothelial cells, not flat ones like we find in regular capillaries. The thickness of the cells somehow makes it easier for lymphocytes, especially T cells, to pass through. This is the main way that T cells get into lymph nodes. It's not all that important to remember this term (HEV) unless, of course, you are studying to be a doctor or nurse and need to pass an exam. More important is to understand that you find specialized structures in exactly the places they are needed. The body exhibits brilliant design.

The spleen is very much like a lymph node in some ways, so much of our learning about nodes will carry over to the spleen. The spleen is wrapped in a connective tissue capsule, it has a hilum, and it contains many of the same immune cells. One obvious difference is that the spleen does not have any lymph vessels going in or coming out. Only blood enters the spleen. The spleen sits to the left of your stomach (from your viewpoint) and right under the diaphragm. It is attached to these organs with connective tissue.

The spleen has two notably different sections called the **red pulp** and the **white pulp**. The white pulp is the tissue that is like a lymph node. White pulp is found wrapped around the lower portions of the blood vessels. In the white pulp we find many T cells, macrophages, and **follicles** containing B cells. We find the same processes going on here, with APCs presenting antigens to T cells, and T cells activating B cells. Additionally, we find something a bit different going on with the B cells in the follicles. Here, the B cells can act like APCs, presenting antigens to T cells. ("Rules" in biology rarely are without exception. Most of the time B cells sit there naive until a T cells activates them, but here in the spleen follicles, we see them playing a different role. It is possible that this goes on at other places in the body, too.) Since there are no lymph vessels going in and out of the spleen, immune cells wanting to come and go from the spleen must use the blood vessels.

The red pulp results from another exception to a rule. The circulatory system (the entire network of blood vessels) is a "closed" system. We mentioned this in the last lesson when we saw that the lymph system is not closed, but has many open-ended vessels where lymph enters. Here in the spleen, we find the one place where arterioles come to an end and let blood leak out. The red pulp is red because it is filled with loose blood cells. These loose red blood cells can only get into the venules that will put them back in circulation if they are healthy and have a normal shape. Diseased and damaged red cells will not be able to return to circulation and will be eaten by macrophages. Because there are so many red cells floating around in the spleen, it also functions as a storage area for an emergency supply of extra red cells. If you have sudden blood loss, your spleen can release some extra cells.

LYMPH NODE

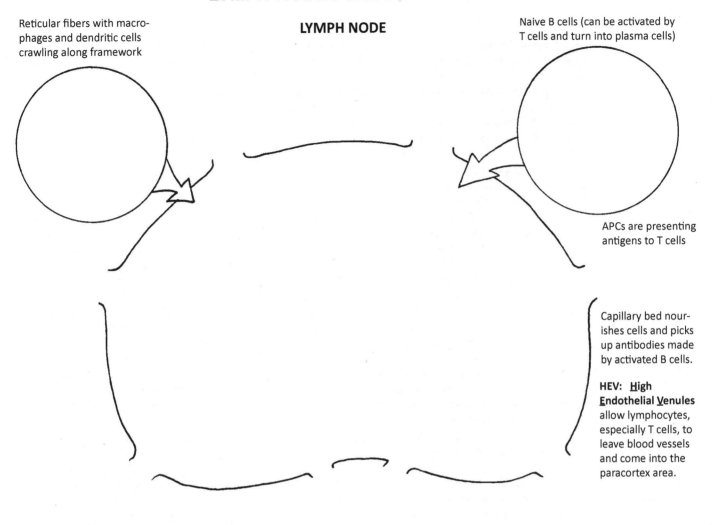

Reticular fibers with macrophages and dendritic cells crawling along framework

Naive B cells (can be activated by T cells and turn into plasma cells)

APCs are presenting antigens to T cells

Capillary bed nourishes cells and picks up antibodies made by activated B cells.

HEV: <u>H</u>igh <u>E</u>ndothelial <u>V</u>enules allow lymphocytes, especially T cells, to leave blood vessels and come into the paracortex area.

SPLEEN

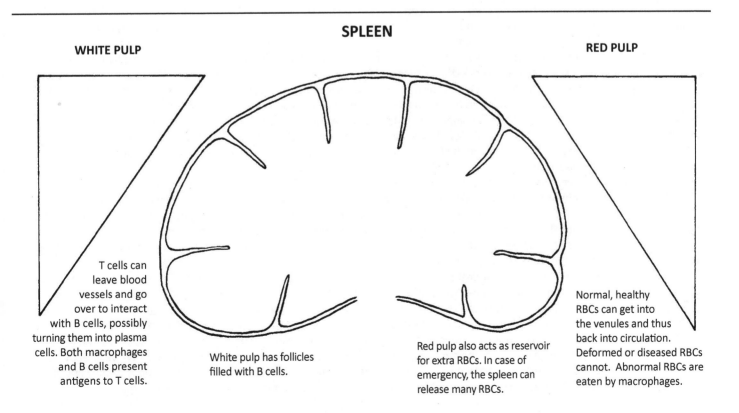

WHITE PULP

RED PULP

T cells can leave blood vessels and go over to interact with B cells, possibly turning them into plasma cells. Both macrophages and B cells present antigens to T cells.

White pulp has follicles filled with B cells.

Red pulp also acts as reservoir for extra RBCs. In case of emergency, the spleen can release many RBCs.

Normal, healthy RBCs can get into the venules and thus back into circulation. Deformed or diseased RBCs cannot. Abnormal RBCs are eaten by macrophages.

57: SKIN

We don't usually think of the skin as an organ of the body, but it is actually the largest organ of our body. Organs are made of tissues, and here in the skin we will see all four tissue types, and a surprising number of connective tissue types (loose, irregular, adipose, blood, lymph).

The skin can be divided into basically three layers: **epidermis, dermis, hypodermis.** "Dermis" is Greek for "skin." The epidermis came from that blue layer we learned about in embryology, the ectoderm. The dermis came from the red layer, the mesoderm. You'd never guess that these two layers came from completely different embryonic sources. The basement membrane is the "dividing line" between the these layers. Remember, the basement membrane is always the bottom layer of epithelial tissue.

The epidermis is made of stacks and stacks of epithelial cells. The bottom layer is called the **basal layer** and it is the only truly active layer. This is the only place in the epidermis that mitosis occurs. Most of the cells in the basal layers are keratinocytes. These cells produce **keratin**, a waxy chemical that gives skin its waterproof qualities. As keratinocytes multiply, the old cells get pushed up. More cells pile up underneath and eventually the old cells get pushed to the surface. On their way up they begin to die. Organelles start to disappear until finally the cells are nothing but empty shells filled with waxy keratin. The surface of our skin is dead cells, millions of which fall off every day. (Microscopic creatures called dust mites love to eat cells that fall off our skin, which is why they like to live in pillows and blankets. Yes, your pillow is probably inhabited by tiny spider-like creatures!)

The very top layer of the epidermis is called the **stratum corneum.** ("Stratum" means "layer," and "corneum," means "hard and waxy.") There are other sub-layers with hard-to-remember names, but they are of minor importance in our overview, so we'll just skip them. If you want to know more, just Google "epidermis layers."

There is only one type of immune cell in the epidermis. We first learned about it back in the lesson on macrophages. It is called a **Langerhans cell,** and it is similar to a macrophage. It moves about, looking for germs or dirt particles. It is the first line of defense up there in the top layer of skin. There is one Langerhans cell for every three dozen keratinocytes.

The basal layer of the epidermis is where you will find **melanocytes**. Melanocytes produce a pigment called **melanin.** ("Melano" is Greek for "black, or dark.") Melanin is what gives color to our skin and hair. The melanocytes make the melanin molecules and package them into vesicles that are exported out of the cell into the surrounding area. The nearby keratinocytes pick up these vesicles and bring them inside. The dendritic shape of the melanocytes helps in the distribution of the vesicles.

There are basically two types of melanin: brown and red. If you've ever seen a tiny red mole on someone's skin, you are seeing the red kind. Brown is the color we are more familiar with. People who have dark brown skin have very active melanocytes. People with light skin have melanocytes that don't produce much melanin. Melanocytes can be encouraged to produce more melanin if exposed to sunlight. This is what causes a suntan. It's important to note that everyone has approximately the same number of melanocytes in their skin. The difference between dark and light skin is how active these cells are. Again, the ONLY difference between dark and light skin is the activity level of the melanocytes. Melanin acts as a natural sunscreen. It "catches" ultraviolet rays from the sun that can damage DNA and destroy Langerhans cells. Sunlight is not all bad, however, as it also seems to stimulate the production of vitamin D. Your body must process vitamin D in order to make it usable by all body cells, and the epidermis of the skin is the primary site for this.

When cell growth in the basal layer gets out of control and too much mitosis happens (because of a mutation), this can result in skin cancer. When a keratinocyte mutates and becomes cancerous we call this basal cell carcinoma (or squamous cell carcinoma). When a melanocytes mutates it causes melanoma. Melanoma is the kind that is more likely to spread to other parts of the body.

The dermis is full of things we've already studied. The top part of the dermis is loose (areolar) connective tissue and the lower part is dense irregular tissue. These connective tissues are filled with fibroblasts that are making collagen, elastin and reticular fibers. There are many immune cells roaming about, such as macrophages and lymphocytes (T and B cells). The dermis also has a rich blood supply. Did you notice that the epidermis has no capillaries? This is a very good design because you wouldn't want to be bleeding every time your skin brushed against something. The damage has to go down to the dermis before blood appears. The dermis also has lymph capillaries, not just blood capillaries. As we learned in the lymph lesson, lymph vessels are open at the ends, not closed like blood capillaries. The lymph vessels recycle the fluid between the cells (interstitial fluid) and remove all the waste products that leak out of the cells into this space.

The dermis also contains hair follicles. We'll learn more about them in the next lesson. The hair is associated with sebaceous glands that produce oily **sebum** that keeps our skin soft and helps it to be water resistant. (Too much sebum on our scalp and we feel the need for a shampoo!) Each hair has a tiny muscle, called the **arrector pili**, that can pull it up straight. ("Arrector" means "stands up straight" and "pili" means "hair.") When the hair goes up straight, the skin bunches together at the top of the follicle and forms what most of us call a "goose bump." Geese (or chickens or turkeys) that are plucked and ready for the oven are covered with little bumps at the places where the feathers were plucked out.

What students usually find most interesting about the dermis are all the strange-looking nerves. There are sensors for pain, light touch, deep touch, hold and cold. The deep touch sensors called Pacinian corpuscles (*CORE-puss-uhls*) are dendrites that will start action potentials when stimulated by touch. The ends of the dendrites are surrounded by protective "padding" that helps them not be too sensitive. There are light touch sensors called Meissner's corpuscles. Hot and cold nerve endings are triggered by temperature. We have these hot and cold nerves on our tongue, too and they can be mistakenly triggered by the chemicals in some foods. Nerve endings also surround the hairs, giving the nervous some feedback about what is happening to the hairs. We've already learned a bit about the function of these neurons, and we know that they are afferent nerves, taking signals from the body to the central nervous system (spine and brain). There are efferent (motor) neurons connected to the tiny muscle that moves the hair.

Though we think of skin as primarily being epithelial cells, it is actually a blend of all 4 tissue types. The dermis and epidermis are separated by the basement membrane (of the epidermis). We have 4 kinds of connective tissue: loose, irregular, adipose and blood cells. We have muscle, and both sensory and motor neurons.

1) basement membrane
2) basal layer (keratinocytes)
3) melanocytes
4) dermal papillae
5) fibroblasts
6) collagen and elastin

7) deep touch sensors (Pacinian corpuslce)
8) light touch sensors (Meissner's corpuscle)
9) free nerve endings (pain sensors)
10) cold sensors
11) heat sensors
12) motor neuron

13) sweat gland
14) arrrector pili muscle
15) sebaceous glands
16) "bulge" (stem cells)
17) lymph vessels

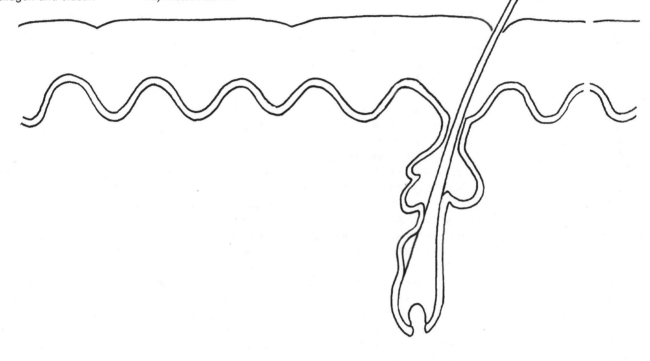

THE EPIDERMIS is made of mostly **keratinocytes**. The keratinocytes in the basal layer are the only ones that go through mitosis. As they divide, the new cells go upwards. As the cells mature, they begin producing a lot of a waxy protein called **keratin** and they also begin to lose their organelles.

By the time they reach the top, they have lost everything, even their nucleus. They are dead cells filled with keratin. We have these dead cells flaking off our skin all the time.

FIBROBLASTS

COLLAGEN AND ELASTIN

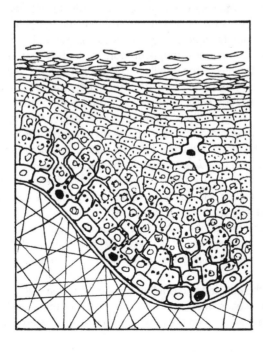

Top layer of dead cells is called the **STRATUM CORNEUM**

KERATINOCYTES: all cells except melanocytes and Langerhans cells

LANGERHANS cells are a type of macrophage and are the only immune cells in the epidermis.

MELANOCYTES produced the pigment **melanin**. Melanin is brown or red (usually brown). Dark skin has more melanin than light skin. Melanocytes release the melanin in little vesicles and these vesicles are taken in by keratinocytes.

BASAL LAYER (of keratinocytes)

BASEMENT MEMBRANE

58: HAIR and NAILS

Hair and nails are considered to be "accessories" of the skin. They are produced by the same type of cells that make skin, so they really are part of the skin, specifically the epidermis. The skin with all its accessories is called the **integumentary system**.

To understand how hair and nails form, we need to review how the epidermis works. As with all epithelial tissue, there is basement membrane on the bottom, with its network of collagen fibers that can bond to tissue beneath. Above the basement membrane we find the basal layer of cells, which act a bit like stem cells. (In fact, some texts will call them stem cells.) These basal cells go through mitosis constantly. A few daughter cells will stay there on the bottom in the basal layer, but most daughter cells will gradually be pushed higher and higher as more cells keep accumulating beneath them. As they rise, two things happen. First, they begin to produce large volumes of the protein **keratin**. Keratin is strong as well as water-resistant, and gives skin these two properties. Second, when the cells are almost full of keratin, the organelles begin to disintegrate and disappear. By the time the cells are close to the top, even the nucleus is gone. The keratinocyte is technically "dead" at that point.

An interesting side note at this point is that right before a keratinocyte dies, it produces enzymes are that capable of chewing through the desmosomes that hold the cells together. If this does not happen, the dead skin cells at the surface will not be able to flake off. Might not sound like a big deal, but it is. People who have a genetic mutation in the genes for producing these enzymes will have a horrible skin condition that causes them much suffering. They must bathe and scrub and scrape their skin daily to try to peel off the dead cells. It's inconvenient and painful.

Hair and nails are also made of keratinocytes that are being pushed up, filling with keratin, and then dying. However, the keratinocytes in the hair and nails know not to make these destructive enzymes before they die. We really don't want our hair and nails flaking apart. Somehow, these keratinocytes know that they are located not in the skin but in hair and nails and they alter their behavior accordingly. The hair shaft has a central core called the **medulla**, a colored **cortex**, and a clear outer layer called the **cuticle**. All of these words are common science words. "Medulla" means "middle," "cortex" means "outer layer," and "cuticle" means "skin."

A hair follicle is basically a bit of epidermis that got pushed down into the dermis. Imagine the epidermis as a piece of stretchy fabric lying on a pan of jello. Put a finger on the fabric, then poke it down as far as your finger will go. You've just created a hair follicle. The inside of the hair follicle is epidermis. (If you look back at your drawing or model of the skin, you will see that this is so). The epidermal cells keep track of where they are in the follicle, though, and the ones on the sides behave differently than the ones at the bottom. The cells at the very bottom are extremely active and they are the ones that generate the hair. You will also find some melanocytes mixed in with the epidermal cells at the bottom of the follicle. Interestingly, melanocytes are not made by the basal cells. Keratinocytes don't differentiate into melanocytes. Melanocytes originated in the neural crest region of the one-week-old embryo. As the embryo developed skin, the melanocytes traveled throughout the body, then lodged in skin and in hair follicles.

At the bottom of the follicle, there is a tall bump where the dermis pokes up into it. This is called the **dermal papilla**. The dermis is where the blood supply is, so the dermal tissue must supply blood to the fast-growing basal cells. The active basal cells form a curve around the top of the papilla. As they go through rapid mitosis, they will create a bulge of cells called the **hair bulb**. A similar process to what happens in the skin will take place. The fast-growing basal keratinocytes will pick up melanin from the surrounding melanocytes. The keratinocytes will get pushed up higher and higher as cells beneath them multiply. As they rise they will begin to produce large amounts of keratin, then they will lose their organelles as they die. As the dead cells rise, they will stay tightly bound together in a column, forming the hair shaft.

The basal cells go through growth cycles, the length of which depends on where they are found in the body. Eyelash hairs have a very short cycle of only a few months. Hairs on the scalp have a cycle of several years. The active part of the cycle is called the **anagen** phase. ("Ana" is Greek for "up.") During anagen, the hairs do what we just described in the last paragraph. Scalp hairs spend several years in this stage. Then comes **catagen**, when the basal cells divide less frequently and the whole follicle shrinks. ("Cata" is Greek for "down.") This stage lasts for a few weeks. Lastly, there is the **telogen** phase. ("Telos" is Greek for "far away.") In this phase, the hair bulb actually detaches from its blood supply and goes "far away" from the dermal papilla. Telogen is the resting phase, where the follicle just sits and rests. Then, after a couple of months, stem cells from "the bulge" migrate down to the bottom and begin to grow a new bulb. The bulb reattaches to the papilla and a tiny hair begins to grow. This new hair pushes the dead hair above it up and out. We lose 50 to 100 hairs on our head each day due to new hairs pushing out old ones. (NOTE: Stem cells in the bulge also help to repair tears in the epidermis above them. They turn into keratinocytes to form temporary patches.)

Humans have about 100,000 hair follicles on their scalp. Each follicle goes through this complete cycle about 20 times in our lifetime. The timing of the phases of the cycle are controlled by "growth factors" produced other places in the body, often in glands. The growth factors are carried by the blood and come into the follicle through the dermal papilla. Significant changes in hair follicles occur during puberty, for example. Skin that was hairless in childhood suddenly begins to grow hair.

Nails grow in much the same way as hairs, except that they don't take up melanin, and the end result is a flat "plate" not a thin shaft. Nails have a basal area with actively growing cells going through mitosis. Like the keratinocytes of the skin, nail keratinocytes produce large amounts of keratin, then gradually die. The dead cells full of keratin are pushed out further away from the basal area, causing nail growth. Nails on fingers grow at a rate of about 2 to 3 centimeters a year. Fingernails grow about four times faster than toenails. (Index fingernails also grow faster than pinky nails.) The actively growing base of the nail is sometimes called the **matrix** (one of the most over-used words in science, along with medulla, cortex, and cuticle). The basal matrix controls the shape of the end product, and in animals can make horns and claws.

The tissue under the nail is called the **nail bed** and is full of capillaries, giving nails their pink color. The white half-moon shape at the bottom the nail is the **lunula**, ("Luna" means "moon.") The lunula is part of the actively growing matrix of basal cells. The lunula is best seen on the thumb and might not be seen at all on the pinky. The edges of skin around the nail are called the **nail folds**.

Hair and nails are part of skin and grow in much the same way. However, unlike skin, the keratinocytes in hair and nails don't produce destructive enzymes that cut the bonds bewteen the cells. The cells stay firmly connected.

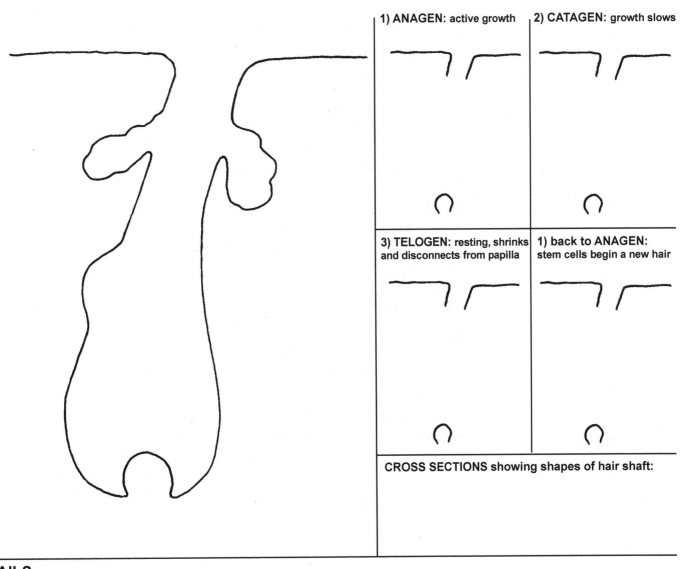

HAIR GROWTH CYCLE

1) ANAGEN: active growth

2) CATAGEN: growth slows

3) TELOGEN: resting, shrinks and disconnects from papilla

1) back to ANAGEN: stem cells begin a new hair

CROSS SECTIONS showing shapes of hair shaft:

NAILS

TOP VIEW: **SIDE VIEW CROSS SECTION:**

59: TEETH and TONGUE

An embryo begins to develop teeth at as early as 6 weeks. That's amazing considering that it doesn't even have fingers yet. Primary (baby) teeth begin to develop at 6 to 8 weeks and permanent (adult) teeth begin to develop at 20 weeks. Fingers develop from 7 to 9 weeks. A baby is born with no visible teeth, but has both sets fully or partially developed already, up in the gums.

Teeth begin as tiny "buds" of cells that have decided to differentiate into teeth. Embryonic growth factor chemicals told these cells to be different from the cells around them, and turn into teeth. The bud grows larger and changes shape, and, as the number of cells increases, the cells differentiate again and become the cells that make the inner and outer parts of the teeth. Cells that form body parts often have names that end in "-blast," and in the teeth we find **ameloblasts, odontoblasts, and cementoblasts**. All of these cells function in a similar way, but to varying degrees. They all secrete protein fibers, such as collagen, as well as chemicals that make mineral atoms stick to the protein fibers. We find this process in bone, too, and will study it more in the first lesson on bone. The protein fibers and the minerals blend together to make a biological substance that is a lot like the reinforced concrete used in the construction industry to make buildings. The steel bars in the concrete are like the protein fibers, and the concrete itself is like the deposited minerals. Like concrete, minerals by themselves tend to be brittle. The protein fibers (often a type of collagen) function like the steel bars, giving a rigid but resilient, and slightly flexible, framework. (Some people might classify this as a type of "biomimicry" where humans have imitated engineering examples found in nature.)

The **ameloblasts** are responsible for making the hard, white **enamel** that forms the surface of the tooth. Enamel is the hardest substance found in the body. Ions of calcium (Ca), phosphate (PO_4) and the hydroxyl group (OH) combine to form a mineral substance called **hydroxy-apatite** (hi-drox-ee-ap-ah-tite), $Ca_{10}(PO_4)_6(OH)_2$. The **odontoblasts** are responsible for making the **dentin**, a hard substance that lies right underneath the enamel. Enamel is mostly minerals with just a little protein fiber. Dentin has more protein fibers and a little less mineral content, so it is not quite as hard but much more resistant to impacts, whether the daily small impacts of chewing or the sudden traumatic impacts of injury. The ameloblasts and odontoblasts start out sitting in a line next to each other. Their products, enamel and dentin, pile up between them, and the cells get farther and farther way. By the time the enamel and dentin are fully formed, the ameloblasts are on the top of the tooth and the odontoblasts are deep in the tooth, on the inside of the dentin. The ameloblast die and disappear after they are done forming the enamel. After your tooth development period is over, you can't replace that enamel because the cells are no longer there. The odontoblasts survive, and will be part of the living tooth.

The **cementoblasts** produce **cementum**, a substance that is about 50% protein fibers and 50% minerals. Cementum is the layer that fastens the tooth to the gums. Unlike enamel and dentin, cementum is constantly produced throughout our life. Surprisingly, it has the highest fluoride level anywhere in the tooth. Fluoride treatments by dentists add fluoride to the outer mineral layer of the enamel.

The inner part of the tooth is called the **pulp**. It contains many blood capillaries, some lymph capillaries, and nerve endings that can sense hot, cold and pain. The pulp narrows and goes down through the roots, in channels called **root canals**.

The teeth are surrounded by the gums, more properly called the **gingiva**. The gingiva tissue is basically the same as the epithelial tissue found in skin. There are basal cells that produce keratinocytes, which gradually produce more keratin and eventually die and become very flat (squamous). Gingiva can be hard and stick to the bone tightly (lower gums), or it can be unattached and free to move around (the little bits of gum between your teeth). We can keep our gums healthy by using dental floss to keep food particles and bacteria from collecting in the space between the tooth and the gingiva. Healthy gums will not bleed at all when flossed. Inflammation of the gingiva is called **gingivitis**. ("-Itis" means "inflammation.")

There are 20 primary (baby) teeth. The permanent teeth number 32, though 4 of those are the "wisdom" teeth than most people have removed during their teen or early adult years. The four types of teeth are **incisors** (central and lateral), **cuspids** (canines), **bicuspids** (first and second premolars), and **molars** (three sets). It is easy to see from their shape and where they are located in the mouth what role they play in biting and chewing.

The tongue is a muscular organ. It is made of muscles and is attached to muscles. It is covered by a type of epithelium called **mucosa** or **mucosal membrane**. We find mucosal tissue lining all body cavities such as mouth, nose, stomach, intestines, lungs and reproductive organs. Mucosa can tolerate being wet all the time, unlike our regular skin. Our entire mouth, except for our teeth, is covered with mucosal tissue. Somehow or other, when the mucosal tissue gets to the edge of our lips, it knows to stop. The mechanism that allows cells to know their exact location has yet to be discovered. On the tongue, the mucosa covers thousands of little bumps called papillae.

Underneath the tongue we find the **frenulum**, which attaches the tongue to the floor of the mouth. We also see bluish veins (though blood is never blue!) and two barely visible pores (at the base of the frenulum) that are openings to salivary ducts. These ducts are like very long pipes that run all the way to the salivary glands that are under the tongue and at the top of the neck.

There are four types of papillae: **circumvallate** (or just vallate), **foliate**, **filiform** and **fungiform**. The diagram shows where each type is found. They all contain taste buds except for the filiform. The filiform papillae only provide friction (for chewing) and sensation (touch, hot, cold, pain). The other three types have microscopic taste buds that can sense sweet, sour, salty, bitter and umami (savory). (NOTE: Those old "maps" of the tongue are outdated. It is now being taught that all parts of the tongue are equal.) The fungiform papillae get their name from the fact that they are shaped like mushrooms.

Taste buds are made of a bunch of **receptor cells**, some of which have cilia protruding out the top. Other cells make a basal layer. Dendrites of neurons go up between these cells, waiting for the cells to stimulate them to begin an action potential. The receptor cells will be triggered by the chemicals present in food particles. The action potential will continue through a series of neurons until it reaches the part of the brain that interprets tastes. The brain will turn the electrical signal into a taste sensation.

Most of a tooth lies below the surface. Teeth have deep roots that go down into the bone. The white enamel is the hardest substance in the body and is non-living. The inner pulp is alive.

The tongue is much larger than it appears. We see the body of the tonuge, but the large "root" is below the surface. The tongue is made of many muscles, and it connects to many others.

Enamel was made by ameloblasts, which disappear once they have done their job.

Odontolbasts made the dentin and are still there on the inside of the dentin, next to the pulp.

periodontal membrane

Cementum is secreted by cementoblasts. It is 50% collagen, and 50% minerals such as calcium, phosphorus, and fluorine. Unlike enamel and dentin, cementum is made throughout our lifetime.

Four types of lingual papillae:

1) circumvallate *We only have 8-12 of these.*

2) foliate *Located along the back sides.*

3) filiform *These are most numerous and we have thousands.*

4) fungiform *We have 200-300 located mainly near tip and sides.*

Those tongue maps are no longer valid. Current research shows that all areas of the tongue can taste sweet, sour, salty, bitter and umami (savory).

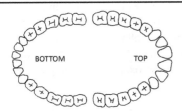

BOTTOM   TOP

<u>Types of teeth</u>:
Incisors
Cuspids ("canines")
Bicuspids
Molars

Underneath the tongue

PAPILLAE

TASTEBUDS

Taste receptors in taste buds are similar to smell receptors in the nose. Both are triggered by chemicals. The cells start an action potential in the nearby neurons, which travels to the brain where it is interpreted as an odor.

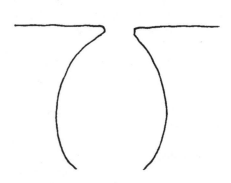

The bumps you see on your tongue are papillae, not taste buds. Taste buds are microscopic. The filiform papillae do not have taste buds. They simply provide friction and sensation. (Animal tongues (notably cats) often have very large and long papillae, making their tongues feel rough.)

Taste buds can only sense sweet, sour, bitter, salty and umami (savory). Most of taste involves smell.

60: MOUTH, NOSE and THROAT

In this drawing we are seeing a cross section of the head as if it was sliced right down the middle. We won't see the ear or the jaw bone or any of the neck muscles. We've even removed the dividing wall in the nose, called the **septum**. The septum is that thing that divides your right and left nostrils. In this view we are seeing the interior of the right nostril and the right side of the sinus area. We'll see a little be of the brain in this drawing, plus the top of the spinal cord.

In this view of the tongue we can see a few of the muscles, but there are others that have been removed because they attach to the head in places near the ear and jaw which have been removed. The tongue is made of 8 muscles all together; 4 of them are **intrinsic**, or completely within the tongue, and 4 are **extrinsic** and attach to places outside of the tongue. To talk or eat you must coordinate the actions of all 8 muscles. These muscles have difficult names that you only need to know if you are studying to be a doctor or a speech therapist.

The surface of the tongue is covered with epithelial tissue. The top of the tongue is called the **dorsum**. The word root "dors" always refers to the back of something. We just need to imagine that the tongue is like a 4-legged animal where its back is actually the top. Under the epithelium is a layer of muscles running parallel to the dorsum. Then under that are some muscles that run perpendicular, going downwards. On the bottom you can see one of the extrinsic muscles that attaches to bone. It is attached to the jaw bone at one end and a tiny bone called the **hyoid bone** at the other end. The hyoid is a bone that most people are not aware of. It a small U-shaped bone that is held in place by muscles stretching out from it in every direction. If the muscles disappeared the hyoid bone would fall out. In this drawing we only see a cross section of the hyoid bone so we can't see its U-shape. However, we do see it attaching to a part called the **epiglottis**. The epiglottis is a piece of soft cartilage covered with epithelium. It is a flap that covers the **trachea** (air pipe) during swallowing so food does not get into your lungs. Nerves running into the epiglottis provide us with our gag reflex.

Right under the epiglottis is the top of the trachea. Behind it is the **esophagus**, which is the food pipe that goes into the stomach. Also at the top of the trachea is the **larynx**, which is the area of the **vocal chords**. The opening between the vocal chords is called the **glottis**. The vocal chords are two pieces of soft cartilage that can be stretched or relaxed to make them short and tight, producing high-pitched sounds, or long and loose, producing low-pitched sounds. (The air you breathe always goes past the vocal chords, but if you don't have them stretched tight, no sound in produced.) The larynx area is protected by the **thyroid cartilage**. (When the thyroid cartilage sticks out noticeably, it is often called the "Adam's apple.") The larynx area has many muscles attached to it at various points, which help with speech and swallowing.

The area right behind the tongue is called the **pharynx**. The palatine tonsils are in this area, but we can't see them in this picture because it is a cross section. Notice the lingual tonsil that we learned about in the last drawing. At the very top of the pharynx, almost in the sinuses, we find the **adenoids**. The adenoids are lymph tissue, just like the tonsils. In this area we also find tiny holes which are the openings of the **Eustachian tubes**. We will learn about the Eustachian tubes in the lesson on the ear.

The top of the mouth area is called the **hard palate**. (PALL-it) The back of the palate is the **soft palate**. You can use your tongue to feel the difference between the hard and soft palate. Both hard and soft palates have lots of collagen, but only in the hard palate do we find the collagen filled with minerals. The soft palate stays soft because it does not fill with minerals. The very tip of the soft palate is called the **uvula**. The uvula gets pressed against opening the sinuses when you swallow. This area around the uvula is sometimes called the **nasopharyngeal** area, using a combination of the words nose and pharynx. (Don't let long words scare you—just look for smaller words inside them.)

The nostrils lead into **nasal passages**. The nostrils are lined with hairs that will hopefully prevent the entry of dust and dirt. Inside the nasal passages are three long folds called **conchae** (singular: **concha**). They have a framework of bone covered in mucosal epithelial tissue. The folds are also sometimes called the **turbinates** because their job seems to be to create turbulence in the stream of air that is passing through. The turbulence slows down the air and makes it come into contact with the mucosal tissue for a longer period of time. Coming into contact with the mucosa warms the air and adds moisture to it, making it more comfortable for our lungs to breathe. Cold, dry air is irritating to the lungs.

The nasal passages have little side passages called **sinuses**. There are four distinct sinus areas. The two that we are familiar with are the largest ones, the **maxillary sinuses** behind our cheek bones and the **frontal sinuses** behind our forehead. When we have a sinus infection it is usually in either the maxillary or the frontal sinuses. The sinuses get infected more easily than the central nasal passages because the sinuses are less open. During a cold or flu, inflammation in the sinuses can cause them to stop draining. The tear ducts also drain into the nasal passages, which is why our nose runs when we laugh hard or cry.

The other two sinus areas are less well known. Right above the adenoids we find the **sphenoid sinus**. It also has small ducts that allow it to drain into the maxillary sinus. The sphenoid sinus provides a little bit of protection for the brain part behind it, the **pituitary gland**. The pituitary gland is a marble-sized brain part that controls the production of many hormones. (We'll learn more about it when we get to the endocrine system.) The sphenoid sinus is in the middle of the sphenoid bone. This bone is part of the bottom of the skull, but when separated from the rest of the skull bones, it looks a bit like a butterfly. You can't see any of the butterfly shape in this drawing because it is a cross section. We'll see the sphenoid bone again in a later lesson.

The **ethmoid sinuses** are located in the **ethmoid bone**. The ethmoid might just be the most complicated-looking bone in the body. It is so complicated that each side of it is called a **labyrinth**. ("Labyrinth" comes from Greek mythology—a story about an impossibly complicated maze that a hero must go through.) A few of the labyrinth bones project into the nasal passages. These are the bones inside conchae. and form the rigid framework for the structures called **conchae**. (kon-kee or kon-kay) The ethmoid sinuses and most of the ethmoid bone cannot be seen in this drawing because it is a cross section and we've removed the entire left side of the head.

Very related to the tongue and teeth are the **salivary glands**, though we can't see them in the main drawing. Salivary glands provide the necessary fluids for the taste buds to function, and for us to swallow comfortably. Saliva contains an enzyme called **amylase**, which begins to break down starches found in bread, rice and pasta. Saliva helps teeth by neutralizing acids. Bacteria are helped by a slightly acidic environment so the presence of acid promotes tooth decay. (Sugary carbonated beverages have both sugar and acid, which is why they are so bad for your teeth.) There are three main areas of salivary glands. The **parotid glands** are located on the upper jaw. These are the glands near your ear that will twinge when you blow up a balloon or taste something super sour. The **sublingual glands** are located right where their name says: under the tongue. The **submandibular glands** are under the mandible (jaw). The submandibular and sublingual glands both have ducts that come out under the front of the tongue, as we saw the last lesson.

The "roof" of the nasal cavity is the part of our nose that can smell. There is a special patch of epithelial skin, about the size of a very large postal stamp, that contains millions of **olfactory receptor cells**. (The word **olfactory** is the technical term used to describe a body part involved in the sense of smell.) These receptor cells are actually a type of neuron. These neurons are unlike any other neurons in the body because they are constantly being replaced. As a general rule, you are born with all the neurons you'll ever have. The neurons in your brain and muscles stay with you for your entire life. If they are damaged they have a limited ability to repair themselves, but they don't regrow like skin cells do. This is why injuries to the spinal cord can be permanent. However, here in the nose, these receptor neurons are being replaced every few months. The new neurons come from stem cells that live at the bottom of this epithelial tissue.

The receptor neurons have cilia that reach out into the nasal cavity. Other cells in the epithelium, called **Bowman's cells**, produce a watery mucus and specialized proteins that help molecules stick to the receptors. When odor molecules are inhaled through the nose, some of them will stick to this mucus and thus come into contact with a receptor neuron. If the shape of the odor molecule happens to be close enough to the shape of a receptor, it will bind to it. That will trigger the neuron and make it begin an action potential (electrical signal). The axons of the receptor neurons are bundled together, and these bundles go up through small holes in the ethmoid bone and then into a brain part called the **olfactory bulb**. The olfactory bulb gathers all the signals coming from the olfactory epithelium and sends them on to other brain parts. Here we find another difference between smell and our other senses. The input signals from our other senses go to a relay center in the middle of the brain where the brain decides what to do with them. The relay center then sends the signals on to other brain parts. However, the signals coming in from the nose don't go to this relay center, but go directly to other brain parts. One of these brain parts interprets the signals as what we could call a "smell." Two other brain parts that get the signal are the memory and the emotional center. They say you never forget a smell, and perhaps this is why. The strong link between smell and memory is a survival mechanism that helps wild animals avoid danger. Also, smells can trigger emotions and reflexes. Delicious smells can make our mouth water. Smells that we first encountered during a negative situation can trigger those same negative emotions even when the situation no longer exists.

Smells are combinations of odor molecules. When you smell pizza, you are sensing a complex combination of molecules coming from tomatoes, spices, cheese and bread. Yes, some of the pizza molecules actually become airborne and float around. (Heat makes molecules move around faster, causing more surface molecules to launch into the air, so a hot pizza will produce more smell than a cold one.) The odor molecules trigger only the neurons that match those molecules. So each smell triggers a unique pattern of receptors. Our brain categorizes those odors, though, so that we can find similarities between smells. A smell can be "fruity," for example, without being identified as specific fruit.

Most of our sense of taste actually comes from our sense of smell. As we chew our food, molecules from the food go back through the pharynx and then up into the nasal cavity. Undoubtedly, you've discovered that if you hold your nose, your sense of taste is much diminished.

We can see a few brain parts in this drawing. We've already mentioned the **olfactory bulb** and how it relays signals from the olfactory neurons to several other brain parts. Above the olfactory bulb we see the main part of the brain, the **cerebrum**. We've also mentioned the **pituitary gland** and how it sits in the little "pit" sculpted out for it in the sphenoid bone. Just behind the adenoids we find the **pons**. The word "pons" comes from the Latin for "bridge." The pons is like a bridge between the upper brain (the part we recognize as "brain") and the lower brain, or brain stem, that controls basic functions like breathing and the pumping of the heart. The pons is primarily involved with sleep and helps to control your waking and sleeping cycles. Research has suggested that our REM sleep—the periods when we dream—originate from the pons.

Behind and underneath the pons we find the **medulla oblongata**, or "brain stem." This is the area that controls breathing and heart rhythm. The medulla oblongata then turns into the spinal cord and goes down through all the vertebrae (back bones). Further behind the pons we see part of the **cerebellum**. "Cerebellum" means "little brain." The cerebellum has a texture very different from the rest of the brain. It sits below the main part of the brain, at the top of your neck. The cerebellum helps with balance and coordination and muscle memory. The cerebellum seems to be where memories like "how to ride a bike" or are stored.

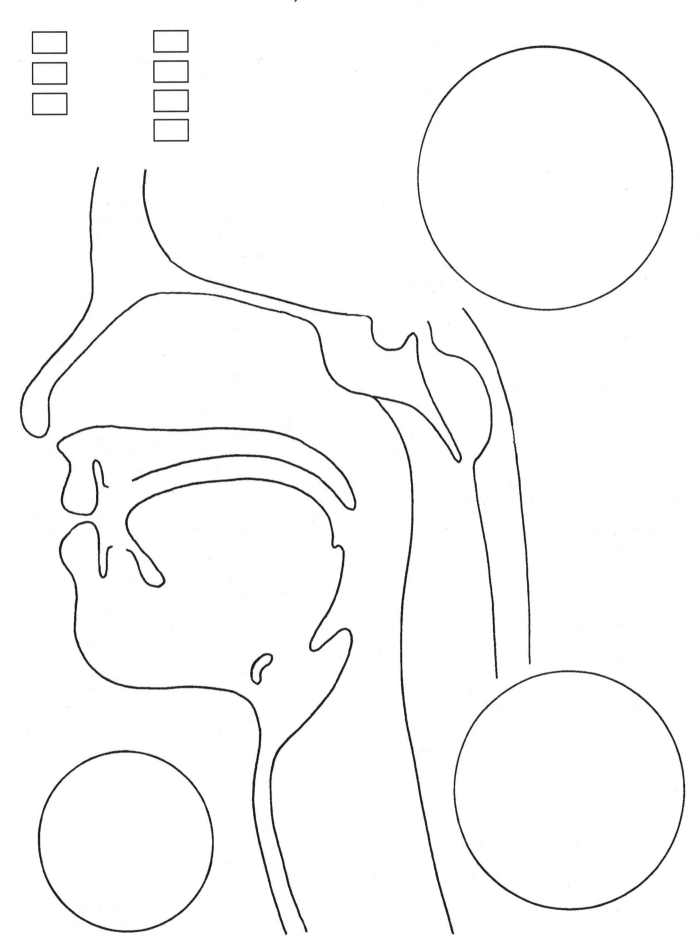

61: THE EAR

The ear can be divided into three main areas: the outer ear, the middle ear, and the inner ear. The outer ear is what we think of as the "ear." Technically, the outer ear also includes the ear canal, or auditory canal, most of which we cannot see. The technical word for just that outer flap of skin we call our ear is the **pinna**. The pinna is a complicated looking shape and we might wonder if there is anything significant about this shape. Scientists think that the shape of the pinna might help to collect and direct sound waves into the ear canal, especially the sound of the human voice. There are names for the various parts of the pinna. The only names we'll learn here are the **lobe** (which you already know), the **helix** (curly outside rim), the **antihelix** (curly inner shape), and the **tragus** (that flap you push on when you want to close your ears and not hear something).

The outer ear ends at the **eardrum**, also called the **tympanic membrane**. The eardrum is a thin, round membrane stretched across the end of the canal. It is not completely flat, but is slightly cone-shaped, bending in to the middle ear. The eardrum vibrates when sound waves hit it. The sound waves are passed along to the parts of the middle ear.

Sound is made of mechanical pressure waves that travel through the air. Very high sounds are made of waves that are short and travel quickly. The highest sound a human ear can hear has a wave speed, or frequency, of about 16,000 waves per second. (Some charts list 20, 000 as the upper limit.) Low sounds are made of waves that are long and travel more slowly. The lowest sounds our ears can hear have a frequency of about 20 waves per second. The unit of measurement for sound is **hertz, Hz**. Hertz means waves, or "cycles," per second. So we can say that humans can hear a range of sound from 20 Hz to 20,000 Hz. Many animals can hear much higher frequencies. Cats can hear sounds that mice make at about 75,000 Hz. Brown bats can only hear high sounds from 10,000 to 90,000 Hz. Don't bother talking to a bat because your voice is so low (100-200 Hz) that it can't hear you!

The middle ear, on the other side of the eardrum, begins with the **malleus** bone, also called the "hammer." ("Malleus" is Latin for "hammer.") The ends of this bone touch the eardrum and pick up vibrations. The vibrations move the malleus back and forth. The malleus is connected to the **incus**, or "anvil." ("Incus" is Latin for "anvil." The anvil is a tool used by a blacksmith. It is heavy block that the hammer strikes.) Connective tissue ligaments hold these bones in place. At the end of the incus we find the tiny **stapes** *(stay-peas)*. The stapes is often called the "stirrup" because it does indeed look like a stirrup on a horse saddle--the part where your foot rests. ("Stapes" is Latin for "stirrup.") These three bones of the middle ear are the smallest bones in the body. They transfer the vibrations from the eardrum to the inner ear where they will be turned into electrical signals.

The space around these bones is not filled with fluid, but is an air space. Problems might arise if the air pressure inside this space is higher or lower than the air pressure outside of the head, so there is a thin tube that connects the middle ear to the outside world. This tube is called the **Eustachian tube**, named after the scientist who discovered it in the 1500s, Bartolomeo Eustachi. The Eustachian tube is only a few millimeters in diameter and is flattened shut most of the time. When you go up in an airplane, the air pressure around you drops, and the high pressure air in the middle ear must escape through the Eustachian tube. We often experience this sudden opening of the tube as a "popping" sensation inside our ears. Divers experience the opposite, with the pressure around them suddenly increasing. They must make a yawning motion in their nasopharynx region in order to open the Eustachian tube and let the pressure equalize.

The end of the stirrup (stapes) is where the inner ear begins. The stirrup touches an oval area called the **oval window**. The oval window is part of a structure that has two parts, the **cochlea** and the **vestibular system**, but looks like it is just one part.

The **cochlea** gets its name from the Greek word for snail shell, "kohklias." The cochlea sits inside a "pocket" of bone in much the same way that the pituitary gland does. The cochlea is a set of coiled tubes that are filled with fluid. This fluid absorbs the vibrations that come in through the oval window. If you straightened out the cochlea and cut a cross section, it would look like two tubes stuck together. Vibrations enter through the top tube and leave through the bottom one. The end of the bottom tube touches the outside of the cochlea at a place called the **round window**. The oval window and round window work together, pulsing in and out as vibrations travel through the fluid. When the oval window presses inward, the round window pops out, and vice versa.

Between the tubes we find the **organ of Corti**. This long and thin little "organ" has cells that vibrate to certain frequencies. These cells are located on the **tectorial membrane**. When the organ of Corti is flexed up and down by the sound vibrations the tectorial membrane rubs "hairs" at the ends of **hair cells**. This causes the hair cells to depolarize and release neurotransmitters that will start an action potential in the dendrites of nearby neurons. (The inner hair cells are the ones that send signals to the brain.) The neurons get bundled as they exit the cochlea and become the **cochlear nerve** that goes into the brain. In the brain these electrical signals will be interpreted as sounds. High sounds are picked up by the first part of the cochlea, closer to the oval window. Low sounds are picked up near the center of the spiral.

The other part of the inner ear is the **vestibular system**. (In Latin, a "vestibule" is an entrance hall.) This is what gives us our sense of balance. The main body of the system is the **vestibule**. There are also three C-shaped tubes called **semi-circular canals**. The canals contain fluid that sloshes back and forth as we move our head. Each canal senses a different motion of the head: 1) up and down in a "yes" motion, 2) back and forth in a "no" motion, and 3) when the head tips down toward a shoulder. All together, the semi-circular canals are sometimes called the **labyrinth**. We saw this word in the lesson on the sinus. (The word labyrinth is a very general word and can be applied to any part that has a complicated, maze-like structure.) Sensor cells on the inside of the semi-circular canals are connected to neurons that send signals to the brain. As the fluid moves around the brain gets information on which way the head is tipped.

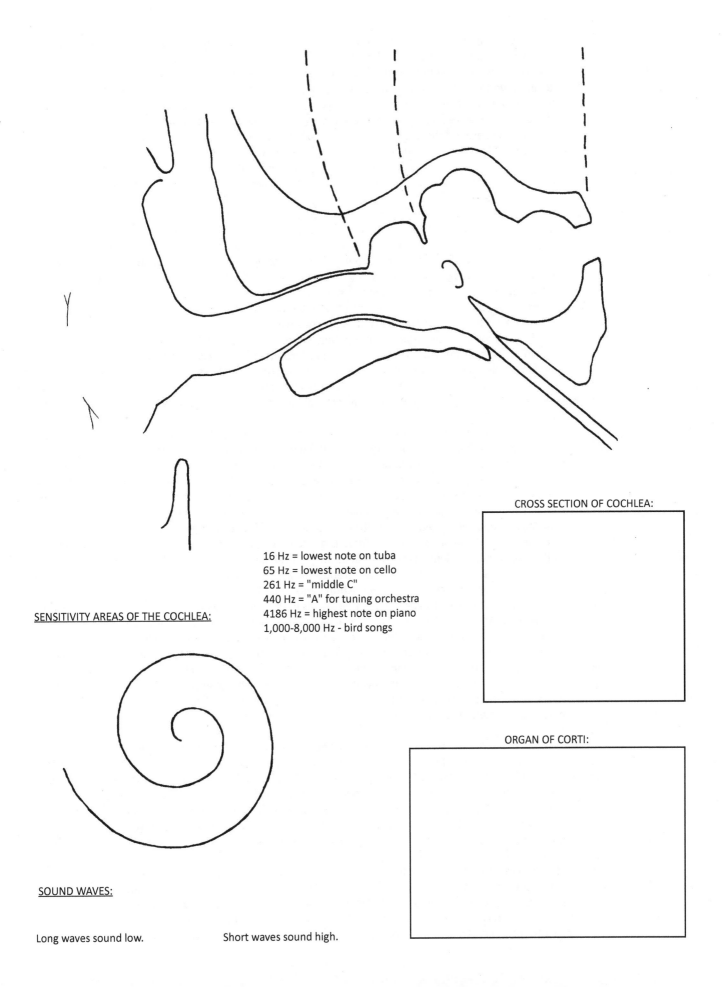

16 Hz = lowest note on tuba
65 Hz = lowest note on cello
261 Hz = "middle C"
440 Hz = "A" for tuning orchestra
4186 Hz = highest note on piano
1,000-8,000 Hz - bird songs

CROSS SECTION OF COCHLEA:

SENSITIVITY AREAS OF THE COCHLEA:

ORGAN OF CORTI:

SOUND WAVES:

Long waves sound low.   Short waves sound high.

62: THE EYE (anatomy)

The eye is so complicated we need to take two lessons to cover it. This lesson is about anatomy: drawing and naming all the parts. The next lesson will be about physiology, which is how the parts work.

The eye is a round sphere about an inch (3 cm) in diameter. We only see the front part of the sphere, peeking out from between our eyelids. The outside covering of the eye is made of white connective tissue (collagen type 1, plus elastin) and is called the **sclera**. ("Scler" means "tough or hard.") The colored part of the eye is called the **iris**. ("Iris" is Greek for "rainbow.") Humans have an iris that is fairly small, so that the white sclera is seen all around it. If you've ever looked at a dog's eye, the colored area is so large that you have peel back the skin around the eye to see any white. The fact that our eyes have a visible iris makes it much easier for us to tell from a distance in what direction someone is looking. The iris is like a donut with a hole in the middle. The hole is called the **pupil**. The pupil's job is to let light into the eye. If there is very little light, the pupil opens and becomes very large so that as much light as possible comes in. If there is too much light, the pupil shrinks and becomes very small. The size of the pupil is controlled by a ring of muscles inside the iris.

The color of the iris is caused by the amount of **melanin** produced by melanocytes. Human eyes contain various amounts of melanin, which is always brown or black. A small amount of melanin will produce light brown eyes and a lot of melanin will produce eyes that are almost black. Blue eyes are caused not by a pigment, but by the scattering of light—the same phenomenon that makes the sky blue. Green and hazel eyes are caused by a combination of the light scattering effect and a very small amount of melanin. If the melanocytes are non-functional or are completely missing, this causes a condition called **albinism**. Albino eyes look pink. People with albinism usually have little or no pigment in their hair and skin, as well.

The iris and pupil are protected by the **cornea**, a clear layer that is connected to the sclera. The cornea is made of connective tissue that is similar to the sclera. The main difference is that the cornea tissue is dryer, causing it to be transparent. Between the cornea and the iris there is fluid called **aqueous humor**. (Remember, "humor" means "fluid.") There is also a very small chamber between the iris and the lens, which is also filled with this fluid. The fluid is produced by the **ciliary body**, which is behind the lens. There are tiny drain holes to drain off old fluid as new fluid comes in to replenish the old. These drain holes are called **Schlemm's canals**. These canals are very similar to lymph vessels, taking away extra fluid and recycling it. If these drain holes become clogged, too much fluid will build up behind the cornea, causing a condition called **glaucoma**. Glaucoma can be very damaging and can even cause blindness.

Behind the pupil is the **lens**. The lens has a central core and a tough outer capsule. If the core becomes cloudy it can cause a condition called **cataracts**. An eye surgeon must drill into the lens and replace the defective tissue. The lens is connected to the ciliary body by thin ligaments called **zonules**. The ciliary body is flat and circular and goes all the way around the lens. It contains a ring-shaped muscle that can change the shape of the lens when it tightens or relaxes. When it is relaxed, the lens is more flat and will focus on things that are far away. When the muscle tightens, the lens gets more round and will focus on things that are close to the eye. The ciliary body is also responsible for making the aqueous fluid that fills the chambers in front of the lens.

The space behind the lens is filled with a "gel" called **vitreous humor**, or the **vitreous body**. "Vitreous" means "glass-like." This gel fills the interior space and keeps all the parts in place. There is a small canal, the **hyaloid canal**, that runs down the middle, from the lens to the back of the eye. When the lens thickens to focus on something close to the eye, its volume increases just a bit. This increased volume could cause problems as it presses on the vitreous body. The fluid in the canal is able to absorb this pressure, so no harm is done. This is a trim detail of design that most people don't know about.

Another lesser-known detail of the eye is the presence of oil glands in the eyelid. Inside the eyelid is a flat piece of connective tissue called the **tarsal plate**. ("Tarsos" is Greek for "flat surface.") This plate makes the eyelid just stiff enough to keep its shape, while still allowing it to be soft and flexible. Inside this plate are microscopic oil glands that make a special kind of oil for lubricating the eye. The oil keeps the water on the surface of the eyeball from evaporating too quickly. It also makes the edge of the eyelid just oily enough to form a slight barrier that keeps normal amounts of water from running down your cheeks. Of course, when you cry, the system is overloaded and tears do roll down your cheek. However, small daily amounts of water are held back. The inside of the eyelid and the visible part of the sclera are covered with a protective layer called the **conjunctiva**. Infections of the conjunctiva are very common and are often called "pink eye."

The movements of the eyeball are controlled by six muscles. Four of them are called **rectus** muscles. "Rectus" means "straight" so the rectus muscles move the eye either straight up and down or side to side. The rectus muscles are **superior** (meaning "on top,"), **inferior** (meaning "on the bottom"), **lateral** (meaning "on the side"), and **medial** (meaning "towards the middle"). There are two **oblique** muscles that rotate the eye clockwise or counterclockwise, allowing for diagonal motion. The superior (top) oblique muscle goes through a little loop called the **trochlea**. The trochlea acts like a pulley, letting the muscle be attached behind the eye instead of to the side. Five of these muscles attach to an area of connective tissue at the back of the **orbit** (the space where they eyeball sits). Only the inferior oblique does not. Also, the muscles that pull the eyelid up and down attach to the back of the orbit. Only the front part of these muscles is shown in the diagram.

The **lacrimal gland** is located above the eye, on the side away from the nose. The lacrimal gland makes tears that keep the eye moist and clean. ("Lacrima" is Latin for "tears.") The little indented place at the "corner" of the eye is called the **lacrimal caruncle**, or just the **caruncle**. The caruncle contains some sweat glands and sebacious glands covered with a layer of protective tissue. Near the caruncle on both upper and lower lids are tiny holes called the **puncta**. (One punctum, two puncta.) The puncta lead to the **lacrimal ducts** that go down into the nose. These ducts are drains for the tears after they have washed the eyeball.

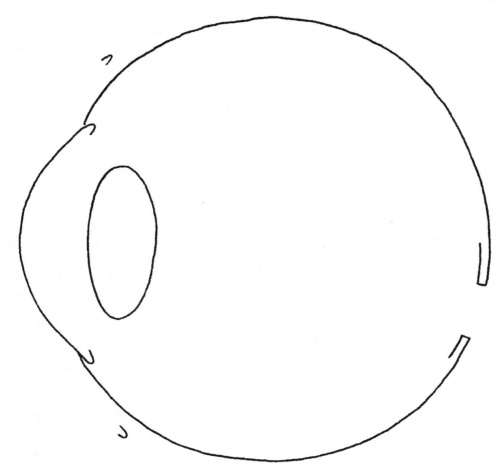

EYE MUSCLES (left eye shown)

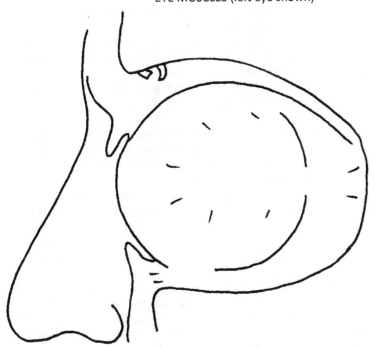

63: THE EYE (physiology)

The eye must constantly adjust its focus from near to far, far to near. All day long we are looking back and forth from distant objects to things that are very close. The light from these objects must be focused onto the **macula** of the retina, and onto the **fovea**, in particular. The fovea is the "sweet spot" of the retina because it contains millions more cones than any other place on the retina. In order to make sure that the light entering the eye hits the fovea, the lens can change shape. When an object is far away, the lens must be more flat. When an object is very close, the lens must become more round. The physics of lenses is the same, whether the lens is made of glass, plastic, or living tissue. Flatter lenses focus on distant object and rounder ones focus on closer objects. However, with glass and plastic the shape of the lens cannot change, of course.. Only living lenses can change their shape to adjust for all circumstances. The process of adjusting is called **accommodation**.

We learned that the ciliary body is in charge of changing the shape of the lens. Inside the ciliary body is a ring of muscle. When this muscle relaxes, it gets larger. This stretches the zonules (those "zonules of Zinn"), pulling the lens into a flatter shape. When the muscles contract, the lens appears to get pushed into a more round shape. Exactly how this happens is still being debated. There are two leading theories. One theory says that the zonules all relax, and the other theory is that only some of them relax. It's important to remember that scientists don't have everything figured out. There is still a lot we don't know.

The ring of muscles in the iris is also contracting and relaxing all the time. There is a feedback mechanism in the eye that automatically controls the size of the iris. If there is too much light, the muscles contract and shrink the pupil. If there is not enough light, the iris will open wider.

The image that hits the macula is upside down. That's just the way light works. The optic nerve will take this information to the brain, and the brain will flip the image and make us think we are seeing it right side up. The vision center in the brain is not right behind the eyes as you might expect, but at the very back. Also, the sides are reversed with the right side of the brain controlling the left eye and the left side of the brain controlling the right eye.

If you look into an eyeball with an ophthalmoscope, you will see two distinct spots. One will be the macula. The other is the **optic disc**, the place where the optic nerve leaves the eye. The optic nerve also contains many blood vessels, so you will see vessels coming out of the optic nerve area. The nerve is to the outside of the eye from this viewpoint.

The receptor cells in the retina are of basically two kinds: **rods** and **cones**. (We must say "basically" because a third kind was discovered in the 1990s and is believed to be the cell that controls the pupillary reflex. However, almost every source you will read in a book or on the web will say there are two kinds.) The rods are very long and thin, and the cones have a cone shape on their ends. They have a similar structure, with a nucleus located in a central area, and the skinny rod or cone shape on one end and a "synaptic" ending on the other. The synaptic endings look a bit like an axon terminal and they do a similar job, but technically these are not nerve cells in the same way that neurons are.

The rod or cone shaped ends of these cells contain about a thousand discs made of phospholipid membrane. The membrane is there to hold a very important molecule in place: **rhodopsin**. ("Rhodopsin" comes from the Greek word "rhodon" meaning "pink," and the Greek word "opsis" meaning "sight.") This is the molecule that responds to light. The rhodopsin molecule contains a smaller molecule called **retinal**, which the cells make from vitamin A. The retinal molecule changes shape when light hits it. This change of shape starts a short cascade of events that leads to the cell's sodium gates being opened up. Strangely enough, the cell is already full of sodium ions (Na^+). The addition of even more sodium ions causes it to become **hyperpolarized**. As we learned in lesson 52, **polarized** means more negative on one side and more positive on the other. This makes a cell ready for an action potential. The action potential happens when the ions rush back to the other side, **depolarizing** the cell. Then the cell has to reset again. In this case, light causes even more polarization. This STOPS the cell from sending a signal. So light stops these cells from sending signals! This seems backwards and certainly must have surprised the scientists who discovered it.

We have a strange system here. When they are NOT being stimulated by light, rod and cone cells are constantly releasing neurotransmitters (**glutamate**, from lesson 9) at their synaptic end. These transmitters, however, are **inhibitory.** (Excitatory transmitters cause (excite) cells to start action potentials, and inhibitory transmitters prevent (inhibit) them from doing so.) When light hits rod and cone cells, they stop releasing their inhibitory transmitters. The cells that they are connected to, the **bipolar neurons**, are then released from inhibition and can send a signal to the next cells in the line, the **ganglion cells**. The ganglion cells are the ones that have axons extending into the surface layer of the retina and then on into the optic nerve. Ganglion cells are the only ones that start a true action potential. The rods, cones, and bipolar cells use what is called a **gradient potential**, more like a dimmer switch than an off/on switch.

There are other nerve cells in and around the bipolar and ganglion cells. **Amacrine** and **horizontal** cells form horizontal connections between cells. (Amacrine comes from "a-" meaning "not," and "macro" meaning "big.") Amacrine cells connect rods and cones in such a way that they can function cooperatively. Horizontal cells connect ganglion cells.

Rods are very sensitive to light and a single rod can detect a single photon of light. Because of their sensitivity, rods can function when there is very little light. They give us our night vision. Rods are found all over the retina, but mainly outside of the macula. Because rods are found very far from the central focal point, they also give us our **peripheral vision**, at the edges of our field of sight. Cones are found primarily inside the macula and need a lot of light in order to function. As you go away from the fovea, the number of cones decreases and the number of rods increases. The total number of rods in the entire retina is about 100 million, and the number of cones is about 6 million. Cones are of three types and can detect one color of light: red, green or blue.

Notice that the rods and cones are in the back layer of the retina. Light must pass through all the nerve cells to reach them. Good thing the nerve cells are transparent! The cells that have pigment (color) in them are **retinal pigmented epithelial (RPE) cells** behind the rods and cones. The pigment absorbs extra light so it doesn't scatter about the retina. The pigmented epithelial cells also nourish and protect the rod and cone cells. These epithelial cells are very unusual because they do not go through mitosis, and must last your lifetime. If they stop functioning you experience **macular degeneration**, which can lead to blindness.

THE EYE (physiology)

63

How does the eye focus? The ciliary body controls the shape of the lens.

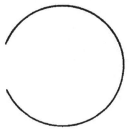

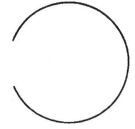

To focus on distant objects, the ciliary body relaxes, causing the zonlues to tighten, making the lens become more flat.

To focus on objects that are close, the ciliary body tightens, causing the lens to become more round.

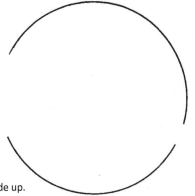

THE RETINA (front view):

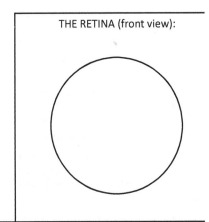

The image hits the retina upside down.
The brain must learn to see the image right side up.

HOW THE RETINA WORKS

Rods and cones are backwards from other receptor cells. They are turned on all the time, constantly releasing neurotransmitters. Light actually turns <u>off</u> rods and cones, and prevents them from being active. It is when they stop "firing" that a signal is sent to the bipolar cells.

The mechanism that starts the turning-off process is a pigment molecule called <u>**rhodopsin**</u>. Rhodopsin is found in the phospholipid membranes in the "pancakes" (discs) in the ends of the rods and cones. It holds a smaller molecule called <u>**retinal**</u>. When light hits retinal, its shape changes and this starts a chemical cascade that results in sodium ions rushing into the cell. The influx of sodium stops the cell from releasing its inhibitory neuro-transmitters. The bipolar cells are then activated.

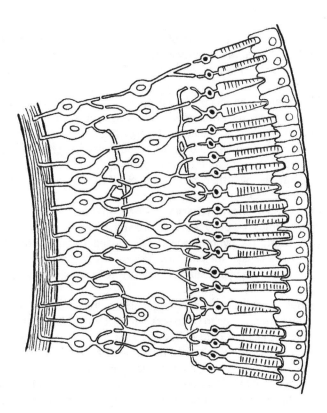

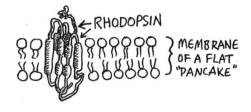

RODS: Cannot sense color, only light/dark. Function in low light conditions.
CONES: Sense one of these: red, green blue. Need lots of light to function.

The fovea has about 150,000 cones per mm^2. Other parts of the retina might have 10,000 or fewer cones per mm^2.

64: THE BRAIN (part 1)

The brain is the most complex object in the universe. There's no way we're going to learn enough about it in just two lessons. Please feel free to learn more about the brain using books, videos and websites. There's lots of good stuff out there!

From the top, the brain might remind you of a walnut. Both are kind of wrinkly and have two halves, or **hemispheres**. In general, the right hemisphere of the brain controls the left side of the body and the left hemisphere controls the right side of the body. The place where the optic nerves (from the eyes) cross over and go to the other side is clearly visible on the bottom of the brain. This is called the **optic chiasm** *(KIE-az-im)*. The crack between the lobes is called the **longitudinal fissure**. (Fissure just means "crack.") The rest of the cracks are called *sulci* (singular: *sulcus*). The lumps and bumps are called **gyri** (singular: **gyrus**). The purpose behind all the wrinkles and bumps is to provide more surface area. Imagine a bath towel lying on a flat surface, then imagine the edges being pushed toward the middle creating many wrinkles and folds. The surface of the brain is called the cortex, and it gets bunched together like that imagined towel so it takes up less flat surface area.

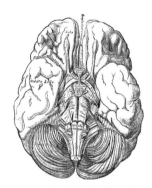

This drawing by Andrea Vesalius in the 1500s shows the brain as seen from the bottom. You can clearly see the nerve chords from the eyes crossing over as they go to the opposite brain hemisphere.

From the side we can see that the brain has three distinct sections The top part, which we saw in the top view, is called the **cerebrum** *(sah-REE-brum)*. It's the part that does all the thinking. However, two other parts are equally important. That little lump under the back of the cerebrum is called the **cerebellum** *(sare-eh-BELL-um)*, which means "little brain." This little piece of the brain enables you to coordinate your movements. Without it you could not walk. It also allows you to remember motions like riding a bike, playing the piano, or tying your shoe. The long thin piece going downwards is called the **brain stem**. It looks like it would be the stem if the brain was a flower. The brain stem is where vital functions are controlled, such as breathing and heart rate.

The cerebrum has a notable lobe at the bottom called the **temporal lobe**. It is located behind your ear. This lobe has many important functions including hearing, smell and speech. We'll learn more about it in the next lesson.

If we cut the brain in half lengthwise, making a **sagittal section** (cross section would be ear to ear), we can see most of the internal features. Some of the parts come in pairs, so we have to cheat just a bit. If we went straight down the middle, we'd miss both members of these pairs, so we'll just assume we went a little to one side in order to see one of these parts. We'll discuss each part number by number.

1) Olfactory bulb: There are two of these, one for each sinus. All the sensory neurons lining the inside of the nose connect to the olfactory bulb. (Olfaction is a fancy word for the sense of smell.) The signals travel from the bulb to a location in the temporal lobe where they are interpreted as odors.

2) Pituitary gland: We saw this in a previous lesson. It's tucked into its own little protective bone cavity. The pituitary, pea-sized as it is, has two parts, the anterior and posterior (front and back). The anterior part produces hormones that control many body parts including the thyroid, the adrenal glands (more on both of these in a future lesson), the growth of bones and muscles, reproductive organs in both males and females, and milk production in females. The back part controls the function of smooth muscles in the reproductive organs, plus it sends hormone signals to the kidneys telling them how much water to absorb. The pituitary is directly tied into the hypothalamus and receives signals from its neurons.

3) Hypothalamus: The hypothalamus is right underneath the thalamus. ("Hypo" means "under.") The most well-known functions of the hypothalamus are body temperature and appetite. Without your hypothalamus making you feel hungry you wouldn't know you needed to eat. The hypothalamus is also linked to the pituitary gland, as we saw in number 2. The neurons of the hypothalamus stimulate the cells of the pituitary. This means that the hypothalamus is part of the regulation of all the things that the pituitary does. Additionally, the hypothalamus plays a role in our sleep cycles and even some of our emotions.

4) Thalamus: The this is often described as the "relay station" for all the incoming signals. It has a central location in the brain, which is just where it needs to be to sort out all the signal traffic that is coming in and going out. All the signals collected by your senses go to the thalamus before they go to the brain parts that will interpret them. The thalamus will decide how important each signal is and whether to relay it on to the upper brain. If you are concentrating on something very important, or watching something very exciting, the thalamus might decide you don't need to be bothered with the minor pain signals coming from your skin. Then, later, you'll notice that you injured yourself and think, "When did that happen?" There are limits, of course, and if the pain is great enough, the thalamus will let the signal go through no matter what you are doing.

5) Intermediate mass of thalamus: This is a connecting part and joins the left and right lobes of the thalamus. In this view we can't see the left and right lobes; we see only one. Most brain parts have a left and right section. (The pituitary is an exception.)

The next four parts form a group called the **limbic system**. This system is your emotional center. Even if you don't think you are a very emotional person, you still have a limbic system. The limbic system also plays a vital role in memory.

6) Fornix: This is a connecting arch that goes up and over the thalamus. (The word "fornix" means "arch.") It connects the mammillary bodies on the front end (anterior) to the hippocampus on the other side (posterior). As with most brain parts, there is a fornix on each side, but they touch in the middle so there can be some communication between left and right sides.

7) Mammillary body: The name of this brain part has nothing to do with its function. It was first discovered well over a hundred years ago, long before anyone had a clue what it did, so they just named it for what it looked like to them, and someone thought that the pair of them looked like mammary glands (i.e. breasts). Yeah, sorry. But they used the scientific name, "mammillary" so it didn't sound so bad. But this part has nothing to do with mammary glands. The clue to its function came from some people who had a nutritional disease that damaged this part specifically. The symptoms of the disease were memory problems. Therefore, it was concluded that the mammillary bodies must play a role in memory. Now the theory is that they act like a relay center, receiving signals from the hippocampus at the other end, and sending some of the signals into a central area of the thalamus. They play an important role in our sense of direction and our memory of places and spaces.

8) Hippocampus: This is at the posterior end of the fornix, and is definitely a huge player in memories. As with the fornix and the mammillary bodies, the name comes from its shape, not its function. Someone thought this part looked a bit like a seahorse, and "hippocampus" is Greek for either "seahorse" or "horse-like sea monster." (Or, maybe they were thinking of "kampe" which means "caterpillar.") Seahorse is actually a good image to use, because it is about the right size and shape.

The hippocampus is the part that transfers short term memories into long term storage. It also helps to retrieve memories when you want to remember them. It's a bit like a librarian who both puts books into storage and also goes and finds them when needed

9) Amygdala: *(ah-MIG-dah-la)* This is near the hippocampus. ("Amygdala" is Greek for "almond.") The amygdala seems to be involved with strong emotions, especially fear and anger. Brain technology now allows researchers to see which brain parts are working in different situations, and they've been able to watch the amygdala light up when certain pictures or ideas were shown or read to the subjects in the experiment. From this, and from dissection of rat brains (and a few human brains) they've been able to determine that this brain part is involved with negative emotions, that it is different in males and females, and that the two lobes of the amygdala are not identical. The left side can experience either happiness or sadness, and might be a key player in conditioned learning, where you receive either positive rewards or negative punishments. The right side is restricted to anger, fear and aggression and makes sure events that produce these emotions get recorded in the permanent memory. It's part of your survival system to make sure you don't repeat behaviors that led to negative consequences. You never forget traumatic events. (Which is also why this brain part is involved in post-traumatic stress syndrome.)

In early childhood, the amygdala seems to be responsible for the fear that babies experience when they see faces they don't recognize. The amygdala also senses people invading your "personal space." Malfunctions in the amygdala seem to play a role in anxiety disorders and alcoholism. In general, males tend to have larger amygdalas, and their amygdalas take longer to develop than they do in females. Males also tend to have the right side of the amygdala be larger than the left. This is one of the lesser known difference between males and females.

NOTE: A few other parts are sometimes listed as being part of the limbic system, because they are connected to it, such as the cingulate gyrus, or the thalamus.

10) Cingulate gyrus: *(SING-gu-late GIE-rus)* Basically, this part connects the cerebrum to the limbic system. However, it is not merely a connecting piece, but adds functions of it own. It connects all those emotional brain parts with the frontal lobe up in the cerebrum, which is the part where conscious decisions are made. The cingulate gyrus makes sure your frontal lobe is involved in any decisions that are very emotional. You might be so angry that you want to throw your friend out the window, but your frontal lobe says no, that would not be a good idea. The cingulate gyrus seems to be involved in positive things like the emotional bonding that happens between mothers and their babies. Also, it connects our language centers in the cerebrum to these emotional centers in the limbic system, making it possible for you talk about how you are feeling.

11) Corpus callosum: ("Corpus" means "body" and "callosum" means "thick.") This is the part that connects the left and right hemispheres. It is made of bundles of axons, so it looks white. People who have very severe forms of epilepsy sometimes have to have a cut made down through their corpus callosum, so that seizures cannot travel from one hemisphere to the other. This has allowed researchers to do very interesting experiments, testing what happens when the two hemispheres can't communicate. These people are surprisingly normal and if you met them you would not immediately suspect anything was wrong. The problems are only apparent during particular tasks, like asking the brain to name something they are seeing with only their left eye. (Since the right side of the brain controls the left eye, and the right brain usually doesn't have a speech center, the person will not be able to talk about what he/she is seeing.) The results of these tests have helped us learn a great deal about the different functions of the left and right sides of the cerebrum.

12) Mid-brain: This is a center for connections and reflexes. The connections from the cerebrum to the cerebellum go through this area. Also, the pathways that carry signals from the cerebrum to the muscles are found here. One part of the midbrain makes a neurotransmitter chemical called dopamine which is necessary for transmission of signals to muscles. If these cells stop producing dopamine, the result can be "Parkinson's disease." The midbrain also has our pupillary reflexes (adjustments for bright or dim light) and the automatic focusing feature of the lens. The midbrain also seems to play a role in maintaining consciousness (being awake and alert). Some people include the colliculi (16) as part of the midbrain.

13) Pons: ("Pons" means "bridge.") This looks like a lump at the top of the brain stem. It has a number of different functions, but the most well-known of these is the sleep cycle, sometimes called the "circadian *(sir-CADE-ee-an)* rhythm." Your pons will wake you up eventually, even if your alarm clock doesn't go off. The type of sleep that produces dreams seems to originate in the pons. Several very large nerves come out of the pons area, and connect to various parts of the face, eyes, and ears.

14) Medulla oblongata: This long (oblong) part is in charge of keeping us breathing and keeping our heart beating. Our breathing and heart rate keep going while we are asleep thanks to our medulla oblongata.

15) Pineal gland: This tiny part produces melatonin, a chemical involved in sleep. It is connected to the eyes and receives information about whether it is light or dark (day or night).

16) Colliculus: *(col-LICK-u-luss)* This area has two sections, the superior (upper) colliculus and the inferior (lower) colliculus. The superior part has important eye reflexes, such as being able to maintain our fix on an object while turning our head. The inferior part is connected to the ears and contains our startle reflex.

NOTE:

There's still one group of brain parts not included on this diagram. There are four places on the inside the cerebrum, kind of around the outside of the thalamus, where there are clusters of nerves, called ganglia. Since these clusters occur generally in the lower half, or underneath, the cerebrum, they are called the **basal ganglia**. They are harder to understand and harder to draw than the 16 brain parts in this lesson, and they are often not shown on brain anatomy diagrams for beginners.. This drawing is pretty full of brain parts already, so I thought it best not to include them.

The basal ganglia are coordinating areas that interconnect different parts of the brain. Some of the ganglia help to coordinate muscle movement. There are several neurological diseases that affect the basal ganglia, including Parkinson's disease and Tourette's syndrome. Both disorders involve muscles moving too much, out of the person's control. Tourette's syndrome also has a behavioral aspect, as people with Tourette's sometimes do things that are viewed as socially unacceptable, such as yelling or biting.

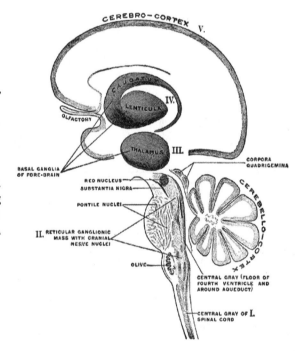

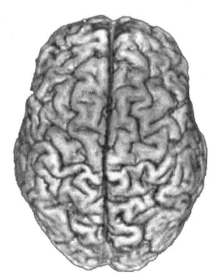

Top view of the cerebrum

BRAIN (part 1)

The brain is extremely complicated. All these drawings and labels have been simplified. If you want more detailed information, the Internet can provide plenty. (There are dozens of small parts and connecting pieces with long Latin names.)

TOP VIEW

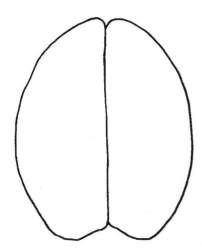

The **LEFT** hemisphere controls the right side of the body.   The **RIGHT** hemisphere controls the left side of the body.

SIDE VIEW

The purpose of wrinkles is to provide more surface area. The surface is where all the neuron cell bodies are and where most of our "thinking" takes place.

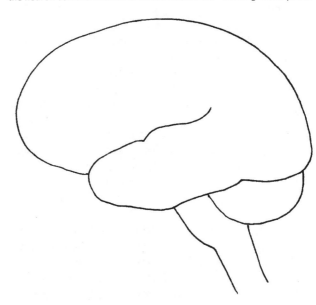

CLOSE-UP of protective layers

SAGITTAL SECTION

CROSS SECTION of cortex
(cells not to scale)

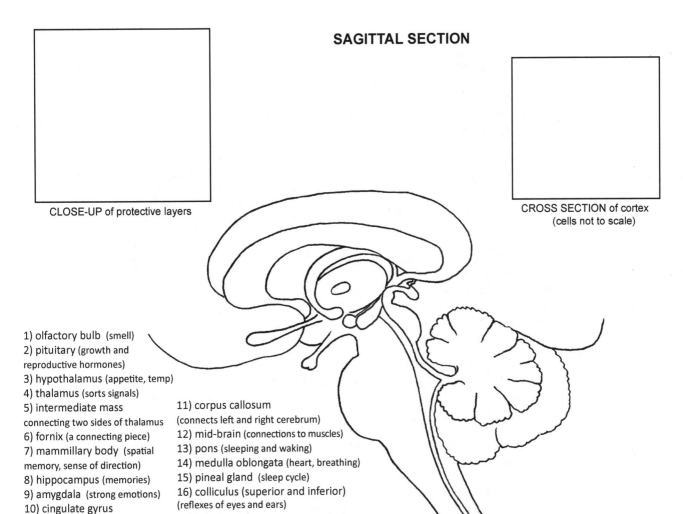

1) olfactory bulb (smell)
2) pituitary (growth and reproductive hormones)
3) hypothalamus (appetite, temp)
4) thalamus (sorts signals)
5) intermediate mass connecting two sides of thalamus
6) fornix (a connecting piece)
7) mammillary body (spatial memory, sense of direction)
8) hippocampus (memories)
9) amygdala (strong emotions)
10) cingulate gyrus (connects top to bottom)

11) corpus callosum (connects left and right cerebrum)
12) mid-brain (connections to muscles)
13) pons (sleeping and waking)
14) medulla oblongata (heart, breathing)
15) pineal gland (sleep cycle)
16) colliculus (superior and inferior) (reflexes of eyes and ears)
17) 4th ventricle (fluid-filled space)

65: THE BRAIN (part 2)

We saw in the last lesson that the cerebrum is divided into two hemispheres, connected by the corpus callosum. Each hemisphere has subsections called lobes. One of the lobes, the *temporal lobe*, is clearly visible, with a sulcus dividing it from the rest of the cerebrum. The other lobes can't actually be seen. We only know about them because of many years of brain research. For over a century, researchers have been carefully documenting brain diseases and making notes about what areas of the brain are affected. Modern brain imaging has added even more information. We can now make a "map" of the brain showing each area and labeling its functions.

1) Temporal lobe: This area has many functions. The centers for **smell** and **hearing** are in this lobe. **Speech and language** are also in this lobe, but usually on just one side. Most people (9 out of 10) have their speech on the left side. The left temporal lobe has sub-sections that specialize in one particular aspect of speech. **Wernicke's area** is where we understand the meaning of sentences. *Broca's area* is where we think of what words we want to put together to make a sentence. The right temporal lobe contributes to speech by adding patterns of rhythm and intonation (voice going up and down).

2) Occipital *(ok-SIP-it-al)* **lobe:** This is the area at the very back of the cerebrum. Oddly enough, this is where your vision center is, at the point farthest from your eyes. We've already mentioned that the optic nerves cross over on the underside of the brain so that the left eye connects to the right occipital lobe and the right eye connects to the left occipital lobe. The place where the nerve cross is called the **optic chiasm** *(kie-az-im)*. The occipital lobe interprets the electrical signals coming in from the retina and forms them into an understandable image. Part of this interpretation process includes turning the upside down retinal image right-side up.

3) Parietal *(par-EYE-it-al)* **lobe:** This is right above the occipital lobe. The lobe is named after the skull bone (the parietal bone) that covers it. This lobe does many things we take for granted. It is concerned with the perception of space and where our body is in that space. This lobe keeps track of where our arms and legs are, and is the reason we can bring our hands together even if our eyes are closed. The parietal lobe understands the 3-dimensionality of objects and lets us rotate objects in our mind. Interestingly enough, the brain region that Einstein seems to have had more of was the parietal lobe. The ability to visualize math and physics ideas seems to be largely in the parietal lobe.

4) Sensory cortex: This is a thin strip at the top of the parietal lobe where the incoming signals from the body connect to the brain. If you cut a cross section of the cerebrum at this point, it might look something like these diagrams. The top diagram just shows the cerebrum sliced in half. The bottom diagram shows this slice with some body parts and words added to show you how much of this strip of cortex is connected to which parts. The important thing to notice is how much space is devoted to parts of the body that need to be very sensitive, like the fingers and the face. We don't need our feet or our back to be as sensitive as our

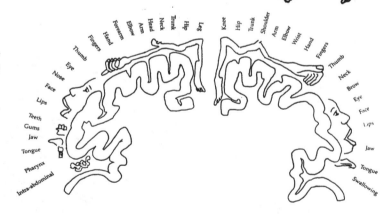

fingertips. This diagram, with the body parts drawn on it, is often called the "homunculus," meaning "little man," because it looks like a body stretched out. Artists sometimes like to have fun with this concept and draw a person proportioned according to these proportions. Their drawings show people with huge lips, face, hands and fingers and very tiny torsos and legs.

5) Motor cortex: This strip of cortex lies right next to the sensory cortex. This is where the signals leave the brain to go out to the muscles. The "homunculus" of this body part would look very similar to the sensory one, with a lot of nerve fibers going out to body parts that require fine controls such as hands, fingers and mouth.

6) Frontal lobe: This is the lobe that does the actual "thinking" that the brain is so famous for. This is where we do mathematical calculations, think of stories to write, decide what to eat for breakfast, solve logic puzzles, and ponder our existence. The left and right frontal lobes have slightly different functions, with the left side being more logical and analytical, and the right side more creative and intuitive. Here are typical lists of left/right brain features:

LEFT HEMISPHERE: logic, analysis, sequencing, keeping track of time, names, symbols, computation, moral decisions
RIGHT HEMISPHERE: creativity, intuition, music, art, sculpting, understanding things as a whole, flashes of insight ("Eureka!")

Many lists also include speaking and writing as a left brain activity, but we already mentioned this in the paragraph on the temporal lobe. (And we need to remember that 10% of the population have their speech on the right side, not left.)

Our personalities are to a large degree determined by our frontal lobes. Scientists first learned this through a famous accident. In the year 1848, an American man named Phineas Gage was working on laying railroad tracks when an explosion caused a long metal bar to be blown right up through his skull, passing through his left frontal lobe. Amazingly, he survived this event, but afterwards his personality changed for the worse. Before the accident he was cheerful, honest and hard-working. After the accident he became angry, irresponsible and lazy. This was a good indication that the left frontal lobe plays an important part in personality, especially our moral decisions. (His skull and the metal pike that went through it have been preserved and are now in a medical museum at the Warren Anatomical Museum in Boston, Massachusetts.)

MEMORY

How are living cells capable of forming permanent memories? What actually happens in the cells? The biochemical mechanisms of memory have not been fully discovered yet. (In other words, we don't know!) Some research seems to indicate that there are permanent changes made in the synapses between some neurons, causing them to form strong connections. Tiny "spines" on the dendrites seem to play a part in these synapse changes. Other research points to changes in the cytoskeletons of neurons, and their possible role in making permanent changes inside the cells. In 2017, researchers in Norway discovered that the dendrites and cell bodies of neurons are covered with a thin protective covering made of sugars. If this covering is removed, long term memories are erased. (Don't worry, they only did this to rats who had long term memories of blinking lights.) These are all clues, but we really don't know the whole story yet.

Though the cellular level remains largely a mystery, we are slightly more knowledgeable about the way memories are shifted around in the brain, from short term to long term. We know that ALL memories start out as very short term impressions brought in by our senses. These impressions of sight, smell, hearing, taste or touch, only last for a few seconds, and your brain can only keep track of about 12 of them at a time. They come into our thalamus area to be sorted, and the important ones are then sent into short term storage. The impressions that the thalamus decides to ignore are erased and disappear forever. We could think of many examples of details about life that our brain often ignores: the color and shape of each car we pass in traffic, the location of all the clouds in the sky above us, the type of shoe each person in our class is wearing, the faces of people we see in crowds, all the individual bird songs we hear while hiking, what last month's news headlines were, etc. Life is full of details that we really don't need to remember.

After sensory impressions are sorted by the thalamus region, the important ones are made into short term memories and stored in the very front of the frontal lobe, in an area known as the **prefrontal cortex.** This short term memory area can only hold about 7 items. (Even if you are a super genius, your short term memory still only holds about 7 items!) It's like a temporary "clipboard" and is constantly erased and reused about every 10 seconds. If a short term memory is important, it needs to be turned into a long term memory. The hippocampus takes the information from the short term storage area and transfers it into the neocortex, that outer edge (gray matter) of the cerebrum. Some short term memories are eventually erased after days or weeks or months, but others persist for years.

Quite unexpectedly, researchers discovered that long term memories are not stored all in one place. Each memory is broken up into its sensory components, such as sight, sound, smell, and emotion, and stored as these individual parts. When you want to recall that memory, the hippocampus finds all those pieces—what you saw, what you heard, what you felt about the situation—and reassembles them back into the whole memory. When you are done remembering, it puts them back again. Unfortunately, this is not a perfect process and each time you recall the memory, small changes occur. After a long period of time, our memories can fade or can change significantly. However, if strong emotion was involved at the time the memory was made, and therefore the amygdala played a role, the memory will be encoded much more strongly and will be less likely to fade or change over time.

As with many brain parts, the role of the hippocampus in memory has been figured out by studying brains with damage to that part. The first person that scientists were able to study was a patient from the 1960s, known by his initials, "H. M." Surgeons had removed his hippocampus and amygdala, hoping to cure him of seizures. Big oops, though, as they did not realize how important the hippocampus is for forming memories. H. M. became severely disabled, unable to form long term memories.

Perhaps the most famous person with hippocampus damage is Clive Wearing, a British man who caught a virus (in 1985) that went to his brain and destroyed his hippocampus. Clive can no longer form long term memories. He still has short term memory and can remember new facts for about 10 seconds. After that, his "clipboard" resets and he completely forgets. Before he got sick, Clive was a professional pianist and symphony conductor. When he is asked to play the piano, he'll tell you he's never seen one or played one before, but when he puts his hands on the keys, he'll start to play tunes that he had memorized before his illness. Those muscle memories are in his cerebellum, not his cerebrum, and are not dependent on the hippocampus. Because of his cerebellum memories, he can also remember how to get dressed, shave and brush his teeth, ride a bike, and things like that. (You can find Clive's story in several documentaries posted on YouTube.)

Memories that involve learning facts are called either *declarative* or *explicit* or *procedural* memories. (The terminology in memory science is not standard. You'll find different names for the same things.) Memories that involve moving body parts (dance steps, sports skills, handwriting, etc.) are called either *non-declarative* or *implicit* memories. The first type is stored in the cortex of the cerebellum. The second type is stored in the cortex of the cerebrum.

VENTRICLES

The brain has "empty" places called **ventricles**. These are fluid-filled spaces that help to cushion and protect the brain, as well as providing nutrients to brain cells around the outside of the ventricles. When we did lesson 51, on nerve tissue in the CNS, we **ependymal** (e-PEND-i-mal) **cells** forming a lining on the inside of these ventricles. These ependymal cells produce **cerebrospinal fluid**, or CSF. This fluid drains from the ventricles in the brain, down through a canal inside the spinal cord. The canal goes all the way down the spine to the "tail bone" at the bottom. Doctors can take a sample of this fluid by putting a needle into the spine, between two vertebrae bones. This is called a **spinal tap**. By testing this fluid they can learn about problems in the brain, especially infections.

There are four ventricles. The first two are called **lateral** ventricles, with one ventricle in each hemisphere. The **third** ventricle is in the middle beneath the laterals. We labeled the **fourth** ventricle in the last lesson. It is down near the cerebellum. The tube that goes from the 3rd third to the fourth ventricle is called the **cerebral aqueduct**. (Aqueducts were stone water troughs built by the Romans to bring a water supply into a city.) The canal that drains down from the fourth ventricle is called the **central canal**. It goes all the way down to the tail bone. At the bottom of the canal, the CSF will be reabsorbed back into the bloodstream.

When the CSF flows into the fourth ventricle, it can take another route, besides going down into the central canal. It can go through canals that will take it up and around the top of the cerebrum. It flows through a space right under the arachnoid layer, called the **sub-arachnoid space**. From there it will be reabsorbed back into the bloodstream, as osmotic pressure pushes it into the tiny veins that run along the outside of the brain. These veins will eventually drain down into the large **jugular** vein in the neck.

Ventricles were first discovered by someone who poured hot wax into a cow brain, then removed the brain to see what shape the wax had cooled into. He found that there were very well-defined areas, with lobes on each side. Further research found that all mammal brains have ventricles, including human brains.

The LEFT hemisphere is known for:

1) _____   4) _____
2) _____   5) _____
3) _____   6) _____

The RIGHT hemisphere is known for:

1) _____   4) _____
2) _____   5) _____
3) _____   6) _____

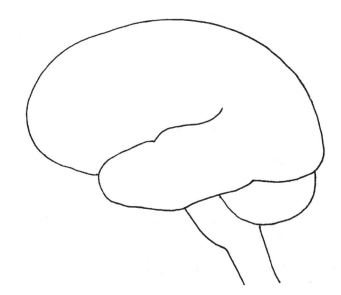
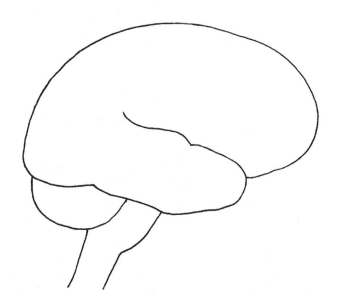

HOW SOME MEMORIES ARE ENCODED:
(This would be declarative/explicit memories.)

Short term involves glutamate crossing synapses.
Long term involves production of new spines on dendrites.

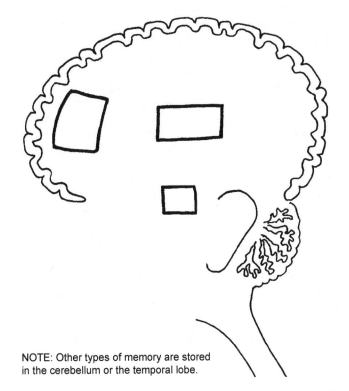

NOTE: Other types of memory are stored
in the cerebellum or the temporal lobe.

VENTRICLES:

Ventricles are filled with cerebrospinal fluid (CSF).
Ventricles and CSF help the brain by providing:

1) _____
2) _____
3) _____
4) _____

CSF is found in the 4 ventricles, as well as under the arachnoid
layer around the exterior. The CSF circulates around these spaces,
then drains either into veins at the top and middle of the brain, or
down into the central canal that goes into the spinal cord.

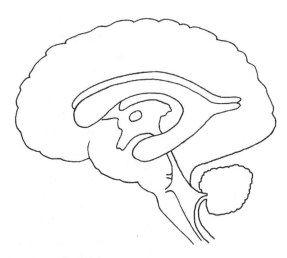

CHOROID PLEXUS: epithelial cells that produce CSF

66: BLOOD VESSELS

We've seen blood vessels in a number of places already, but this lesson will focus on the vessels themselves. There are three types of vessels: arteries, veins, and capillaries. Arteries are defined as any vessels that carries blood away from the heart. Veins are vessels that carry blood toward the heart. Microscopic capillaries are the "in between" places where arteries turn into veins.

Both veins and arteries have the same three-layer structure. The inner layer is called the **tunica intima** and is made of another three layers: a layer of epithelial cells (the endothelium), basement membrane (which is under all epithelial cells), and then an elastic layer. The middle layer is called the **tunica media** and is made of smooth muscle surrounded by an elastic layer. The outer layer, the **tunica externa**, or "adventitia,: *(ad-ven-TISH-ah)*, is made of elastic connective tissue that is also strong. Since fibroblasts make collage and elastin, you will find fibroblasts in this outer layer. Macrophages also can be found in this outer layer.

Arteries are built for high pressure. Since they carry blood away from the heart, they must withstand the forceful pumping of the heart's strong muscles. The muscle layer is very thick, allowing the artery to contract a bit every time the heart pumps. This contraction compensates for the increased pressure in the vessel. The pressure needs to be high, though, so that the blood will keep flowing in all the vessels. (If you laid all your blood vessels end to end, you could circle the globe several times!) Arteries get more muscular as they get smaller. Sometimes large arteries are called "elastic arteries" and smaller arteries are called "muscular arteries." When arteries get very small they are called **arterioles**. When arterioles get microscopically small they are called **capillaries**. Small veins are called **venules**.

Capillaries are only slightly larger than a red blood cell. In fact, some capillaries are small enough that the red cells must squeeze to fit through. The red cells can release a chemical messenger called nitric oxide (formula: NO) that causes them to relax and flex so the red cells can squeeze through.

Veins are built for lower pressure. After the blood leaves the bed of capillaries, it is flowing much more slowly. The pressure is low enough that veins have one-way valves to keep the blood from going backwards. (If these valves stop working properly you can get **varicose veins**, a condition where sections of veins will enlarge and become very visible under the skin.) We met similar one-way valves in the lymph system. Veins have a larger **lumen**, or interior space, than arteries do. Veins have less smooth muscle, but they still have the same basic three layers: the tunica intima, the tunica media and the tunica externa.

Macrophages in the tunica externa of vessels can play a role in the development of a condition called **atherosclerosis**. When damage occurs in endothelial cells, macrophages are attracted to that area to help clean up. In the clean up process they eat fats that are in the area and acquire a foamy texture, giving them the name **foam cells**. Chemical changes result in more macrophages coming to the area and they eat even more fat molecules. Gradually the macrophages die, leaving their remains and all the fat molecules they ate lodged in the wall of the vessel, in the tunica intima. We call this a **plaque**. As long as plaques stay small and stay where they are, nothing too terrible happens. But if they become large they can block blood flow, and if they tear away from the wall of the vessel, they eventually get stuck in a smaller vessel causing a blockage. If the blocked vessel is in the brain we call this a **stroke**. If the blocked vessel is in the heart, it can cause a heart attack. If the blockage is in an arm or leg it can cause a lot of pain and a need for immediate medical treatment. Everyone has atherosclerosis to some degree, as we start forming these plaques even during late childhood. The goal is to keep them to a minimum. Things that help prevent atherosclerosis are not eating too much sugar, eating the right kind of fats (more omega 3 than others), exercising, keeping our weight under control, and not smoking.

Body cells must be near a capillary in order to receive enough oxygen and food. They don't need to be touching the capillary, but they need to be close to it. As red blood cells come through the capillaries, oxygen can be released from the hemoglobin molecules that are carrying them. Oxygen will diffuse to areas that have a low oxygen level. Water can also diffuse out, as well as glucose. Other molecules will go out, also, such as hormones or other chemical messengers. On the other side of the capillary bed, wastes can enter the blood. Once again, simple diffusion is often the way they go in. Wastes that need to be carried away include CO_2, urea, creatinine, and lactic acid. More on these in a future lesson.

Small arterioles can use their smooth (involuntary) muscles to regulate how much blood goes to various parts of the body. If you are digesting food, arterioles in the digestive system can relax and allow maximum blood flow to digestive organs. If you are exercising, arterioles in the muscles will open and arterioles in the digestive tract will shrink. When you exercise, your face gets red because tiny vessels are opening up to help get rid of excess body heat.

There are three types of capillaries. **Continuous** capillaries are the most abundant type and are found most places in the body. Endothelial cells can take nutrients from the blood stream (through endocytosis) and pass them out the other side (through exocytosis) to the waiting body cells. Sometimes things leak through tiny cracks, but the cracks are soon mended. **Fenestrated** cells have tiny holes in them. ("Fenestra" is Latin for "window.") The holes allow a much faster rate of leakage from inside to outside the capillaries. Fenestrated capillaries can be found in the intestines, pancreas, kidneys and in glands. The most leaky type of capillary is the **sinusoidal**, and it is found only in the liver and spleen. These large gaps allow blood cells to exit the capillaries.

BLOOD VESSELS

66

There are 3 main types of vessels: **arteries** (away from the heart), **veins** (toward the heart) and **capillaires** (microscopic).

ARTERIES : built for high pressure

The heart is a very strong pump. When blood leaves the heart, it does so under high pressure. Arteries must be able to withstand high pressure. Smooth muscles in the vessels contract with each pump.

VEINS: built for low pressure

Veins experience much less pressure because they are farther away from the heart. In fact, they have one-way valves to ensure that blood does not flow the wrong way.

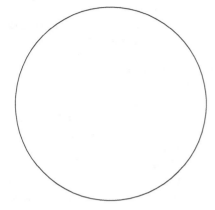
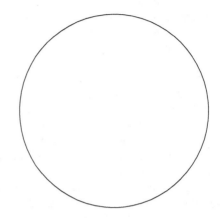

CAPILLARIES form "beds" (networks)

ARTERIOLE (small artery)

The smooth muscles of arterioles control how much blood goes to which parts of the body.

VENULE (small vein)

TYPES of CAPILLARIES:

1) **Continuous**
Where? _____

2) **Fenestrated** ("fenestra" = "window")
Where? _____

3) **Sinusoidal**
Where? _____

67: THE HEART

The heart is a living pump. It beats about 70-80 times per minute round the clock, not needing to rest like your other muscles do. A heart can pump up to 2,000 gallons (7,000 liters) of blood each day.

When we look at a diagram of the heart we need to remember that the left and right parts are labeled from the perspective of the owner of the heart. Therefore, the left side of the diagram is actually the right side of the heart, and the right side of the diagram is the left side of the heart. (Just imagine the heart inside your own chest think of it from that perspective.)

The heart has four main chambers. The two smaller ones on the top are the **atria** (plural of **atrium**) and the larger ones on the bottom are the **ventricles**. Ventricles have sort of a "V" shape, which makes it easy to remember their name, since ventricle begins with the letter "V." There are valves between the atria and the ventricles. The valve on the right side is called the **tricuspid valve** ("tri" means "three" and this valve has three flap pieces) and the valve on the left side is called the **bicuspid valve** ("bi" meaning "two"). Just to confuse you and make things harder, you also need to know that the bicuspid valve is also commonly called the **mitral valve**. The flaps of the valves are anchored to the ventricles by strings made of connective tissue. (This is perhaps where the term "heart strings" comes from. You may have heard the phrase, "tugging at someones' heart strings," meaning appealing to their emotions.)

The muscle tissue of the heart is called the **myocardium**. ("Myo" means "muscle" and "cardium" means "heart.") This muscle tissue needs it own blood supply so there are arteries and veins on the outside of the heart that go down into the muscles. The muscle fibers of the heart work very hard all the time and need a constant supply of nutrients. If one of these exterior **coronary arteries** or veins gets blocked, those muscle cells will not receive enough oxygen and the result will be a "heart attack."

The heart is actually two separate pumps sitting side by side. The right side receives "used" blood from the body (which contains very little oxygen and a lot of carbon dioxide), then pumps it into the lungs where it gets rid of the carbon dioxide and picks up fresh oxygen. The left side receives the re-oxygenated blood from the lungs and then pumps it out into the body. Blood vessels that are taking blood towards the heart are called **veins**. Blood vessels that are taking blood away from the heart are called **arteries**. Most often we see arteries and veins colored coded with red and blue indicating whether they have a lot of oxygen (red) or very little oxygen (blue). Here in the heart, we find the exception to this color coding rule. We will see veins bringing in blood from the lungs which is rich in oxygen.

The large veins coming into the heart are the **superior vena cava**, the **inferior vena cava**, and the **pulmonary veins**. (The words "superior" and "inferior" mean "top" and "bottom" in this case.) The arteries leading out of the heart are the **aorta** and the **pulmonary arteries**. The aorta has three branches that go off the top, the brachiocephalic ("brachio" means "arms" and "cephalic" means "head"), the carotid (which we saw in the head drawing), and the subclavian ("sub" means "under" and "clavia" refers to the clavicle bone that runs between the neck and the shoulders). After bending over the top of the heart, the aorta then goes down behind it and continues downward, going past the liver and kidneys and then into the legs. (This bottom part is called the **descending aorta**.)

The tip (very bottom) of the heart is called the **apex**. (The word apex is a very common word in science and always refers to the tip of something.) The entire heart is wrapped in a membrane "bag" called the pericardium. ("Peri" means "around" and "cardium" means "heart.") There is fluid inside the pericardium, to allow the heart to move around inside the bag without creating any friction.

When we look at the interior view, we can see these two sides of the heart. The wall between the sides is called the **septum**. The diagram shows the blood from the body coming in through the superior and inferior vena cavas and entering the right atrium. When the right atrium contracts, the blood is squeezed down into the right ventricle. As it flows into the ventricle the tricuspid valve opens. After the contraction is over, the valve goes shut, and this is the "lub" sound in the "lub dub" of the heartbeat. The ventricle then contracts and the blood is forced out into the pulmonary arteries where it then goes to the lungs. The valve at the top of the ventricle is called the **semilunar valve**. ("Semi" means "half, or part" and "lunar" refers to the moon. This valve must have flaps that reminded the discoverer of a half moon or perhaps a crescent moon.) When the semilunar valve snaps shut, this creates the "dub" of the "lub dub" sound.

The left side of the heart receives the oxygenated blood from the lungs through the pulmonary veins. (Again, this is why it is helpful to think of the definition of veins as being vessels that go towards the heart, because here we have veins filled with oxygenated blood which would be colored coded as red instead of blue.) The blood comes into the left atrium and then is squeezed through the bicuspid valve (also known as the mitral valve) into the left ventricle. When the left ventricle contracts, the blood goes up through another semilunar valve and into the aorta. The aorta branches off at various points so that blood is equally distributed to all parts of the body. A complete set of contractions—atria and ventricle—is called the **cardiac cycle**.

The steady rhythm of the heartbeat is controlled by special nerves in the right atrium. (These specialized cells start working when an embryo is only a few weeks old!) Neurons in an area called the **SA** (sinoatrial) **node** initiate an electrical signal that travels through specialized nerve fibers across to the left atrium. Thus, when the SA node gives the signal, both atria contract together. The signal also travels to another little spot of specialized cells called the **AV** (atrioventricular) **node**. When the AV node receives the signal, it relays the signal through nerve fibers that run down through the septum and out to the walls of the ventricles. This signal causes the muscles of the ventricles to contract. The timing of all of this works out just perfectly, and the ventricles contract a split second after the atria, resulting in a steady pumping motion. These nodes form the **intrinsic conduction system**. (Intrinsic just means internal.) The heartbeat is also influenced by the **extrinsic** (external) system, with the nodes being influenced by signals from the brain or by hormones released by various body parts. When you exercise, for example, your heart rate goes up as the brain signals it to speed up. When you sleep, your heart rate slows down, as your brain stem gives it signals to relax. A sudden jolt of adrenaline from the your adrenal glands will cause your heart to start thumping fast and hard. (We'll take another look at these "speed up" and "slow down" control systems when we do an overview of the entire nervous system.)

THE HEART

67

ARTERIES go away from the heart.   VEINS go toward the heart.

EXTERIOR ANATOMY

INTERIOR ANATOMY

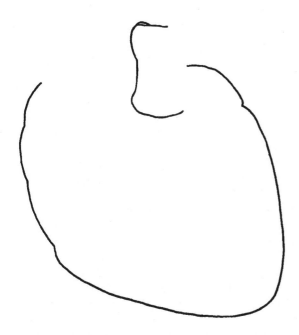

The **pericardium** is a membrane "bag" that goes around (peri) the heart (cardi).

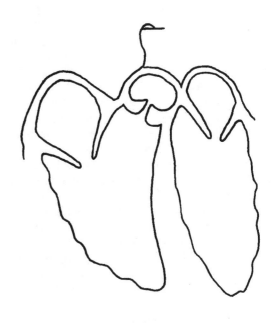

VALVES:
1: Tricuspid
2: Bicuspid (a.k.a. mitral valve)
3: Semilunar valves

"LUB DUB"   (the cardiac cycle)

The familiar "lub dub" sound of a beating heart is made by the valves opening and closing.

The first sound, the "lub," is when the cuspid flaps close.

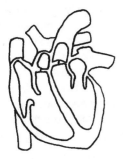

The second sound, the "dub," is when the semilunar flaps close.

Intrinsic Conduction System

(how the heart beats in rhythm)

Extrinsic stimuli (such as adrenaline) can override the nodes' normal rhythm.

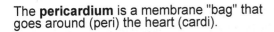
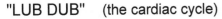

The SA and AV nodes began to function very early in embryonic development. An embryo's heart cells begin beating in rhythm long before the shape of the heart has been completed.

The two phases of rhythm:

1) _____ : _____

2) _____ : _____

68: THE LUNGS

The lungs surround the heart. (The pulmonary arteries and veins of the heart don't have far to travel to get to the lungs!) The heart leans to the left just a bit, so the left lung is slightly smaller than the right lung in order to make space for the heart. The lungs are made of large sections called **lobes**. The right lung has three lobes, the superior, the middle and the inferior, and the left lung has just two lobes, the superior and inferior. Each lobe is made of many smaller lobes called lobules. Each lobule consists of a group of tiny hollow spheres called **alveoli** *(al-VEE-oh-lie or al-vee-OLE-ie)*. (Alveoli is plural, alveolus is singular.)

The tube that leads down into the lungs is called the **trachea** *(TRAY-kee-uh)*. The trachea has rings of stiff cartilage that keep it from flattening. If the trachea was soft and pliable like the esophagus, we would be in constant danger of it collapsing and closing. The cells that line the inside of the trachea are columnar epithelial (lesson 31) and many have cilia on the top. These cilia beat in rhythm to move mucus upward, helping to clean out dust and dirt. The mucus is made by goblet cells. When you cough or clear your throat, the mucus that has been brought up will then be swallowed and go down into the stomach where strong acid will kill most germs.

The trachea branches into two **bronchi** *(bron-kie)*, one for each lung. These two tubes are called the primary bronchi. Each primary bronchus then branches into many smaller bronchi, called the secondary bronchi. The smaller secondary bronchi branch into even smaller **bronchioles** *(BRON-kee-oles)*. The bronchioles get smaller and smaller until they are almost microscopic. At the end of each tiny bronchiole is a lobule made of alveoli. The alveoli are wrapped in microscopic capillaries, and this is where gas exchange happens. The endothelial cells of the capillaries lie right next to the very flat squamous cells of the alveoli. In this way, the blood gets so close to the air inside the alveoli that oxygen and carbon dioxide can diffuse in and out of the blood. This is classic diffusion at work (lesson 6). Gases will diffuse from places where there is more to places where there is less.

The blood in the capillaries has come to the lungs to pick up oxygen, so it has a low level of oxygen. Oxygen from the air will pass right through the alveolar epithelial cells and the endothelial cells, and into the red blood cells. In the red cells, the oxygen attaches to a heme molecule that is being held by a hemoglobin molecule (lesson 39). The oxygen will ride on the heme until it reaches a place where there is less oxygen outside the capillary, and then it will be released to into the tissues that need it.

Carbon dioxide is brought into the lungs as waste that needs to be disposed of. (CO_2 was created in the cells as a by-product of cell processes, and it passed into the blood via diffusion.) CO_2 is carried in the blood in three ways. First, a small amount can simply dissolve into the water of the blood plasma. Second, another small amount can attach to the globin (protein) part of hemoglobin. Third, most of the CO_2 is carried in the plasma not as CO_2 gas but as bicarbonate ions: HCO_3^-. How does this happen? The CO_2 goes into the red blood cells where a special enzyme combines CO_2 and H_2O to make carbonic acid, H_2CO_3. You may remember from previous lessons that when you see "H_2" at the beginning of a molecule, it is likely to be an acid. You might also remember that one of the H's often goes wandering off, leaving its electron behind. The negative sign at the end of the molecule represents this electron. Wandering H's are hydrogen ions, shown as H^+. The hydrogen ions bind to the hemoglobin but the bicarbonate ions (HCO_3^-) diffuse out of the red blood cell and into the plasma. All these negative ions diffusing out could create pH problems for both the red blood cell and the plasma, so an exchange is done where Cl^- ions diffuse into the red cell to replace the bicarbonate ions that are diffusing out. The bicarbonate ions ride along in the blood until they reach the lung, when this process is reversed and the bicarbonates are turned back into carbon dioxide gas that can be exhaled by the lungs.

The alveolus is lined not only with very flat squamous cells, but also cuboidal cells that secrete a special fluid. This fluid consists of water with some very particular proteins and phospholipids. The tails on these phospholipids are both palmitic acid, which we saw in a list in lesson 3. This fatty acid is just the right one for this job. The phospholipid lies on the surface of the water with one tail in the water and one tail sticking up into the air. When you get a whole bunch of these phospholipid head and tails floating on the water, the water molecules are not able to hang together as tightly as they usually are. The result is that gases, like oxygen, can dissolve into the water much more easily. As chemical that reduces surface tension of water is called a surfactant. If the lungs did not produce surfactant along with the secreted water (which must be there to keep the lungs moist), we would not be able to get enough oxygen. Also, the surfactant affects the ability of the alveoli to pop back open after the collapse a bit during exhaling. So there are two reasons we must have this surfactant in that fluid. Developing babies do not begin to produce large quantities of surfactant until the end of the 8th month of pregnancy. This is the most common complication with premature births. The babies must be on breathing equipment that supplies not only extra oxygen but also artificial surfactant.

Special macrophages called **dust cells** (lesson 42) crawl around on the inside surfaces of the alveoli, cleaning up dust, dirt, debris from dead cells, bacteria, fungi and viruses. They do a great job of keeping the lungs clean. Problems arise, however, when they are faced with things they cannot digest, such as asbestos.

Creating a "wall" underneath the lungs and heart is the **diaphragm**. This muscle is shaped like an upside down bowl, and forms a barrier between the thoracic *(thor-ass-ick)* cavity above, and the abdominal cavity below. There are only three places where things connect these two cavities: 1) the inferior vena cava, 2) the descending aorta, and 3) the esophagus. The diaphragm muscles are connected directly to the medulla oblongata in the brain stem (lesson 64). The medulla is able to sense when the CO_2 levels are too high and it sends a signal to the diaphragm (and to the intercostal muscles between the ribs) to contract. When these muscles contract, the result is that the chest area expands. The expansion causes the air pressure inside the lungs to fall, and more air rushes into to fill this area of low pressure. The **serous membrane** "bags" that surround the lungs (**parietal pleura** *(plur-uh)* on the outside and **visceral pleura** on the inside) have a thin layer of liquid between them (serous fluid) that creates suction (like when two plastic cups get stuck together) and this section plays a role in the action of inhalation. The outer (parietal) layer is attached to the wall of the thoracic cavity so when the chest expands with the action of the diaphragm, the inner (visceral) membrane gets pulled along, too, thus opening the lungs.

Although the lungs are the main organs of the respiratory system, the action of the lungs is actually called **ventilation**, not respiration. Respiration is what happens in the cells ("cellular respiration" from lessons 18 and 20 in module 1). When the lungs take in air, this is called **inhalation**. When the lungs expel air, this is called **exhalation**.

The **THORACIC CAVITY** contains the lungs, the heart, the trachea and the esophagus. The **diaphragm** separates the thoracic cavity from the abdominal cavity below.

At the end of each bronchiole is a **lobule** made of microscopic **alveoli**. Each alveolus is covered with a bed of capillaries.

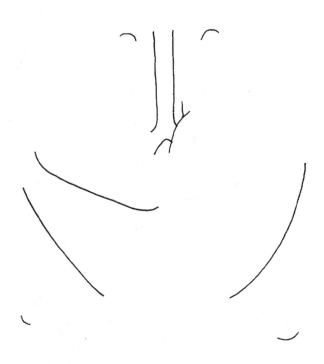

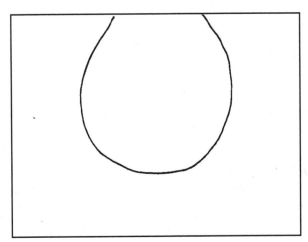

The diaphragm is shaped like an upside down bowl.

Three holes in diaphragm:
1) inferior vena cava, 2) descending aorta, and 3) the esophagus.

Contraction of the diaphragm causes the chest to expand, causing air to rush in.

1) ultra-thin squamous epithelial cells
2) cuboidal epithelial cells that secrete **surfactant**
3) macrophages called "dust cells"
4) layer of water that contains surfactant (phospholipids that lower the surface tension of water so that O_2 can pass through)
5) endothelial cells of the capillaries
6) red blood cells that pick up the O_2

WHAT CAUSES US TO BREATHE:
The diaphragm is "wired" to the medulla oblongata in the brain. The medulla is very good at sensing small changes in the CO_2 level in the blood and sends signals for the diaphragm and the intercostal muscles (between the ribs) to contract when CO_2 gets too high.

HOW O_2 IS TRANSPORTED IN THE BLOOD:

RBCs contain billions of hemoglobin molecules.

HOW CO_2 IS TRANSPORTED IN THE BLOOD:

1) A small amount is carried by the globin part of hemoglobin, or is dissolved directly into the plasma.

2) Most CO_2 is combined with water to form carbonic acid (H_2CO_3), then bicarbonate ions (HCO_3^-) and hydrogen ions (H^+). HCO_3^- diffuses out into the plasma. To keep the pH even, Cl^- ions diffuse in to replace HCO_3^-.

69: THE LIVER and GALL BLADDER

The liver is the largest gland in the body. A gland is defined as something that secretes one or more substances that are useful to the body. As we will see, the liver is the ultimate gland, as it secretes more different types of chemicals than any other gland in the body. The liver of an average-sized adult weighs about 3 pounds (1.5 kg). Human livers are the same color as those chicken livers you see in the grocery store (a reddish brown) and have a rubbery texture.

Both the liver and the stomach lie right below the diaphragm. In the last drawing we saw that the diaphragm has three holes in it: for the inferior vena cava, the descending aorta, and the esophagus. The esophagus leads to the stomach, and we will learn more about it in the next lesson. Both the inferior vena cava and the descending aorta connect to the liver. The aorta brings oxygen-rich blood into the liver. The blood that the vena cava transports into the heart is low in oxygen but it is rich in nutrients, as we will see.

The blood supply to the liver comes from two sources: from the descending aorta, as we just mentioned, and also from veins called the **hepatic portal veins**. (Words that begin with "hep" or "hepa" always have something to do with the liver. "Hepatic portal" might be translated as "doorway in the liver.") The blood in the hepatic portal veins comes from several sources: from the spleen, the pancreas, the stomach, and the intestines. Basically, all of the blood from the abdominal cavity (everything below the diaphragm) comes into the liver through these portal veins. This means that before any nutrients from food are distributed to the body, they must go through the liver first. The liver keeps track of how much sugar and fat are in the blood and will do its best to prevent an overload of either one. For example, if the glucose level in the blood is too high, the liver can start catching and storing the extra glucose molecules, putting them into long strings called **glycogen**. (We also find glucose stored as glycogen in muscle tissue.) About 5 percent of the liver's weight is glycogen.

Between meals, your liver is constantly checking the blood's glucose level and when it begins to fall below about 70 grams per deciliter (a deciliter is one tenth of a liter) the liver will begin to break down some of the glycogen and turn it back into glucose, and release it into the blood. If the liver runs out of stored glycogen and the body still needs more glucose, the liver cells will begin turning amino acids into glucose. Ultimately, all forms of energy (sugar, fats, proteins) must be converted into glucose so that it can be fed into the Krebs cycle and then into the Electron Transport Chain. If the liver runs out of space to store glucose as glycogen, it will then begin turning the glucoses into triglycerides and send them out into the body to be picked up by adipose cells that will store them as fat. The process can go the other way, too, as the liver can turn protein and fat into glucose. This is how carnivores who eat only protein and fat survive. Their livers convert the protein and fat into glucose. This process is called **gluconeogenesis**. ("Gluco" = sugar, "neo" = new, "genesis" = to make) In the process of turning aminos into glucose, the nitrogen part of the amino acid must be discarded, as there isn't any nitrogen in either sugars or fats. The nitrogen from the amino acid is discarded as a molecule of NH_3, ammonia, a substance that is very poisonous to the body. But the liver has a way to deal with ammonia—it quickly turns it into a molecule of urea, which still contains that nitrogen but is much less poisonous. The urea can safely float around in your blood until it reaches the kidneys where it will be filtered out and then sent out of the body with the urine.

The liver can also turn sugars and fats into amino acids. There are still those 6 "essential" amino acids that your liver can't make and that you must get from your diet, but the rest can be manufactured by liver cells. If you eat too much of one kind of amino and not enough of another, the liver has enzymes that know how to transform one amino into another. Pretty amazing!

Another major function of the liver is to produce **bile**. Bile is a liquid made of water, bile salts (also called bile acids), cholesterol and phospholipids. Bile acts very much like the dish soap you use to cut the grease on your dirty dishes. Bile "emulsifies" fats, meaning it breaks them up into very small pieces. Each tiny bit of fat is surrounded by phospholipids that have their water loving heads facing outwards so that the fat molecules can travel safely in the blood. (The job of the salts/acids is to act as a surfactant.)

The gall bladder is the storage area for bile. When fatty foods reach the **duodenum** (the first part of the small intestine, pronounced either *du-ODD-den-um* or *du-oh-DEE-num*), chemical signals are sent to the gall bladder telling it to release some bile. (In people who have had their gall bladder removed, the bile just flows directly from the liver into the duodenum.)

Other functions of the liver include making and transporting cholesterol using "shuttles" called LDL and HDL. The LDL's are the delivery system taking cholesterol out to the cells. The HDL's are involved in collecting leftovers and returning them to the liver for recycling. (Doctors tells us we want more HDL's than LDL's.) The liver also stores iron (Fe) and fat-soluble vitamins: A, D, E K, plus B12.

A microscopic view of the liver shows that it is made of individual units called **lobules**. Each lobule is approximately in the shape of a hexagon. At the center is a **central vein** that will eventually lead into the hepatic vein that goes up into the inferior vena cava. At each vertex of the hexagon you find a **triad** of three vessels: 1) a hepatic artery, 2) a portal vein, and 3) a **bile canaliculi** (tiny canal). The hepatic arteries contain blood that came in from the descending aorta so they are rich in oxygen. The portal veins contain blood that came in from the veins in the stomach and intestines so they are rich in nutrients. Both the hepatic arteries and the portal veins lead into the central vein. (As with all other blood vessels in the body, they are made of endothelial cells.) The fluid in the bile canaliculi comes from all the surrounding cells, and goes out through larger and large bile vessels until it reaches the **bile duct**. The large "common" bile duct connects to both the gall bladder and the duodenum.

The blood from these tiny portal veins and hepatic arteries joins together in a space called the **sinusoid**. On the walls of this sinusoid space, macrophages called **Kupffer cells** crawl around looking for bacteria to destroy or red blood cells that need to be recycled. It is in the Kupffer cells that hemoglobin is broken down into heme and globin.

The large cells (that are not endothelial cells) are called **hepatocytes**. The hepatocytes have three notable features: 1) a large amount of smooth ER, 2) microvilli (that look like fringe), and 3) occasionally a double nucleus. When cells have more DNA than normal, this is called **polyploidy**. Why many liver cells are polyploid is not yet fully understood. Notice the bile canaliculi between the cells. The hepatocytes do numerous jobs including breaking down toxins such as alcohol, making and monitoring glucose, fats and amino acids (as we've already mentioned), making clotting factors for the clotting cascade (from lesson 38), making blood proteins such as albumen, and making all the immune system proteins that begin with "C" (C-reactive protein and C1 through C9, from lesson 48).

The space between the hepatocytes and the endothelial cells is called the **space of Disse**. Plasma from the blood leaks out into this space, interacts with the hepatocytes along their microvilli, then drains into the lymph system. Stellate cells live in this space; they store vitamins and make the collagen network that holds all the lobule cells in place.

The liver is the largest gland in the body. It weighs about 3 lbs (1.5 kg). It is the ultimate "mulit-tasker" and by some counts does as many as 500 jobs! The gall bladder is simply a storage bag for one of the products that the liver makes.

MICROSCOPIC VIEWS:

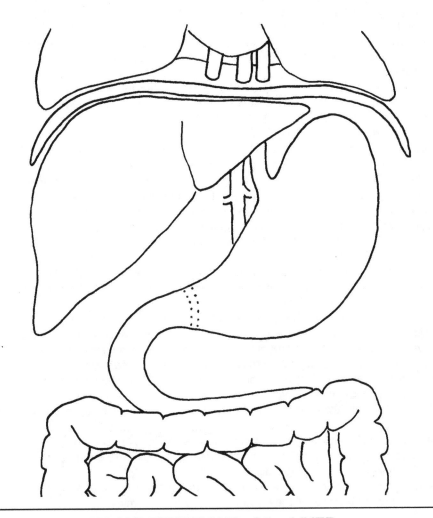

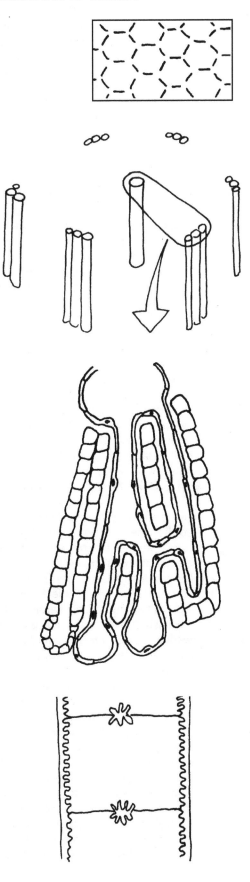

MAJOR FUNCTIONS of the LIVER:

A: _____ : _____
_____

B: _____ : _____
_____

C: _____ : _____
_____

C: _____ : _____

C: _____ : _____
_____

D: _____ : _____

E: _____ : _____

F: _____ : _____

G: _____ : _____

70: STOMACH and duodenum

We've already seen how the esophagus pokes down through the diaphragm. In this drawing we'll add a nerve that travels down with it. The **vagus nerve** is the main nerve of the digestive system. The vagus has two branches, left and right, that come down from the brain stem. They go through the diaphragm alongside the esophagus and then they branch off to connect to most of the organs in the abdominal cavity. There are both afferent and efferent nerves in the vagus. Remember, the afferent are sensory neurons and the efferent are motor neurons. The motor neurons control the muscular motions of the digestive organs, most notably **peristalsis**. In peristalsis, the circular and longitudinal muscles coordinate their movement in such a way that the **bolus** (food glob in esophagus) or **chyme** (mush in intestines) is pushed along through the tubes.

Right where the esophagus meets the stomach, there is a ring of muscle called the **esophageal sphincter.** A sphincter (*sfink-ter*) muscle is a special type of muscle that works backwards from regular muscles. In most muscles, the "normal" state of the muscle is relaxed and uncontracted and the action of the muscle is to contract. In a sphincter muscle, the "normal" state is contracted and to use the muscle it must be relaxed. Sphincters come in very handy in places in the body that require keeping something in. If we did not have a sphincter at the very end of our digestive system, for example, our feces (poop) would just plop right out all the time. The sphincter holds everything in until we are able to visit a bathroom. The stomach has two sphincter muscles, one at the top and one at the bottom. The esophageal sphincter prevents the highly acidic stomach fluid from splashing up into the esophagus. When stomach acid gets up into the esophagus we call this **acid reflux**, or "heartburn." The sphincter at the bottom of the stomach keeps food inside until it is adequately churned into a soft mush that can safely go through all the bends in the intestines. Strong muscles in the **pyloric** *(pie-LOR-ick)* region of the stomach push chyme through the **pyloric sphincter** and into the **duodenum**. The duodenum is shaped like the letter C, and connects the stomach to the small intestine. The bile duct is attached here as we saw in the last lesson. The pancreas also has a duct that empties into the duodenum, often connecting to the bile duct.

Food is pushed down through the esophagus by muscle movement called peristalsis. This motion is done by involuntary muscles controlled by the automatic mechanisms in the brain stem. Even if you stand on your head while eating, the food would still go down into the stomach, against gravity, because of the action of the muscles of the esophagus.

The stomach, as you probably know, is a very acidic place. Cells in the lining of the stomach produce hydrochloric acid, HCl. The pH level (on the scale of 1 to 14) is about 2. The acid is what activates a protein digesting enzyme called **pepsin**, also made by the cells in the lining of the stomach. The cells actually make an inactive form of the enzyme called **pepsinogen**, and the acid turns it into pepsin. Remember how fibrinogen was turned into fibrin? A molecule that will turn into an active enzyme is called a **zymogen** *(ZIE-mo-jen)*. ("Zym" means "enzyme," and "gen" means "to make.") The body does this to prevent the enzyme from acting too soon or in the wrong place. You don't want pepsin digesting your own cells before it gets to your food, and you don't want fibrinogen turning into fibrin while it is floating around in your blood. Pepsin's job is to break down protein chains into amino acids that can be used by your cells' ribosomes to make their own protein structures.

If we peeled back the outer tissue layer of the stomach we would see three layers of muscles: **oblique** (meaning diagonal), **circular** and **longitudinal**. The combined action of these directional layers produces a churning, mixing motion. Only the stomach has the outer oblique layer. The rest of the digestive system has only circular and longitudinal.

If we look at the inside of the stomach we see deep folds and wrinkles. These folds allow the stomach to stretch and expand as it fills with food. As the stomach stretches, signals are sent to the brain to tell the muscles layers to start churning.

If we take a really close-up look at the stomach lining (under a microscope) we see that the surface is made of tiny finger-like things called **villi**. These are not microvilli, just villi. The cells that form the top of the villus are called **pit cells**, because the spaces between the villi are called **gastric pits**. The pit cells are basically goblet cells that make large amounts of mucus all the time. They coat the surface of the stomach lining with mucus to protect it from all the hydrochloric acid. The bottom half of these spaces between the villi are called gastric glands and they are made of several types of cells. **Parietal cells** are at the top of this gland area, dividing it from the mucus-producing top area. (Parietal means "wall.") Parietal cells make the hydrochloric acid. **Chief cells** make the pepsinogen zymogen that will be activated into pepsin. Chief cells also make a chemical called IF (intrinsic factor) that binds to vitamin B12 to protect it from being destroyed by the HCl. **G cells** make a hormone called gastrin that stimulates the parietal cells. There are also a few **enteroendocrine** *(en-TARE-oh-EN-do-krin)* cells that secrete a number of different hormones that control various aspects of digestion. All these hormones allow for your body to maximize digestive processes and also to adjust to unusual situations such as fasting, over-eating, and illness.

The villi and all their various cell types make what we call the **mucosa layer**. Under the mucosa layer we have the **submucosa**. The submucosa is basically loose connective tissue, with a network of collagen, blood and lymph vessels, and immune cells. We have not drawn the lymph vessels or the immune cells in this drawing, but we will definitely see the lymph vessels in our drawing about the intestines because they perform an important job in digestion. Under the mucosa is the **muscularis** layer, with its bands of smooth muscles going in various directions. Of course, these are not voluntary muscles like we find in our arms and legs. These smooth, involuntary muscles operate all the time with no thought required.

You will notice that there are capillaries running up into the villi. In the stomach, these capillaries can absorb a few things, such as alcohol, but they are not able to absorb most nutrients. Absorption primarily takes place in the small intestines. We will also see in the lesson on the intestines that the villi have microvilli, greatly increasing their surface area.

The digestive organs, including the stomach, are surrounded by an outer layer called the **serosa**. (We saw a serosa layer round the lungs, also.) The serosa is a layer of epithelial cells with a bit of connective tissue underneath. These epithelial cells secrete **serous fluid** which helps to lubricate the outsides of all the organs so that friction is not created when we move and bend our torso.

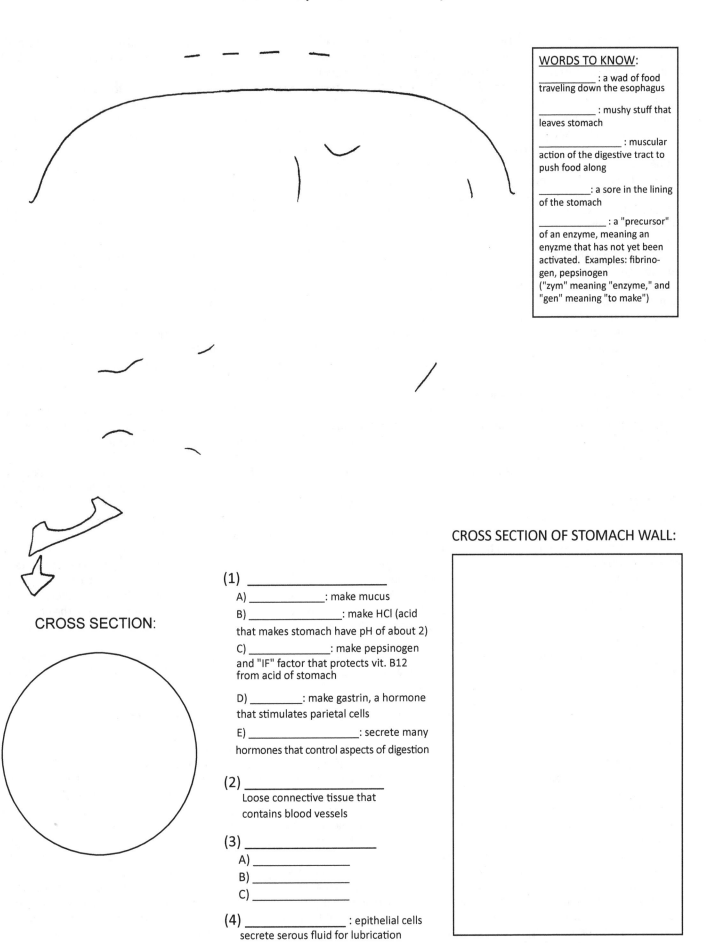

CROSS SECTION OF STOMACH WALL:

CROSS SECTION:

(1) _____

A) _____: make mucus

B) _____: make HCl (acid that makes stomach have pH of about 2)

C) _____: make pepsinogen and "IF" factor that protects vit. B12 from acid of stomach

D) _____: make gastrin, a hormone that stimulates parietal cells

E) _____: secrete many hormones that control aspects of digestion

(2) _____

Loose connective tissue that contains blood vessels

(3) _____

A) _____

B) _____

C) _____

(4) _____ : epithelial cells secrete serous fluid for lubrication

71: PANCREAS (and digestive enzymes)

The pancreas is classified as a "mixed gland." This means that it functions as both an exocrine gland and an endocrine gland. An **exocrine** gland secretes its products to the outside ("exo" means "out") through some kind of duct or physical opening in the gland. An **endocrine** gland secretes its products directly into the blood via capillaries that come into the gland.

The exocrine part of the pancreas is made of groups of cells called **acini**. These cells cluster together to make sort of "mini-glands" that have ducts in the center. These cells make vesicles filled with zymogens that will go through the duct system and be dumped into the duodenum. The zymogens will be turned into their active forms (enzymes) as they go through the small intestine. Some people have a pancreatic duct that looks like a Y, with each branch of the Y making an opening into the duodenum. Other people have a duct that does not branch and has just one opening into the duodenum. This is because of the way that the pancreas forms during the embryonic stage. It starts out as two separate buds, each with their own duct. At a later stage, these two buds merge together to make one whole pancreas. Sometimes the ducts merge and sometimes they don't.

The most well-known of these enzymes are **pepsin, trypsin** and **chymotrypsin**, which all are proteinases that break down proteins. These exocrine cells also make **lipase**, an enzyme that breaks down lipids (fats), and **amylase**, an enzyme that breaks starches down into **maltose**, a disaccharide made of two glucoses stuck together. The cells of the intestines will make maltase, the enzyme that breaks maltose into its two glucose units. Amylase is also produced by salivary glands so that starches begin to be broken down while still in the mouth. Plants also make amylase. For example, when a banana ripens, starch molecules are being broken down into glucose. You can taste this difference if you compare a green banana to a ripe one.

That cover proteins, fats and carbohydrates (starches). What's left? Some food also contains cells that have DNA and RNA in them (plant cells, for example). The pancreas makes nucleases to deal with these. A **nuclease** is an enzyme that works on nucleic acids and takes apart the "rungs" of the DNA and RNA "ladders."

These cells of the acini make another very important product that is not an enzyme. They make **sodium bicarbonate**, $NaHCO_3$, which is the chemical we use in our kitchens as baking powder. Sodium bicarbonate is alkaline (basic), the opposite of an acid, so it can neutralize stomach acid. The stomach has that special mucus coating to protect it from its own hydrochloric acid, but duodenum does not have that protection. The acidity of the chyme must be neutralized as it comes into the duodenum. Fortunately, the pancreatic duct opening is right there, squirting in sodium bicarbonate.

The endocrine function of the pancreas is found in little "islands" or "islets" in the middle of the "sea" of acini. The first person to discover them was Paul Langerhans, so they were named **islets of Langerhans**. These little islets are made of about half a dozen types of cells but there are only three that are well-known. The **alpha (α)** cells make **glucagon**, a hormone that tells the liver and muscles to release some of that glycogen they have been storing. (Remember, glycogen is made of strings of glucose molecules.) Glycogen is released into the blood when the level of glucose gets too low (like several hours after a meal). The **beta (β)** cells make **insulin**. Insulin is the hormone that is released when the blood sugar level is too high; it signals the liver and muscles to start pulling glucose out of the blood and start storing it as glycogen. **Delta (δ)** cells make **somatostatin**, a hormone that is made by other places in the body, also, including in the brain (by the hypothalamus, to suppress the pituitary functions). Somatostatin produced by the pancreas affects the digestive organs by telling them to stop making their digestive hormones (insulin, glucagon, gastrin, secretin).

The digestive system has quite a few hormones that send signals to the various organs telling them when to increase their activity. When chyme and stomach acid enter the duodenum, cells lining the duodenum sense this and they begin secreting substances such as **secretin** and **CCK** (never mind what the letter stand for-- it's a very long word!). These hormones go into the bloodstream and eventually (in only a few minutes) end up in the gall bladder and pancreas where they cause these organs to increase production. The stomach actually sends signals to itself, as odd as this might seam. Cells in the stomach produce **gastrin** which goes into the blood and then ends up back in the stomach. The cells in the gastric glands have receptors that sense the presence of gastrin and respond by increasing their production of acid and enzymes.

Processes in the body are never simple! We take eating and digesting for granted and are unaware of the massively complicated chemistry that is occurring inside us. Digestion is actually a lot more complicated that what we have described here. Researchers are still discovering new things about these processes.

****

You have probably heard about "type 1" and "type 2" diabetes. What's the difference?

Type 1 diabetes is an autoimmune disease caused by a malfunction of the immune system. Those T cells that are supposed to tag foreign invaders begin tagging body cells by mistake. Macrophages automatically eat anything that is tagged, so if body cells are tagged they get destroyed. In the case of type 1 diabetes, the T cells begin tagging the beta cells inside the islets of Langerhans. (Why? We don't know for sure, but we are suspicious that viruses could be the initiators of many autoimmune conditions.) Once the beta cells are destroyed, the body has no way to make insulin. The level of glucose in the blood will be very high, which will cause all sorts of other problems. The treatment for type 1 diabetes is to take artificial insulin. It is tricky to monitor the level of glucose in the blood and get just the right amount of insulin injected. If you get too much insulin that is also a dangerous situation. People often come down with type 1 diabetes quite suddenly, almost overnight in some cases, and very often the disease begins in childhood.

Type 2 diabetes is something that happens more slowly, often as we get older and possibly as a result of eating too much sugar for many years. It often begins as "insulin resistance," a condition where your cells fail to respond to insulin and don't get all the extra glucose out of your blood. Sometimes type 2 can be managed without taking insulin shots, simply by controlling the diet and getting more exercise. Other times, type 2 may get severe enough that you need to take insulin in order to normalize blood sugar levels.

The pancreas is a "mixed gland" meaning that it performs both **endocrine** and **exocrine** functions. The exocrine products go into the duodenum. The endocrine products go into the blood.

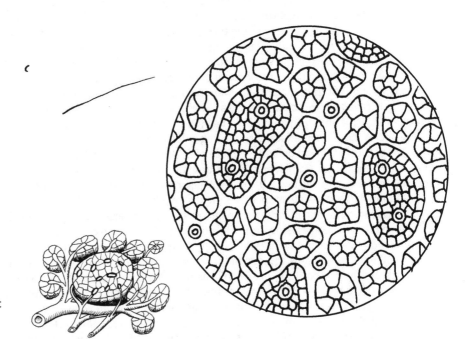

Organs communicate using hormones. When the duodenum detects protein, fats and stomach acid coming into it, its cells start making **secretin** and **CCK**. These hormones go into the blood and eventually reach the pancreas and gall bladder, causing them to increase their output.

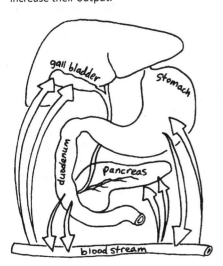

EXOCRINE: secreted to OUTSIDE

The exocrine cells are called: _____

Exocrine products of pancreas:

1) _____
(NaHCO$_3$ that neutralizes stomach acid
(In kitchen we call it: _____)

2) _____ break proteins

ex: _____

ex: _____

ex: _____
These are made as zymogens which are activated by an enzyme in the intestines.

3) _____ breaks apart fats

4) _____ breaks starch into units of maltose (disaccharide)

5) _____ break RNA and DNA

ENDOCRINE: absorbed (by the blood) while INSIDE the gland

The endocrine cells are organized into:
_____ of _____

Endocrine products of pancreas:

1) ___ cells make _____
that_____

_____

2) ___ cells make _____
that_____

_____

3) ___ cells make _____
that_____

_____

72: INTESTINES

The intestines are divided into two very distinct regions: the small intestine and the large intestine. The small intestine is subdivided into three regions: the **duodenum** (du-ODD-den-um, or du-oh-DEE-hum) the **jejunum** (jeh-JU-num) and **ileum** (ILL-ee-um).

We've already seen that the duodenum is the part that connects the stomach to the intestines. The pancreas and bile duct are attached here, and the duodenum secretes hormones that help to control the rate of digestion. One of the main functions of the duodenum is to neutralize the acidic chyme coming from the stomach. The pancreatic juice helps with this, but the duodenum also has Brunner's glands that make an alkaline (basic) secretion.

The jejunum gets its name from the Greek word for "fasting." It seems that whenever the ancient Greeks would open a dead body to dissect it, they would always find this part empty, as if the person had not eaten. This mystified them. We now know that the jejunum has the most active peristalsis, and food is moved along through it very quickly. Perhaps in the minutes during and after death, peristalsis pushes the chyme along into the ileum. The jejunum is about 2 meters (6 feet) long, but this is approximate because there is no firm dividing line between the jejunum and the ileum. The pH is somewhere between 7 to 9, so it is slightly alkaline. The diameter of the jejunum is slightly greater than the ileum. The inside has many circular folds called **plicae** (plee-kay) **circulares**. The plicae are covered with tiny "fingers" called villi. The villi give the interior a soft, kind of velvety look.

The ileum is about 6-7 meters (18-20 feet) long. The diameter is slightly smaller than the jejunum, the walls are thinner, and peristalsis not quite so fast (possibly because the thinner walls mean less muscle). The pH is still slightly alkaline, between 7 and 8. The ileum is where bile salts are reabsorbed and sent to the liver for recycling. The ileum is also the site for reabsorbing vitamin B12. A notable feature of the ileum, and one of the things that sets it apart from the jejunum is the presence of **Peyer's** (pie-ers) **patches**, which are little lumps of lymph tissue. The immune cells in these patches help to fight harmful bacteria.

Both the jejunum and ileum have the four basic layers of tissue: **mucosa, submucosa, muscularis and serosa**. Here around the intestine, the serosa is also called the **visceral peritoneum**. (Visceral means organs.) The visceral peritoneum is called the mesentery when it extends out from the organs. More on that in the next lesson. Note that the mucosa layer has its own very thin layer of smooth muscle called the **muscularis mucosa**. These tiny muscles are moving around all the time, making the villi move and come into contact with the maximum amount of chyme. (Can you imagine the massive tiny network of neurons required to control them?)

The end of the ileum is attached to the beginning of the large intestine (also called the **colon**) at a place called the **cecum** (see-kum). The cecum is the part behind the place of attachment, kind of like a dead end. Attached to the cecum is the **appendix**. (The appendix is basically a piece of lymph tissue, and seems to play a role in recovery from intestinal infections.) Right where the ileum attaches to the cecum you find a band of sphincter muscles that control the flow of digested food into the large intestine.

The colon (large intestine) is about 1.5 meters long (5 feet) and is divided into areas using words that describe their location and position: **ascending** (going up), **transverse** (across the top), **descending** (going down), and **sigmoid** (from the Greek letter S: sigma). The thing that looks like a stripe running along the length of the colon is a thin band of muscle. The colon is where water is reabsorbed. The digested food is sloppy when it enters the colon, but when it leaves it should be fairly hard. (But not too hard! Dry feces are difficult to push out, causing a condition called constipation.) When the chyme enters the colon we then refer to it as **stool**.

No actual digestion occurs in the colon. The pH of the colon ranges from 5.5 to 7, so it is slightly acidic. This is a good environment for the helpful bacteria that live there. These bacteria are able to digest some of the fiber (usually plant material) we eat but can't break down. Our bodies don't make the enzymes necessary for breaking down the cell walls around plant cells. The bacteria eat the fiber and then release chemicals we can use, such as vitamin K, some B vitamins, and small fatty acids (which happen to be anti-inflammatory). Eating plenty of fiber ensures that these good bacteria will be healthy. Having lots of "good" bacteria also helps to prevent "bad" bacteria from multiplying.

The very last part of the large intestine serves as a holding area for feces and is called the **rectum**. The ring of muscles at the end of the rectum is called the **anus** (which means "little ring"). When the rectum gets full and begins to stretch, signals are sent to the brain to tell you that you need to use the bathroom. The correct name for solid waste is **feces** (fee-sees), and the act of releasing the feces is called defecation.

If we take a microscopic look at the villi, we see five main types of cells.

1) **Enterocytes** take nutrients out of the chyme and put them into the blood or the lymph system. (Words that start with "entero" have something to do with the digestive system.) They produce digestive enzymes such as proteases, lipases, amylase and nucleases. They do the final, final work of digestion, such as chopping those maltose molecules in half, making two glucoses. They also perform a very important function for lipids. They take the tiny bits of fat that the bile emulsified and package them chylomicrons which are then absorbed by tiny lymph vessels called **lacteals**. The chylomicrons will ride in the lymph fluid and eventually get dumped into the blood, along with all the lymph fluid, into a vein right under the collar bone. The enterocytes are the masters of endocytosis and diffusion, constantly taking in nutrients from the lumen, then pushing them out the other side into the space that contains the blood and lymph vessels. (This space is called the lamina propria and is made of loose connective tissue. Many macrophages and lymphocytes (T and B cells) also live here. The edges of the enterocytes have **microvilli**, making a "brush border" that looks fuzzy. The function of both the villi and microvilli is to increase surface area.

2) **Goblet cells** that make mucus.

3) **Enteroendocrine cells** that make hormones used for communication with other organs.

4) **Paneth cells** that make antibiotic chemicals which are toxic to microbes that might make you sick.

5) **Stem cells** that divide constantly, providing a fresh supply of all these other cells. These other cells only live for about 5 days, so these stem cells are very active. If the stem cells get too active, though, they could form a tumor.

The intestines consist of two distinct regions: the small intestine (divided into **duodenum**, **jejunum**, and **ileum**), and the **large intestine** (also called the **colon**).

What do the jejunum and ileum have in common?

1) _____

_____

2) _____

Differences between jejunum and ileum:

<u>Jejunum</u>   <u>Ileum</u>

1) _____   1) _____

2) _____   2) _____

3) _____   3) _____

4) _____   4) _____

The colon's function is to reabsorb:

_____

MICROVILLI

"brush border"

The enterocytes are the masters of endocytosis and diffusion! Nutrients pass through these cells and out the other side. The cells package triglycerides into chylomicrons that go into the lacteals (lymph vessels). Glucose and amino acids go into the blood capillaries.

VILLI

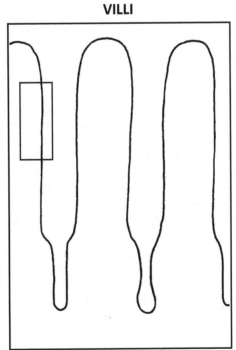

crypts/glands

= enterocytes   = goblet cells

= enteroendocrine cells   = stem cells

= Paneth cells (antibiotic chemicals)

PLICAE CIRCULARES

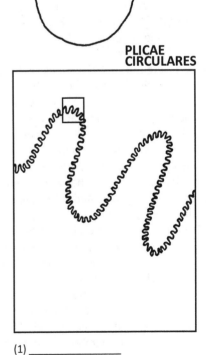

(1) _____

a) Epithelium
b) Lamina Propria (loose connective)
c) Muscularis Mucosa (very tiny muscles)

(2) _____

(3) _____ (circular, long.)

(4) _____

73: BODY CAVITIES and MESENTERY

We've already mentioned that all the organs in the body are wrapped in "bags" made of thin layers of connective tissue. These bags are often inside of other bags, or even inside of several bags. The largest bags define areas that we call body cavities. There are three main cavities: the thoracic cavity, the abdominal cavity and the pelvic cavity.

The **thoracic cavity** is subdivided into three smaller cavities: the **pleural cavity** (containing the lungs), the **mediastinum** (in the middle), and the **pericardial cavity** (containing the heart). The pleural cavity is easy to understand because it simply contains the lungs. The pericardial cavity contains not only the heart itself, but the large blood vessels that come out of it, such as the aorta. The pericardial cavity is inside of the third cavity, the mediastinum. ("Media" means "middle" and "stinum" can refer to the "sternum," which is the bone that runs down the center of the ribcage.) The mediastinum contains not only the pericardial cavity but also part of the esophagus and trachea, the thymus (remember this organ from module 2 on the immune system?), and quite a few nerves and lymph nodes. The thin layers of connective tissues that surround these cavities are made primarily of collagen and elastin (produced by fibroblast cells) but they also contain some very tiny nerves and blood vessels, as well as cells that make serous fluid. As we've seen in the past few lessons, the diaphragm separates the thoracic cavity from the abdominal cavity.

The **abdominal cavity** contains the stomach, the spleen, the tail of the pancreas, the last half of the duodenum, the small intestines, most of the large intestines, and the mesentery (thin layers of connective tissue that anchor the intestines to the back wall of the abdominal cavity). There would also be many blood vessels and nerves all through the abdominal cavity.

The **pelvic cavity** contains the urinary bladder, the sigmoid colon, the rectum, and the reproductive organs. As with the other cavities there would also be many blood vessels and nerves in this cavity. Oddly enough, the kidneys and most of the pancreas are not in any of these body cavities. They sit behind the abdominal cavity and are called **retroperitoneal** organs. ("Retro" means "behind.")

All three of these cavities, the thoracic, the abdominal and the pelvic can be categorized into one very large area called the **ventral cavity**. ("Ventral" means on the stomach side.) There is also a dorsal cavity, which consists of the cranial cavity, containing the brain, and the spinal cavity, containing the spinal cord. ("Dorsal" means on the back side.)

The mesentery wasn't considered to be an organ until 2012 when anatomists at the University of Limerick, in Ireland, began studying electron microscope images of it. These images allowed them to examine the tissues in great detail. They realized that all the seemingly unconnected sheets of tissue were actually all connected. Since they were all connected and they were all doing basically the same job, this allowed the mesentery tissues to be classified as an organ. The definition of an organ is "a group of tissues that work together to perform a certain function." They felt that the mesentery satisfied this definition and should be added to the list of organs of the body. They presented their research to other scientists and finally everyone agreed to officially declare the mesentery to be an organ. In the past, young students were never taught anything about the mesentery and were thus unaware of its existence. This will change. Now, even elementary students will have to learn the word "mesentery" along with the words "stomach" and "intestines."

What does the word "mesentery" mean? "Meso" means "middle," and "entery" is a variation of "entero" which refers to the digestive system, particularly the intestines. The mesentery isn't all in one place like the stomach or the liver or the pancreas. It is spread throughout the abdominal cavity.

The mesentery looks like very thin sheets of strong connective tissue. They are translucent, meaning clear enough to let light through but not perfectly clear. They have two main jobs. The first is to anchor the intestines to the back wall of the abdominal cavity. Just imagine what would happen if nothing was keeping the intestines in place. All those squiggles and loops of intestine would slide around and get tangled. Gravity would also tend to pull them down and eventually all your guts would sag down to the bottom. Yuck. Good thing the mesentery keeps them in place! The second thing the mesentery does is to provide a surface through which nerves, blood vessels and lymph vessels can run. We'd have the same tangling and sagging problem with all these vessels if they were not anchored in place.

The mesentery is made of two thin layers of serous membrane stuck together. We've already met serous membrane in several places. We saw it around the around the heart, forming the pericardium, and we also saw it in the lungs, forming the parietal and visceral pleura. Serous membrane is made of a layer of **mesothelium** (a single layer of simple squamous epithelial cells that secrete serous fluid) stuck to a layer of connective tissue (collagen and elastin) made by fibroblast cells. It's the connective tissue layer that contains the blood vessels and nerves.

Serous membranes are called by different names, depending on where they are located. When the serous membranes line the walls of the body cavities they are called **parietal peritoneum**. (Remember, "parietal" means "wall.") When the membranes wrap around organs they are called **visceral peritoneum**. ("Visceral" means "organs.") When two membranes are stuck together to form a sheet, they are called **mesentery**. Serous membrane around the lungs is called **pleura**, and serous membrane around the heart is called **pericardium**.

There are names for various parts of the mesentery. The part that holds the colon in place is called the **mesocolon**. There is a sheet of tissue between the liver and stomach that is called the **lesser omentum**. There is a large sheet of tissue called the **greater omentum** that acts like an apron, covering the front of the small intestines.

BODY CAVITIES are large sections of the body that are enclosed by membranes.

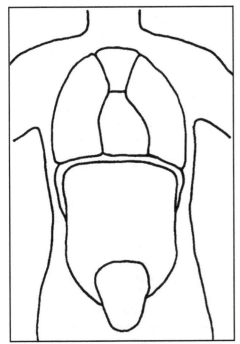

There are 3 main body cavities:

1) _____

   a) _____ (_____)

   b) _____ (_____)
     (is inside mediastinum)

   c) _____
     (superior part shown)
     (includes superior vena cava, aortic
     arch, thymus, part of esophagus,
     part of trachea

2) _____
   (contains stomach, spleen, liver,
   gall bladder, part of pancreas, all of
   small intestines, transverse colon)

3) _____
   (contains urinary bladder,
   reproductive organs such as uterus,
   and sigmoid colon)

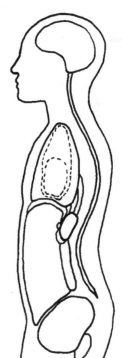

Just FYI, there are 2 other "official" body cavities:
-- CRANIAL
-- SPINAL

A few organs lie outside of the cavities.
They are called **retroperitoneal** ("retro" means "behind").

-Aorta (descending)
-Vena cava (inferior)
-Kidneys
-Part of duodenum
-Part of pancreas
-Ascending colon
-Descending colon
-Rectum

THE MESENTERY is a very thin membrane that holds all the organs in place.   The mesentery also provides a surface for nerves, blood vessels, and lymph vessels. Mesentery is made of serous membrane (which is made of a layer of simple squamous epithelial cells stuck to a layer of connective tissue).

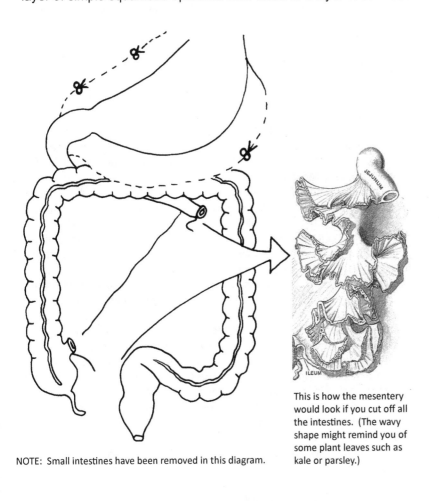

NOTE: Small intestines have been removed in this diagram.

This is how the mesentery would look if you cut off all the intestines. (The wavy shape might remind you of some plant leaves such as kale or parsley.)

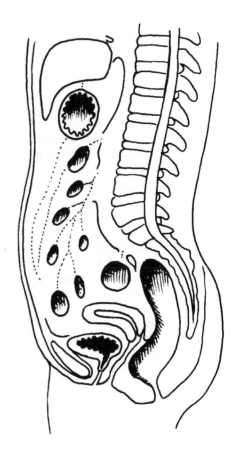

74: KIDNEYS and BLADDER

The kidneys and bladder (and their various tubing) make the urinary system. The goal of this system is to gather what is called "metabolic wastes" and excrete them in a liquid we call urine. Solid wastes go out in the feces. The type of wastes the kidney deals with are tiny molecules made of less than 20 atoms. The waste molecule called **urea** is what gives urine its name. Urea is made from **ammonia**, NH_3, a waste that is formed when amino acids are torn apart. Carbohydrates and fats don't have any nitrogen in them, so they can be broken apart and burned in the Krebs cycle. When amino acids are broken apart, one of the broken pieces is a molecule of ammonia, NH_3. Ammonia is toxic to the body, so the body has to have an immediate way to deal with it. The liver takes two NH_3's and one CO_2 and combines them to make $CO(NH_2)_2$, or urea. Urea is still a waste but it is not that toxic and can safely wait a while to be expelled. Another metabolic waste that helps to give urine its name is **uric acid**. Uric acid is formed as a by-product of the breakdown of nucleic acids (rungs of DNA and RNA). A third type of nitrogenous waste is **creatinine**, which is a by-product of the breakdown of creatine, a molecule in the muscles that acts a bit like ATP, with phosphates that supply energy. (We'll learn more about that in the muscle drawing.) Urine might also contain other small waste molecules containing phosphates or sulfates, as well as various small molecules from food or medicines.

The kidneys are located behind the peritoneal membrane, as we saw in the last drawing. Blood is supplied to the kidneys through the the **renal artery** and the **renal vein**. (Anything starting with "renal" will be related to the kidneys.) There are little organs sitting on top of the kidneys called the **adrenal glands**. We will learn more about them in the lesson on the endocrine system. The tubes that come out of the kidneys and go down into the bladder are called the **ureters** *(YOUR-eh-ters)*. These tubes are lined with small muscles that use peristalsis to push the urine down into the bladder. If gravity was the only force that pulled urine down the tubes, you'd have problems at night when you are lying flat. Urine would collect inside the kidneys and cause major problems. Speaking of night and sleeping, you also get help from your pituitary gland, which secretes a substance called ADH (anti-diurectic hormone) while you are asleep. ADH tells the kidneys to reabsorb as much water as they can, making less urine. Sometimes you still have to get up in the middle of the night to go, but you go less than you would during the day.

The bladder is a waterproof, stretchy bag that collects the urine and holds it there until you get a chance to get rid of it. We saw bladder epithelial tissue in lesson 31. The cells that line the bladder are of a special type that isn't found anywhere else. As the bladder fills and these cells stretch, they send signals along nerves to the brain. The result is that feeling that your bladder is full and you have to find a bathroom. There are sphincter cells at the top and bottom of the **urethra** *(you-REETH-ra)*, the tube that leads to the outside. The top sphincter is made of smooth muscle and is involuntary. The bottom sphincter is made of skeletal muscle and is under voluntary control; it is this bottom sphincter muscle that you relax when you urinate. In males, this lower sphincter is not right at the opening, as it is in females. In males the sphincter is closer to the bladder, just under the prostate gland that sits below the bladder. In females, the urethra is only about 4 cm long, whereas in males it is bout 20 cm long. In females, the urethra is completely separate from reproductive parts, but in males the urethra is part of both the urinary and reproductive systems. More on this in a later lesson.

There are also tiny sphincters where the ureter tubes attach to the top of the bladder. These muscles open anywhere from 1 to 5 times per minutes, allowing a little spurt of urine to enter the bladder each time. When our systems are healthy and working as they should, we are completely unaware of these little spurts going on all the time.

If we look at a cross section of a kidney, we see it divided into a number of distinct areas called **pyramids**. The top portion of these pyramids belongs to the outer layer, called the **cortex**. The bottom of the pyramids belongs to the central core called the **medulla**. There are tubes coming out of the pyramids, feeding into a large tube which will become the ureter.

If we take a close up look at a pyramid, we see that it is made of thousands of individual units called **nephrons**. The nephron is the functional unit of the kidney. Each nephron has a capillary bed around it, called the **peritubular capillary network**. If we following the incoming arteriole, we see that it goes into a tiny sphere called a **Bowman's capsule**. Inside this capsule, the arteriole branches into very tiny capillaries, creating a tangle called the **glomerulus** *(glom-AIR-u-lus)*. The narrowing of the diameter of these capillaries causes the pressure inside of them to go up. (When you want to create more pressure in a garden hose, you twist the nozzle so that it becomes smaller.) This pressure in the glomerulus causes tiny molecules to be squeezed out-- molecules such as water, salts, glucose, amino acids and urea. Large things like blood cells, platelets, globulins, and clotting proteins, etc., are too large to escape. The filtration process is aided by special cells called **podocytes**, that have many feet ("podo" means "feet") that create tiny cracks. (Think of standing in muddy water and having fresh, clean water squeezed up from between your toes.)

This liquid "filtrate" then goes into the walls of the capsule, and then into a long twisted tube called a **convoluted tubule**. The walls of the tube are made of cuboidal cells with many microvilli that increase surface area. Some of the water and nutrients that were taken out are now reabsorbed by these microvilli and then transferred back into nearby capillaries. The cuboidal cells lining the tubes have special carrier proteins that recognize molecules that the body should recycle. The number of carriers for each type of molecule (nutrients, salts, vitamins, etc.) corresponds to how much should actually be in the blood. There are a lot of carriers for some things and less for others. For example, there are lots of carriers for glucose, because all of the glucose should be reabsorbed at this point. (In the case of diabetes, there is so much glucose in the blood that there are not enough carriers to pick it all up and some glucose goes right through the tubes and ends up going into the urine. Testing for glucose in the urine is a key test for diabetes.) The fact that there are only a certain number of carriers for each type of nutrient molecule is what keeps our body from being overwhelmed with too much of something. We can eat out-of-balance meals and our kidneys will strain out the excess amounts of most things. If you eat an entire bag of pretzels or chips, all that excess salt will be filtered out and go into the urine. If you take a vitamin tablet and your body doesn't need all those vitamins, the excess can be filtered out at this point.

The tube twists around, then dips down to create a long U-shaped loop, called the **loop of Henle**. The first part of the loop, the descending portion, is very narrow and is lined with leaky epithelial cells that allow water to seep out. The ascending portion is narrow at first, then gets larger. This upward part of the loop is watertight so water can't leak out here. In this part of the loop, Na^+ ions are actively pumped out and go back into the nearby capillaries. (We saw sodium pumps in the nervous tissue lesson. You can find them in many places in the body.) When Na^+ ions leave, Cl^- (chlorine) ions will naturally follow them because the ions are attracted to each other, being +1 and -1. This puts a lot of NaCl (salt) outside of the tubes, and creates an imbalance in water concentration. The process of osmosis will cause water to flow to areas that have a lesser concentration of water (and a higher concentration of solutes like NaCl). In this case, osmosis will make water flow out of the tubes and back into the blood.

NOTE: The water and salts can hang out for a while in the medulla tissue. They don't always go immediately back into the capillaries. Some diagrams have arrows pointing out into the tissue, not into capillaries. This can be confusing. The answer is that the water does go back into the capillaries but sometimes not immediately.

The very last part of the convoluted tubule (the distal part) has cells that can actively transport certain molecules out of the blood and into the urine. For example, molecules of penicillin and creatinine are pulled out in this way. This last part of the tubule is the last chance the nephron has to pull things out of the blood or to exrete more water.

So what is left in the end? What is urine made of? Mostly water, with small amouts of salts, urea, uric acid, ammonium and creatinine. (More about creatine and creatinine in a future lesson.)

About 99 percent of the salt in our blood is filtered out by the glomerulus, then returned to the blood at various points in the nephron. 67 percent is returned right away in the proximal (near) part of the convoluted tubule. Another 25 percent is pumped out in the upward (ascending) part of the loop of Henle. The last few percent is reabsorbed in the last part (the distal part) of the tubule, before it feeds into the ureter. The salt balance in the blood is very important to maintaining the correct volume of blood. Blood volume is an important factor in blood pressure. Too much salt in the blood will cause water to want to come into the blood vessels to equalize the concentration of water inside and outside the vessels. This is why doctors might tell someone with high blood pressure to decrease the amount of salt in their diet. Salt causes an increase in blood pressure. The kidneys try their best to maintain balance, but sometimes systems struggle to keep up and we can ease the situation by adjusting our eating habits.

There is a special place in the nephron that is devoted to monitoring blood pressure—right where the arteriole comes into the Bowman's capsule. The incoming (afferent) arteriole touches (and is actually connected to) the far (distal) portion of the convoluted tubules. After the tubule dips down and forms that loop of Henle, it goes up and touches the incoming arteriole. This touch point has a special name that looks very complicated: the **juxtaglomerular apparatus**. ("Juxta" means "crossing point" and "glomerular" refers to that tangle of capillaries we call the glomerulus. The word "apparatus" seems a bit too fancy for a simple-looking touch point, but oh well.) The cells of this touch point can sense when there is not enough pressure in the glomerulus. Not enough pressure means that things are not being filtered out. So the cells of this "apparatus" begin making an enzyme called **renin**. The renin turns a molecule called angiotensinogen into its active form, angiotensin. ("Angio" means something related to blood vessels. "Tensin" is related to the word "tension" meaning "tight.") Angiotensin causes blood vessels to tighten. Think about that hose again. Making the diameter of a hose or tube smaller will cause the pressure inside to increase. So angiotensin will increase blood pressure. Angiotensin also acts as a messenger molecule to the adrenal glands that sit on top of the kidneys. It tells the adrenals to start producing **aldosterone** *(al-DOST-er-one)*. Aldosterone is a hormone that promotes the reabsorption of Na^+ ions. When salt is reabsorbed into the blood, water also flows in and blood volume and blood pressure increase. But what if blood pressure gets too high? The sensor mechanism for this is located in the aorta. As blood volume increases, the aorta will experience more stretching. There is a tiny sensor spot in the aorta that senses this stretching and begins to secrete ANH (atrial natriuretic hormone), a hormone that stops the secretion of renin and aldosterone.

Just in case you are still interested in knowing more, a recent research project (2009-2011) was done using men who were training to be astronauts to Mars. While sealed into their spacesuits for many days, researchers were able to control exactly the amount of salt the men took in, and to measure exactly how much salt came out in the urine. They fed the men a diet very high in salt for a while and watched what happened. The results were not what they expected. Extra salt only came out in urine something like once a week or once a month. It appeared that the rest of the time, the kidneys were finding other ways to manage the extra salt. The kidneys got better at conserving water. The men did not get extra thirsty or need to drink extra water, which was quite the opposite was what was expected. The urinary system is "smart" and can adapt to unusual situations.

Another function of the kidneys is to help keep the pH of the blood at a constant 7.4. Part of pH balance occurs in the lungs, as the balance of H^+ (hydrogen) ions and HCO_3^- (bicarbonate) ions is constantly changing. If the blood becomes too acidic (too many H^+ ions in it) the breathing center in the medulla oblongata will tell you to start breathing faster in order to rid the body of hydrogen ions. (The hydrogen ions will end up as H's in water molecules expelled along with CO_2 when you exhale.) The kidneys can also reabsorb or excrete H^+ and HCO_3^- ions. The kidneys have a trick that the lungs don't have. They can also use ammonia, NH_3, as a way to remove H^+ ions from the blood. NH_3 comes into the kidneys as the waste product made by cells that tore apart amino acids. The NH_3 molecule can take on an extra hydrogen and become NH_4^+ (ammonium). This ammonium can be sent out with the urine. This ammonium is what makes the nose-tingling smell found in diaper pails and cat litter boxes.

Lastly, the kidneys do something completely unrelated to all the water and salt balance stuff. They monitor the level of oxygen in the blood and secrete the appropriate amount of **erythropoietin**. (Remember, "erythro" means "red" and "poie" means "to make.") Erythropoietin (ee-RITH-ro-po-EE-tin) tells the hematopoietic stem cells in the bone marrow to make more red blood cells. This exact place this happens is a little hard to determine, but it seems to be mostly in cells of the cortex that are close to the capillaries that surround the tops of the tubules.

The functional unit of the kidney is the **NEPHRON**

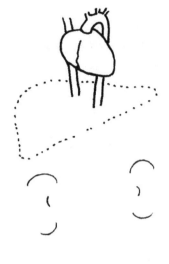

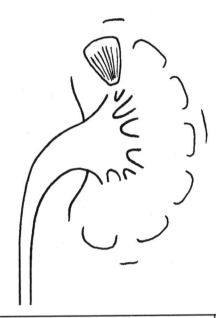

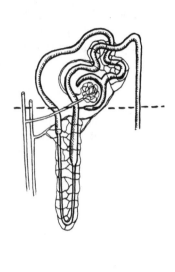

LENGTH OF URETHRA

WALL OF BLADDER

In the **GLOMERULUS**, the blood is under pressure. Very tiny molecules are pushed out: water, salts, ions, glucose, amino acids, uric acid and urea. Big things like proteins and blood cells stay in the capillaries.

In the **CONVOLUTED TUBULES**, water is reabsorbed and goes back into capillaries, but only a certain number of small molecules go back.

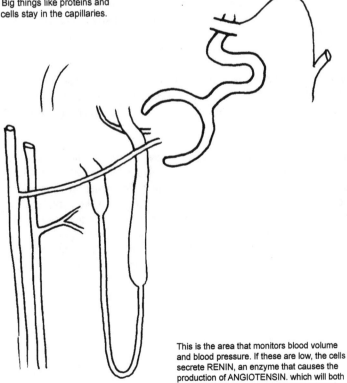

This is the area that monitors blood volume and blood pressure. If these are low, the cells secrete RENIN, an enzyme that causes the production of ANGIOTENSIN. which will both constrict blood vessels and tell the adrenal cortex to produce ALDOSTERONE, which makes more sodium go into the blood.

SUMMARY OF KIDNEY FUNCTIONS:

1) _____

Nitrogenous wastes come from the breakdown of amino acids, which have a nitrogen atom. (The liver turns ammonia into urea.)

2) _____

This is achieved through a balance of water and salt in the blood. The more salt in the blood, the more water goes into the blood, and that means greater blood volume and greater pressure. To increase blood pressure, the kidneys secrete renin, which activates angiotensin, which tells the adrenals to make aldosterrone which causes more sodium to go into the blood.

3) _____

The kidneys can excrete or reabsorb both H^+ ions (which make things acidic) and HCO_3^- ions (alkaline).

4) _____

If the kidneys sense that there is less oxygen in the blood, they will begin to produce more erythropoietin, which tells the hematopoietic stem cells to make more red blood cells.

5) _____

Vitamin D from the diet must be converted to a more active form that the digestive system can use to absorb calcium ions.

NOTE: The pituitary gland (in the brain) secretes a chemical called ADH (anti-diuretic hormone) at night, which causes more water to be reabsorbed, making less urine.

75: BONES (as organs)

We won't be learning the names of bones in this lesson. This lesson is just about how a bone functions as an organ. Like all organs, a bone is a group of tissues that work together for a common goal. As with the other organs, we will see many types of tissue in bones, but connective tissue will predominate. The goals of a bone are:
1) be a framework for muscles (and/or a protector of organs)
2) produce blood cells
3) store minerals, especially calcium

Our drawing shows the humerus bone, though a middle section had to be taken out for it to fit on the page and still be wide enough to show the internal structure. It is one of the "long bones" in the body, connecting the shoulder to the elbow. All long bones have a layer of **hyaline cartilage** on the end. You may have seen this shiny white layer on the end of a chicken bone. The knobby end of the bone is called the **epiphysis** *(ee-PIFF-eh-sis)*. The long, straight shaft is called the **diaphysis** *(die-AFF-eh-sis)*.

Inside the epiphysis is **spongy bone,** also called **cancellous bone** or **trabecular bone**. The spongy bone is filled with red marrow, so it looks red. The red comes from the red blood cells that are being produced. The red marrow is where the hematopoietic stem cells live, so the red marrow also produces platelets and all types of white cells. (See lesson 36 for review of hematopoietic stem cells.)

There is another layer of cartilage in long bones, located between the epiphysis and the diaphysis. This layer is called the **epiphyseal plate** *(ep-i-FIZZ-ee-al)*, or the growth plate. While children are growing, this plate is the site of bone lengthening. (A disorder called achondroplasia is caused by a malfunction in this layer of cartilage. The result is very short arms and legs, a form of dwarfism.) During puberty, levels of the hormone estrogen rise in both males and females, and this is what tells the chondrocytes in this plate to stop reproducing. Thus, bones stop growing. A very thin plate remains, even in adults, although it no longer causes bone growth.

Inside the shaft (diaphysis) there is a **medullary cavity** filled with yellow marrow. In children, this cavity actually still has red marrow in it, but as they grow older the red marrow will be replaced by yellow marrow. Yellow marrow is yellow because of the great number of adipose (fat) cells in it. The yellow color of the fat cells comes from natural food colors such as carotene which makes carrots orange. Fat cells can hold on to carotene.

Bones have tiny holes at various points that allow blood vessels to enter. These holes are called **nutrient foramens**. Arteries and veins bring in nutrients and take away wastes as well as picking up all those new blood cells that are being created.

Bones are covered with a thin layer of serosa-type membrane called **periosteum**. ("Peri" means "around," and "osteum" means "bone.") The periosteum continues off the end of the bone and will help to connect the bone to other connective tissue such as tendons and ligaments.

The bone around the medullary cavity (the bone that makes the shaft) is called **compact bone**. Compact bone is made of a network of collagen fibers that is filled with minerals such as calcium and phosphorus. We can think of compact bone as being a bit like concrete that is reinforced with re-bar. The re-bar rods are like the collagen fibers and the concrete is like the minerals. Cells called **osteoblasts** make the compact bone by secreting collagen and then filling it with minerals (mostly calcium and phosphorus but some magnesium and possibly other ions as well). Osteoblasts are made from stem cells in the outer layer of the bone, right under the periosteum. (You can review information about osteoblasts in lesson 35.)

If we look at a close-up view of compact bone, we see that it is made of osteons. (See lesson 35 for review of osteons.) Blood vessels go into the bone from outside, and create a network of vessels inside the bone, mainly going through the central canals. The periosteum around the outside has two layers. The outer layer is tough and fibrous. The inner layer has many cells in it, including stem cells, fibroblasts that make collagen, osteoblasts and osteoclasts.

Bones are constantly being remodeled. When the calcium level in the blood drops too low, a signal is given to cells called **osteoclasts** to start dissolving bone and releasing the calcium ions into the blood. Calcium is important for many body processes. Hopefully, you remember that the clotting cascade uses calcium, and that it plays a vital role in muscle contraction. Osteoclasts are a very different type of cell from osteoblasts. The osteoclasts do not come from stem cells in the outer bone like osteoblasts do. The osteoclasts might actually be a type of macrophage. (Research is still being done on this.) The osteoclasts secrete H^+ ions (protons) into the bone to dissolve the minerals. They also secrete an enzyme called **collagenase** to dissolve the collagen fibers. When the osteoclasts have been going a while, they will start to produce chemicals that will help to call osteoblasts over to make repairs. Osteoblasts come along and fill in the holes that the osteoclasts made. So you have this continual, ongoing process of remodeling, in order to keep the level of calcium in your blood constant.

You will want to go back and review drawings 35 and 36 before doing this drawing.

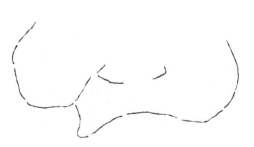

BONE REMODELING   (Blasts VS Clasts)

- Osteons are constantly being torn apart and rebuilt.

- When calcium levels in the blood get too low, **OSTEOCLASTS** dissolve bone to release calcium ions (Ca2+).

- When calcium levels in the blood are high, the **OSTEOBLASTS** put calcium back into bone by building up the osteons.

CLAST = tear down   BLAST = build up

Osteoclast is secreting acid (H+ ions) to dissolve the mineral content.
It will secrete an enzyme called **collagenase** to dissolve collagen.
Osteoblasts might be a type of macrophage.

76: THE SKELETAL SYSTEM

Finally, we get to name the bones! You probably already know some of these names. We are just going to label the major bones of the body, not every bone. If you want to know the names of smaller bones (or the names of the parts of the bones, as every lump and bump has a name) you can easily search the Internet and find them.

Scientists think of the skeleton as having two parts. The **axial** *(AX-ee-el)* skeleton consists of the skull, the spine and the ribcage. The **appendicular** *(ap-en-DIK-u-lar)* skeleton refers to the arms and legs. ("Appendage" means something that sticks off something else.)

The arms hang from the "pectoral girdle," which is a framework made by the **clavicle** (collarbone) and the **scapula** (shoulder blade). Right where the clavicle and the scapula meet at the shoulder, there is a smooth, cup-shaped depression that forms a "socket" for the smooth head of the **humerus** (upper arm bone). This type of joint is called a ball-and-socket joint and we'll look at it again in the next lesson. Below the humerus we find two bones, the **ulna** and the **radius**. Both of these bones are connected to the wrist, which is made of a clump of 8 bones called **carpals**. (Connective tissue keeps all these lumpy bones in place, of course.) The longest bones of the hand, found in the palm and in the thickest part of your thumb, are called the **metacarpals**. The bones of the fingers are called **phalanges** *(fal-AN-geez)*. The phalanges that are closer to the metacarpals are called the **proximal** phalanges. ("Proxi" means "close.") The phalanges at the tips are called the **distal** phalanges. ("Distal" means "distant.") The distal phalanges are the ones that are the farthest away from the wrist.) The ones in the middle are called the middle phalanges. The thumb does not have a middle phalange.

The legs are hung from the "pelvic girdle," which is made of the **coxal bones** and the sacrum. The wide back part of the coxal bone is called the **ilium** (not to be confused with the ileum). The ilium is the bone that we feel when we put our hands to our hips. The frontal part of the coxal bone is called the **pubis**. The bottom of the coxal bone is called the **ischium** (ISS-key-um) and this is the part of the bone we sit on. The two coxal bones are joined in the back by the sacrum, and in the front by a piece of fibrocartilage called the **pubic symphysis** *(SIM-fuh-sis)*.

The rounded top of the **femur** fits perfectly into a bowl-shaped depression in the **coxal bone**. This joint is a ball-and-socket joint, just like the shoulder joint. The bottom of the femur is attached (at the knee) to two smaller bones called the **tibia** and the **fibula**. The bone on the front of the knee (the knee cap) is called the **patella**.

Right under the tibia and fibula you find the 7 tarsal bones. The top tarsal bone is called the **talus** *(TAY-lus)*. The tarsal that forms the heel is called the **calcaneus** *(cal-KANE-ee-us)*. The foot has long bones similar to the metacarpals in the hand. These long foot bones are called the **metatarsals**. The bones of the toes are called the **phalanges**, just like the finger bones.

The names of the bones of the skull are very similar to the names of the lobes of the brains. For example, the **frontal** bone covers the frontal lobe. Thus, we also have the **temporal** bone, the **parietal** bone and the **occipital** bone. You may already know that the correct name for the jaw bone is the **mandible**. We met the **sphenoid** bone and the **ethmoid** bone in lesson 60. (The sphenoid is that butterfly shaped bone on the bottom of the skull. The ethmoid is in the sinuses.) The bone that holds the top row of teeth is the **maxilla**. The **nasal** bone connects the two maxilla bones and forms the top part of the nose. The **zygomatic** bone is the framework for your cheek. The top part of this bone, under the outside of the eye, is called the **zygomatic arch**.

The spine is made of 33 vertebrae. The top 7 vertebrae are called the **cervical vertebrae** and they form the neck. The first cervical vertebra, right under the skull, is the **atlas**. (In Greek mythology, Atlas was the god who held the earth (or the celestial sphere) on his shoulders. If you imagine your skull to be a globe, then your atlas is Atlas.) Your atlas isn't like any of the other vertebrae. It has a special shape that fits with the skull on top and with the second cervical vertebra below it, called the **axis**. The atlas is the bone that allows you to nod your head up and down and tip it side to side, and the axis allows you to look left and right. Working together, these two bones allow you to move your head in any direction. (As an important side note, if these two bones get knocked out of alignment, they can pinch the nerves coming out of your brain stem and cause quite a variety of health issues, such as headaches, digestive disturbances, neurological problems with arms or legs, brain problems like depression or "brain fog," and even (in a few documented cases) nervous system diseases such as MS. Most chiropractors don't deal with these two vertebrae. Special chiropractors, called upper cervical chiropractors, make precise measurements of the misalignment and tap the bones back into place. (If you've had head or neck injury get your atlas checked and spare yourself future problems.) The rest of the cervical vertebrae are simply known by their numbers: C3, C4, C5, C6, and C7.

The next 12 vertebrae are called the **thoracic vertebrae**. These are the vertebrae that are attached to your ribs. You have twelve pairs of ribs and ten of these pairs are fastened to the vertebrae in the back and the sternum bone in the front. The two bottom pairs of ribs, number 11 and number 12, are called "floating ribs" because they don't attach to the sternum, only to the spinal vertebrae in the back. None of the thoracic vertebrae have special names. They are called T1, T2, T3, etc.

Below the 12 thoracic vertebrae are 5 **lumbar vertebrae**. The nerves that come out of the lumbars go into the organs in the lower part of the torso, such as bladder and rectum, and down into the legs. Numbers are used for the lumbars: L1, L2, L3, etc.

Between all the vertebrae you find thick pads made of dense fibrocartilage. This padding allows the vertebrae to move around a bit, but hopefully not slip out of place too much.

Below L5 we have the **sacrum** *(SAY-crum)*. The sacrum looks like one piece but is actually 5 vertebrae that are fused together. The sacrum sits between the two coxal bones that form the hips. The last tiny bit of the spine is called the **coccyx** (KOK-siks) and it appears to be 4 tiny vertebrae fused together. The common name for the coccyx is the "tail bone." The word coccyx comes from the Greek word for the cuckoo. The namers of this bone thought that the coccyx resembled the beak of this bird.

BONES of the SKULL:

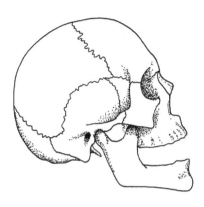

BONES of the SPINE:

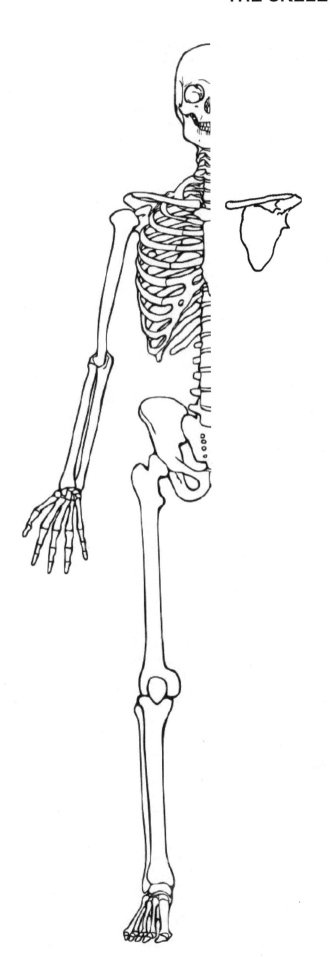

This drawing is by Andreas Vesalius, first published in 1543. His book, "De Humani Corporis Fabrica" is considered to be the first modern anatomy textbook.

77: JOINTS

There are three types of joints. Joints that are freely movable (like the knees, elbows, wrists, etc.) are called **synovial joints**. Joints that are very firm and only slightly movable are called **cartilaginous**. Joints that don't move at all are called **fibrous**. Let's begin with the fixed, immobile joints and end with the movable ones we are most familiar with.

Fibrous joints are places where two bones come together and are firmly attached, as if they are glued together. The **sutures** in the skull are the most well-known fibrous joints. The cranium bones were once separate, when the body was still in its embryonic state. The separations between the bones made it easier for the head to go through the birth canal. After birth, the gaps begin to narrow, and after a few years the cranium is basically one solid piece. The place at the bottom of a tooth where it is held into its socket also counts as a fibrous joint. The pairs of bones in the arms and legs (ulna and radius, and tibia and fibular) are connected at both ends by a fibrous joint.

Cartilaginous joints are not completely fixed, but are only slightly movable. These joints are made entirely of shiny, white hyaline cartilage, which is why they are called cartilaginous. The vertebrae are connected by this type of joint. There are solid discs (intervertebral discs) of hyaline cartilage between all the vertebrae. These discs allow the vertebrae to move just enough so that you can twist your back to the left and right and bend over, but the joints are stiff enough to give the back the strength it needs to be a framework for the rest of the body. (If these discs slip out of place, the result can be very painful.)

The pubic symphysis, which we met in the last lesson, is also a cartilaginous joint. The symphysis joins the two coxal bones but allows just a tiny bit of flexibility when necessary. The costal cartilage that connects the ribs to the sternum also counts as a cartilaginous joint, though we don't tend to think of it as a joint.

The **synovial joints** are the ones that immediately come to mind when we think about joints: the knee, hip, elbow, wrist, shoulder, ankle. Synovial joints have some special characteristics that the first two categories don't have. The bones of the synovial joints are held together by ligaments. **Ligaments** are like biological steel cables or ropes, and are incredibly strong for their size. Since they are made of collagen, they easily connect to the collagen framework found in bones. The bones also have tendons connected to them. **Tendons** connect bone to muscle. Remember that muscles are surrounded by bags of connective tissue, so it is this connective tissue that connects to the bone.

Between the bones of the synovial joint there is a space called the **synovial capsule**. The capsule is made of a thin membrane very similar to serous membrane. Some of the cells in the membrane secrete a fluid called **synovial fluid**, similar to serous fluid. This synovial capsule acts as a shock absorber and keeps the bones from rubbing against each other. In the large synovial joints such as the knee and hip, there are additional pads and cushions such as bursae (fluid filled sacs) and pads of fat.

There are six types of synovial joints:

Hinge: elbow, knee
Ball and socket: shoulder, hip
Pivot: atlas/axis in neck, radius at elbow

Saddle: base of thumb
Plane (or "gliding"): between carpals/tarsals, between sternum and clavicle
Ellipsoidal (or "condyloid"): between metacarpals and phalanges

NOTE: Sometimes intervertebral joints (bewteen the vertebrae) get classified as plane joints. The ellipsoid category can also be a bit loose, with some disagreement about which joints belong in that category.

Let's look at a cross section of the knee. The knee is where the bottom of the femur is joined to the top of both the tibia and fibula. The ends of the bones are covered with slippery, white hyaline cartilage. The slippery nature of the cartilage helps the joint to be almost friction-free. Friction is also reduced by slippery pads called **menisci** (singular: **meniscus** *[meh-NISS-cuss]*), and by about a dozen fluid-filled capsules in and around the joint, called **bursae** (singluar: **bursa**). The bursae *(burs-ay, or bur-see)* are filled with fluid, just like the synovial capsules. The bursae help ligaments and tendons to slide past each other easily as you bend your knee. Inflammation of the bursae is fairly common, and is called **bursitis**. The cruciate *(CRU-she-ate)* ligaments get their name from the fact that they cross each other. ("Cruc" or "crux" means "cross") Cruciate injuries are common in sports, especially football.

The **patella** (knee cap) is held in place from above by the **quadriceps tendon** and from below by the **patellar ligament**. The patella not only protects the knee joint, it also acts as a fulcrum if you think of the leg bones as levers. The patella pushes the tendons out from the knee a bit, acting to increase leverage, giving mechanical advantage to your leg muscles. (Meaning you get more action for your effort.)

The hip is obviously a ball and socket joint. The head of the femur is the ball, and a concave surface on the pelvic bone makes the socket. ("Concave" means that it curves in, like the entrance to a cave. Convex is a curve that protrudes outward. The ball is the convex surface.) In this simple hip diagram, we see ligaments connecting the bones, and a synovial capsule filled with fluid.

The shoulder is a more comlicated ball and socket joint, and it is even more complicated than this diagram. The socket is made of several bones and is not as deep as the pelvic socket. We see tendons that go around the head of the humerus, anchoring the top of the biceps muscle into the joint. The muscles are anchored in just the right places to give us maximum mechanical advantage. There are additional layers of muscle and tendon that form something called the **rotator cuff**, which helps to keep the head of the humerus in place. It was not possible to show the rotator cuff in this diagram. (Notice that we see another cartilaginous joint, also, where the acromion is joined to the clavicle.)

There are three kinds of joints: 1) FIBROUS, 2) CARTILAGINOUS, 3) SYNOVIAL

<u>FIBROUS:</u> (don't move at all)

Ex: sutures in skull, teeth in sockets, ends of ulna/radius, tibia/fibula

<u>CARTILAGINOUS:</u> (move only slightly)

Ex: discs between vertebrae, pubic symphysis, ribs/sternum

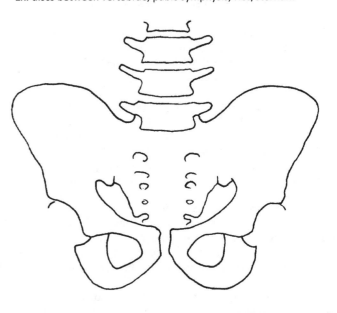

<u>SYNOVIAL:</u> (very flexible)

Synovial joints have fluid-filled capsules in and around the joint to decrease friction. They also have slippery (white) hyaline pads.
There are 6 types of synovial joints: hinge, ball and socket, pivot, saddle, plane and ellipsoidal.

HINGE: the knee (shown) and the elbow

☐ hyaline cartilage   ☐ synovial cavity   ☐ fat
☐ ligaments, tendons   ☐ bursa (fluid filled sac)

BALL AND SOCKET: hip and shoulder

HIP:

SHOULDER:

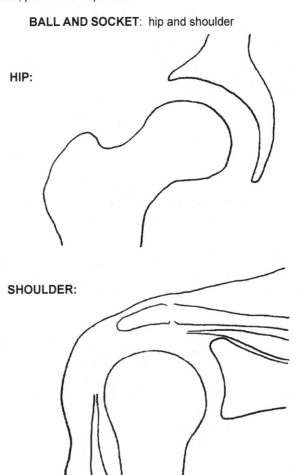

78: MUSCLES (as organs)

All muscle cells have these traits in common:
1) They are controlled by nerves.
2) They contract using use actin and myosin. (See lesson 54 for review of actin and myosin.)
3) They use cellular respiration as a source of ATP.

There are three types of muscle tissue:
1) **Skeletal muscles** are the ones we control voluntarily to move our body parts. The fibers run parallel in long lines, and they have bands going across them, which are the ends (Z lines) of the sarcomeres. (Remember from lessons 53 and 54 that the sarcomere is the tiny unit that contracts, shortening the muscle. It does so by using little molecular paddles called myosin that push against microfilaments called actin. You can review those lessons if you can't remember how this works.) Voluntary muscles are made of fibers, not individual cells, and each fiber has many nuclei. It's like many cells merged together, and they all donated their nuclei to the fiber.

2) **Smooth muscles** are involuntary, which means we have no conscious control over them. The cells are spindle-shaped, and are individual cells, each with a nucleus. There aren't any sarcomeres. We are usually unaware of the action of our smooth muscles. Smooth muscles line the digestive tract and cause the churning in the stomach and peristalsis in the intestines. Very tiny smooth muscles surround blood vessels causing constriction or relaxation. Tiny smooth muscles also control the opening and closing of various glands or other micro-structures. Smooth muscles don't operate like skeletal and cardiac. They do have actin and myosin, but they don't have sarcomeres.

3) **Cardiac muscles** are found only in the heart. They are individual cells, but they are connected end to end to form long tubular fibers. The pattern made by these fibers if often described as looking "branched." The fibers have stripes, like skeletal muscles, because they have sarcomeres. The places where the cells connect are called **intercalated discs**. The discs have **gap junctions** made of little "tubes" that allows ions to flow from cell to cell very quickly, carrying an action potential that causes contraction. (You can review gap junctions in lesson 25.) The heart muscles are especially amazing because they get all the rest they need between beats.

Skeletal muscles are made of microscopic muscle fibers. In lesson 53 we saw how muscle fibers are grouped together into bundles. Look back at the top drawing on the lesson 53 page. We drew a neuron traveling along with this bundle and then going down into the bundle to attach to each of the muscles fibers in the bundle. We drew a circle around them and defined this as a motor unit. Motor units are groups of muscle fibers that all work the same way at the same time because they are all stimulated by the same nerve. However, there are millions of motor units in your body, so even with some muscle fibers doing the same thing as their neighbors, you brain still has many options of which ones to combine to be able to move just the way you want to. Think of how amazing it is for your brain to be able to control your muscles so finely that bodies can do things like get a basketball into a hoop, or dance ballet.

Muscles can only do one thing: contract (get shorter). Because of this, muscles must work in pairs. We see this most clearly in skeletal muscles. The muscle that is contracting is called the **prime mover**. The muscle that pulls the other way is called the **antagonist**. The example most often given is the relationship between the biceps and triceps muscles. The biceps is the muscle that contracts when you "make a muscle" in your arm. On the back side of your arm you have the triceps muscle, and it pulls the lower arm back again, straightening the arm.

If we consider the motion of a prime mover muscle (ex: biceps), the place where it attaches to the bone that is staying still is called the **origin**. The place where the muscle attaches on the bone that is moving is called the **insertion**. In our biceps example, the origin of the biceps is up inside the shoulder (as we saw in the last lesson) at a point partly on the scapula. The insertion of the biceps is on the radius bone. Think about it-- when you curl your arm to make a muscle, it is your radius/ulna that move, not your humerus. The insertion point is on the bone that moves.

Muscles also work in groups. The biceps muscle isn't the only one pulling when you curl your arm. The other smaller muscles that help out are called the **synergists** *(SIN-er-gists)*. ("Syn" means "with.") Synergy means working together. Synergistic muscles are found all over the body, but are especially well studied in the legs and pelvic areas. Having many muscles working together gives the body the ability to move gracefully with a wide range of motions.

Large scale motions, like curling your arm or lifting your leg, can be classified into groups. A movement that decreases the angle of a joint (like curling your biceps) is called **flexion**. A motion that increases the angle is called **extension**. When you move a body part away from the midline of the body, this is called **abduction**. ("Ab" means "away") When you move a body part towards the midline, it is called **adduction**. ("Ad" means "towards.") Rotation is twisting around an axis, like turning your head. You can also rotate a hand or an arm. Circumduction is when you move an arm or leg in a circle, sort of tracing out a cone shape in the air.

Smooth and cardiac muscles get almost all of their energy from cellular respiration, and the ATPs that are made by the electron transport chain in the mitochondria. (If you need to review, you can go back to lesson 20.) Cardiac muscles prefer to burn fatty acids (not glucose) in their Krebs cycles, but smooth muscles almost always use glucose. Skeletal muscles only use ATPs from cellular respiration after their reserve of **creatine phosphate** is gone. Creatine is a molecule similar to ATP. It can hold onto a phosphate. When the phosphate pops off, energy is released. While creatine is holding a phosphate, it is called creatine phosphate, then after the P pops off, it is called just creatine. Creatine molecules are made in the liver and kidneys, then transported through the blood into the muscles. Muscles have enough creatine phosphate stored up to provide energy for about 8 seconds. After 8 seconds, the muscle will begin relying on the electron transport chains in the mitochondria to provide ATPs. As long as there is enough oxygen in the cells, the ETC will keep making ATPs. If exercise is intense and the lungs can't keep up the supply of oxygen, muscles will begin burning pyruvates and making them into lactic acid. When lactic acid is present, the muscle will feel a burning sensation. Once the exercise stops, the muscles begin clearing out the lactic acid and it is gone in a few minutes. The soreness you feel the day after exercise is NOT from lactic acid, as many people believe. Lactic acid is broken down fairly quickly. Soreness is more likely due to inflammation in tissues.

MUSCLES (as organs)

There are 3 kinds of muscles: 1) SKELETAL (voluntary), 2) SMOOTH (involuntary), 3) CARDIAC (heart)

SKELETAL:	SMOOTH:	CARDIAC:

1) _____

2) _____

3) _____

4) _____

1) _____

2) _____

3) _____

4) _____

1) _____

2) _____

3) _____

4) _____

SKELETAL MUSCLES WORK IN PAIRS

Muscles can only do one thing: CONTRACT.
A prime mover and its antagonist work together.

MOTIONS can be classified

Abduction: body part moves away from midline
Adduction: body part moves toward midline
Extension: joint angle increases
Flexion: joint angle decreases
Rotation: rotates around axis
Circumduction: cone shape is outlined

WHERE DO SKELETAL MUSCLES GET THEIR ENERGY?   Here they are, in order of preference.

1) CREATINE PHOSPHATE

2) CELLULAR RESPIRATION (the ETC)

3) FERMENTATION

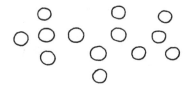

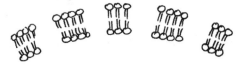

CREATINE is first choice, but can be sustained for only 8 seconds. Creatine holds onto a phosphate. An enzyme can take the P off, and then put it onto an ADP, making ATP. No oxygen is needed.

After 8 seconds, cellular respiration kicks in. Oxygen is needed for the Electron Transport Chain to turn ADP back into ATP. Glucose from glycogen is the preferred fuel for the ETC in skeletal muscles.

Lactic acid fermentation is the third and last choice for energy. This must be used if oxygen is not available. Lactic acid gives that burning sensation in muscles when they are fatigued.

79: THE SKELETAL MUSCLES

Before we start learning the names of the skeletal muscles, we need to do an overview of their fuctions. Skeletal muscles do more than just move your arms and legs.

Functions of skeletal muscles:
1) Make bones move. (the obvious function)
2) Make skin move. The muscles in your face allow you to express your emotions and convey feelings to other people.
3) Help to maintain a constant body temperature. Contraction of skeletal muscles causes ATP to release energy that causes heat to be produced and helps to warm the body.
4) Assists movement in cardiac and lymphatic systems. The pressure of skeletal muscles contraction keeps blood moving in veins (arteries flow because of pumping action of heart), and the lymph fluid moving through the lymph vessels.
5) Help to protect internal organs. Bones are covered with muscles, but so are the internal organs. The muscles of our torso help our body to move, but they also cover the stomach and intestines. (You many have seen feats of strength where someone tenses their abdominal muscles so tightly that they can take blows to their torso without any damage to their organs.)

When learning the names of muscles, it helps to understand that the names aren't designed to be difficult, though they often look pretty scary. The names use Latin and Greek word roots which may be unfamiliar to us, but these words roots aren't difficult in their meaning. Common word roots mean things like straight, circular, near, far, triangular, back, front. Easy to understand. Muscle names will always incorporate one or more of these descriptions:
1) Size (examples: "maximus" and "major" both mean "biggest")
2) Shape (examples: deltoid means shaped like the Greek letter D, which is a triangle, "orb" means "round")
3) Location (examples: "frontalis" means "front," "dorsi" means "back," "anterior" means "front," "posterior" means "back.")
4) Action (examples: "extensor" means "to extend," "masseter" means "to chew")
5) Attachment points (example: the sternocleidomastoid is attached to the sternum, the clavicle, and the mastoid process)
6) Number of attachments (example: "bi" means "two" and the "biceps" has two attachment points)

The muscles are all listed on the drawing page, so there is no need to list them all here. Instead, we'll list the word roots that these names use. (Please note that these word roots are not intended for linguistic use. If you happen to be an expert in Latin or Greek, you'll probably be tempted to nit pick about the endings. The only purpose of this list is to help you decode the meaning of anatomy words, not to learn Greek or Latin.)

Abdo: related to the abdomen (stomach area)
Ante: in front
Bi: two
Brachii: arms
Carpus: wrist
Ceps: head
Cleido: related to the clavicle
Cnem: knee
Deltoid: shaped like the Greek letter D (delta)
Digit: fingers
Dors: back
Extensor: action of increasing angle of joint
Ex: outside, or on the outside
Flexor: action of decreasing angle of joint
Gastro: stomach or belly
Gloutos: buttocks
Latus: wide or broad

Major: large or important
Masseter: to chew
Musculus: "little mouse" (The Greeks thought the biceps muscle was shaped like a mouse!)
Obliquus: slanting
Occiput: back of skull
Ocu: eye
Orb: round (spherical)
Ori: mouth
Pectoral: related to the chest
Quad: four
Rectus: straight
Sartor: tailor (someone who sews clothes)
Sterno: related to the sternum
Trapezium: "little table" (shape with 4 sides)
Tri: three
Zygomatic: related to the zygomatic arch (cheek bone)

There is one part in our diagram that is not a muscle: the Achilles tendon. It is so large, and so visible in these drawings, that it seemed best to go ahead and label it.

PLEASE NOTE THAT this drawing shows only the major muscles of the body. There are muscles along the spine, for instance, (deep under the back muscles we drew), muscles in the eye (which we saw in the eye drawing), muscles in the groin, and more muscles in the face. There are about 600 individual muscles in the body!

In this lesson, we will be using drawings made by famous anatomist Andreas Vesalius in the year 1555.

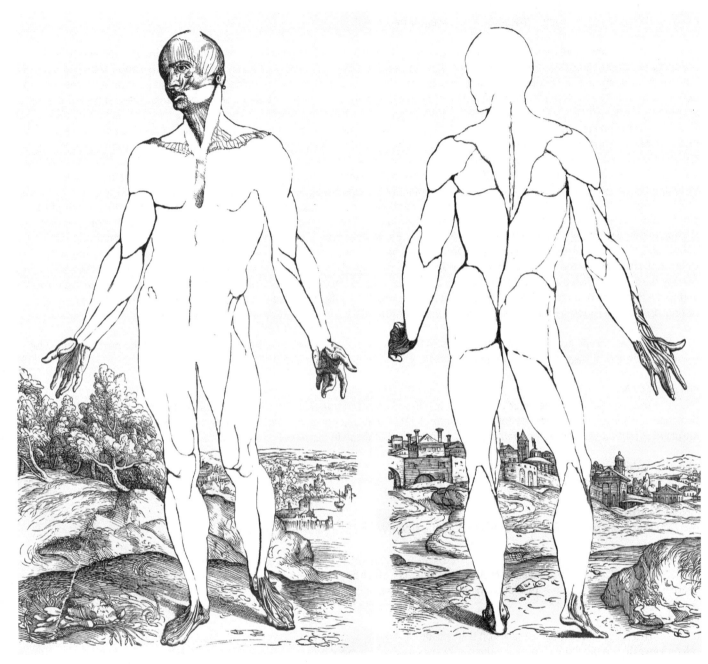

HEAD and NECK

1) **Frontalis:** *wrinkles forehead and moves eyebrows.*

2) **Orbicularis oculi:** *closes eyes*

3) **Zygomaticus:** *smiling*

4) **Masseter:** *closes jaw*

5) **Orbicularis oris:** *closes and protrudes lips (like a kiss)*

6) **Occipitalis:** *moves scalp backwards*

7) **Sternocleidomastoid:** *turns and twists head*

UPPER LIMBS

8) **Deltoid:** *raises arm at shoulder joint ("delts")*

9) **Triceps brachii:** *straightens arm*

10) **Biceps brachii:** *bends arm at elbow*

11) **Flexor carpi group:** *bends hand down at wrist*

12) **Extensor carpi:** *pulls hand up at wrist*

13) **Flexor digitorum:** *closes hand*

14) **Extensor digitorum:** *opens hand*

TORSO

15) **Trapezius:** *moves head, shrugs shoulders ("traps")*

16) **Pectoralis major:** *("pecs") pulls arm across chest*

17) **Rectus abdominis:** ("abs") *"sit-up" muscles*

18) **Latissimus dorsi:** *("lats") pulls arm across back and extends shoulders*

19) **External oblique:** *rotates torso*

20) **Teres major and minor:** *pulls arm down and back*

LOWER LIMBS

21) **Gluteus maximus:** *going from sitting to standing*

22) **Quadriceps group:** *straightens leg*

23) **Hamstring group:** *bends leg at knee*

24) **Sartorius:** *rotates thigh (so you can sit cross-legged)*

25) **Gastrocnemius:** *points toes ("calf")*

26) **Tibialis anterior:** *pulls toes up, and inverts foot*

27) **Achilles tendon**

80: THE ENDOCRINE SYSTEM (an overview)

The endocrine system is the name we give to the major glands of the body. Many of these glands cooperate with each other or with other body parts to keep the body in balance (**homeostasis**).

First, we should review the difference between endocrine glands and exocrine glands. The **exocrine glands** secrete their products into ducts. Exocrine glands include sweat and oil glands, salivary glands, milk glands, and digestive glands in the stomach and intestines. **Endocrine glands** don't have ducts, and secrete their products right into the blood stream. The hormones secreted by endocrine glands act as messengers, carrying instructions to other cells, often cells that are quite far away in a different part of the body.

A **hormone** is a messenger molecule. These messenger molecules can be made of amino acids (**peptide hormones**) or cholesterol (**steroid hormones**). Peptide hormones never enter a cell. Their shape fits into a receptor molecule on the surface of a cell. The peptide molecule fits into the receptor like a key fits into a lock. Once locked in, the receptor causes an ATP to lose 2 phosphates, turning it into a molecule called **cyclic AMP** (**A**denosine **M**ono**P**hosphate). "cAMP" is now the new messenger molecule, since the hormone has to stay outside the cell. The hormone molecule will eventually be torn apart and its atoms recycled.

Steroid hormones are nonpolar (hydrophobic) because they have cholesterol as part of their structure. The interior of the plasma membrane is also nonpolar (hydrophobic) and even has cholesterol molecules floating in it. So the steroid hormones have chemistry that gets along well with the interior of the plasma membrane. This means that steroids can just slip right though the membrane, with no channel protein needed. Once inside the cell, they will then bind with some kind of receptor molecule, either in the cytoplasm or inside the nucleus. The final result is that this "complex" (receptor and hormone stuck together) will then bind to a certain place on the DNA and allow that section of DNA to be copied into messenger RNA. The mRNA will then go out of the nucleus and find a ribosome that can translate it into a protein. This takes a while, so the effects of steroid hormones are slower but longer lasting. In contrast, peptide hormones act quickly and strongly.

This lesson will just give a list of the glands in the endocrine system, then we'll go into more detail in the next lessons. Also, some of these glands have been covered in past lessons.

PITUITARY GLAND

This has two parts, the anterior and the posterior. The anterior makes TSH (thyroid stimulating hormone), ACTH (adrenocorticotropic hormone), FSH (follicle stimulating hormone), LH (luteinizing hormone) and GH (growth hormone). The posterior pituitary makes ADH (antidiuretic hormone) and oxytocin. More on all of these in the next two lessons.

HYPOTHALAMUS

Makes hormones that affect the pituitary gland.

PINEAL GLAND

Makes melatonin, which helps to regulate the sleep/wake cycle (circadian rhythm).

THYROID

Makes T_3 and T_4 for regulating metabolism (how fast the body burns energy), and calcitonin for lowering calcium in blood by decreasing the activity of the osteoclasts (the cells that dissolve bone).

PARATHYROID

This is actually 4 separate spots on the outside of the thyroid. They make PTH (parathyroid hormone) which increases the calcium level in the blood by stimulating the osteoclasts to dissolve more bone.

THYMUS

Trains T cells during childhood. (Lesson 46)

PANCREAS

The Islets of Langerhans make insulin and glucagon that control glucose level in blood. (Lesson 71)

ADRENAL GLANDS

They have two parts: an inner medulla and an outer cortex. The medulla makes epinephrine (adrenalin) and norepinephrine. The cortex makes aldosterone (which affects the kidneys) and cortisol which affects glucose levels and is also anti-inflammatory.

OVARIES and TESTES

Make sex hormones (estrogen, progesterone, and testosterone) that control development in puberty as well as reproduction in adulthood. More on these in future lessons.

Endocrine glands secrete hormones into the blood. Hormones are messenger molecules.

PEPTIDE HORMONES are made of _____   STEROID HORMONES are made using _____

Peptide hormones never enter a cell. They bind to external receptors.   Steroid hormones enter the cell and bind to a receptor inside.
Usually, ATP is turned into cAMP, which starts a cascade reaction.   That receptor molecule will attach to DNA and cause a certain
Cascades allow for rapid manufacturing.   part to be copied into mRNA, which will then build a protein.

THE ENDOCRINE GLANDS

_____ gland has two parts.

_____   _____
-TSH   -ADH (for kidneys)
-ACTH   -oxytocin (females)
-FSH, LH
-GH

_____ makes hormones
that affect the pituitary gland.

_____ gland makes
melatonin, which helps to regulate
sleep cycle.

_____ glands have
two parts.

_____ (inside)
-epinephrine (adrenalin)
-norepinephrine

_____ (outside)
-aldosterone (for kidneys)
-cortisol (raises blood glucose,
anti-inflammatory)

_____ makes
-insulin (lowers blood glucose)
-glucagon (raises blood glucose)
(lesson 71)

_____ gland makes
-T3, T4 for metabolism and growth
-Calcitonin for lowering blood calcium

_____ gland
consists of 4 spots on the thyroid.
It makes PTH (parathyroid hormone) for taking calcium out of
bones and putting it into blood.

_____ gland is most
active during childhood.
It trains T cells (lesson 46).

_____ in females
-estrogen
-progesterone

_____ in males
-testosterone

81: HYPOTHALAMUS and PITUITARY

In the next few lessons, we will take a closer look at some of the major endocrine glands. If you find these lessons on the endocrine system confusing, it's not you, it's the endocrine system. Organs such as the hypothalamus and pituitary bring together all of the other body systems in one way or another. That's a guarantee that it's going to be complicated. The main point in this lesson isn't to memorize all this stuff, but to gain an appreciation for how all the body's organs communicate with each other and work together. You'll see lots of information from previous lessons here in this lesson!

The **hypothalamus** isn't a minor brain part. It's the hub where everything comes together. It receives information from both afferent (in-coming) nerves and from the blood. Most of the brain is out-of-touch with the rest of the body because of the Blood Brain Barrier (BBB) which keeps just about everything except food and oxygen from going into brain tissue. However, in the hypothalamus region we find places where that barrier is leaky. Receptor cells in this area need to sample the blood to find out what is currently in it. If there is too much or too little of something, hormone messengers will need to be sent out to correct the imbalance. The hypothalamus uses hormones to control a number of other organs. Often, these correction operations happen automatically, without us needing to do anything. Blood pressure and heart rate go up and down without any thought on our part. Other corrections need us to do something, in order to make the needed adjustment.

The hypothalamus connects our conscious, "thinking" brain to our chemical needs. It receives signals from many sensory cells in the body, so it is informed about things like body temperature, blood pressure, heart rate, level of glucose in the blood, saltiness of the blood, time of day, activity level of the immune system, any many other things. The hypothalamus also has connections to our rational brain-- the parts that control our actions. The behaviors prompted by the hypothalamus are often are things that we "feel like " doing. When our body is hungry we "feel like" eating. When thirsty, we feel like drinking. We might even feel like having one food and not another. If we are very hungry, we may feel grouchy. We can feel like curling up in a blanket, or sitting in front of a fan. The hypothalamus prompts us to act a certain way, in order to maintain homeostasis.

The influence can go the other way around, too, with our behaviors or thoughts affecting our body chemistry. If we are in the midst of sad or upsetting circumstances, our appetite may be affected. Anger can make our blood pressure go up. Scary sights or scary thoughts can make our heart rate go up. When a mother hears her baby cry, the hypothalamus and pituitary can cause the mammary glands to be activated. And so it goes both ways, with our chemistry affecting our behaviors, and our thoughts and behaviors affecting our body chemistry. The hypothalamus is right at the crossroads, where all this back and forth communication takes place.

The **pituitary** is almost like an extension of the hypothalamus. It hangs below it, connected by a thin "stalk." (In lesson 60 we noticed that the pituitary lies inside a protective "pocket" made of bone.) The pituitary has extensive networks of capillaries so it is well connected to the blood stream. Some hormones, like adrenalin, will need to go into the blood very quickly. Specialized neurons (called neuroendocrine cells) reach down from the hypothalamus into the pituitary and direct all its functions.

The pituitary is divided into two parts: the front part (**anterior**) and the back part (**posterior**). Their functions are completely separate, so we will discuss them separately, as though they are two different glands.

The **anterior pituitary** makes peptide hormones that affect many other glands and organs. It's stunning how much it controls, considering that it is only the size of a very small pea. It makes:
-- **ACTH (adrenocorticotropic homrone)** which affects the adrenal glands. (More about this in a future lesson.)
-- **TSH (thyroid stimulating hormone)** that tells the thyroid to secrete its hormones. (More about this in a future lesson.)
-- **GH (growth hormone)** which does more than just make children get taller. Adults still need growth hormone because it also stimulates the immune system, helps bones to stay strong, affects how the liver processes glucose, helps to maintain the Islets of Langerhans in the pancreas, affects the thyroid's hormones to function properly, helps us build muscle, promotes the breakdown of fats, and more.
-- **PROLACTIN**, which causes female mammary glands to produce milk. ("Pro" means "for," and "lact" means "milk.")
-- **LH (luteinizing hormone)** which makes the testes secrete testosterone and the ovaries secrete estrogen. Release of eggs (ovulation) in the ovaries is caused by a sudden increase in LH. (More on this in a future lesson.)
-- **FSH (follicle stimulating hormone)** causes eggs to mature and sperm to be produced. (More about this in a future lesson.)

The **posterior pituitary** makes these peptide hormones:
-- **ADH (antidiuretic hormone)** which affects the kidneys. ADH molecules stick to receptors in cells in the collecting ducts in the medulla of the kidneys and cause these cells to put more aquaporin channels onto the inner sides of their membranes (the sides that form the inside lining of the duct). This increase in aquaporins allows more water to go out of the ducts and back into the kidney tissue. Basically, water gets recycled a lot more, and less urine is produced. AND, since the signals are being sent out by a part of the brain that is also in contact with the pineal gland (keeping track of night and day) and the thalamus (one of the parts that regulates consciousness), extra ADH is secreted while you are sleeping so that you makes less urine and hopefully don't need to get up in the middle of the night to go to the bathroom (at least not too often).
-- **OXYTOCIN** which affects the smooth muscles in both male and female reproductive organs. (More on this in future lessons.)

Review of location of hypothalamus and pituitary

The **HYPOTHALAMUS** is a very important control center. It receives input, from both the senses (afferent nerves), and from the conscious mind. It also samples the blood to find out if there is too much or too little of various chemicals.

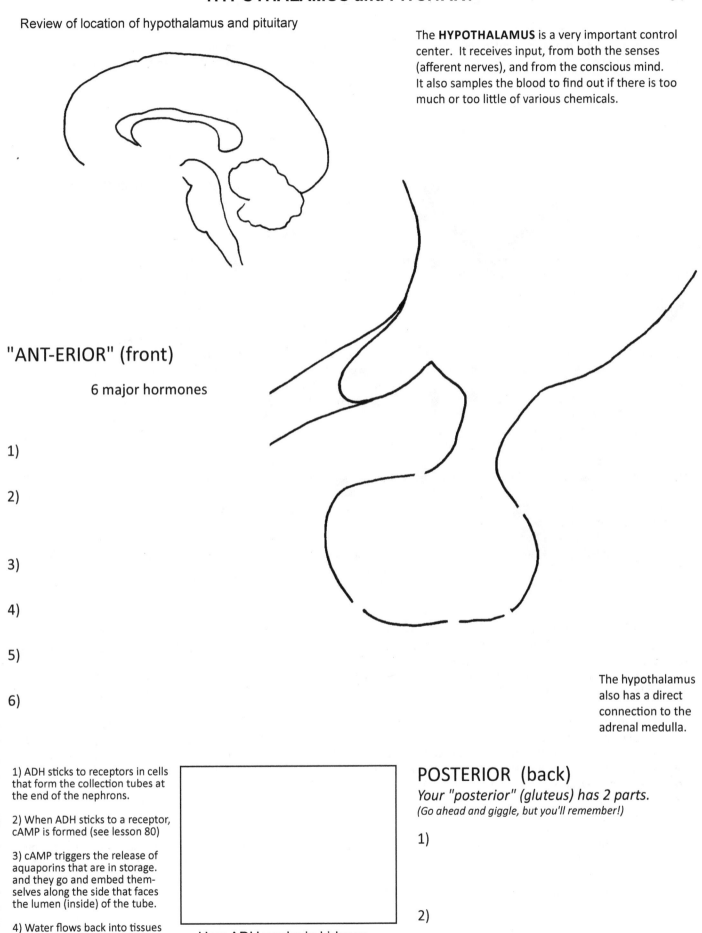

"ANT-ERIOR" (front)

6 major hormones

1)

2)

3)

4)

5)

6)

The hypothalamus also has a direct connection to the adrenal medulla.

1) ADH sticks to receptors in cells that form the collection tubes at the end of the nephrons.

2) When ADH sticks to a receptor, cAMP is formed (see lesson 80)

3) cAMP triggers the release of aquaporins that are in storage. and they go and embed themselves along the side that faces the lumen (inside) of the tube.

4) Water flows back into tissues and is conserved (less urine).

How ADH works in kidneys

POSTERIOR (back)
Your "posterior" (gluteus) has 2 parts.
(Go ahead and giggle, but you'll remember!)

1)

2)

82: THYROID and PARATHYROID

The **thyroid** is the main gland in your throat. It is butterfly-shaped and wraps around your trachea. The thyroid is told what to do by the pituitary gland (which, in turn, is regulated by the hypothalamus). When stimulated by TSH from the pituitary, the thyroid secretes three hormones:

-- **Calcitonin**
-- **T3** (triiodothyronine)
-- **T4** (thyroxine)

Calcitonin inhibits (stops) the osteoclasts in bone tissue from tearing it apart. (Remember, osteoclasts are those little macrophage-like cells that are constantly dissolving bone matrix. The osteoblasts built it back up again.) When osteoclasts are very busy, a lot of calcium and other minerals are released into the blood. By slowing down the osteoclasts, calcitonin helps to reduce levels of calcium in the blood.

Thyroid tissue is filled with tiny **follicles** (sacs) that make T3 and T4. (Sometimes you will see these written as T_3 and T_4.) These molecules are very similar, and body cells can often turn T4 into T3. They are both made of the amino acid tyrosine with some iodine atoms attached. The membranes of the follicle cells have special portals that bring in iodine ions from the blood. If a person's diet doesn't contain enough iodine, the follicle cells will not be able to manufacture the T3 and T4 molecules, which will create a deficiency disease. (Iodine is very compatible with salt molecules, so you will often see "iodized salt" sold as table salt.) After the T3 and T4 molecules are made, they are exported outside the cell into the blood stream, where most of them will latch onto a globulin protein taxi. Over 99% of these thyroid hormones will attach to a taxi, and only a very small amount will end up floating around freely in the blood. (The "free" T3 and T4 molecules are usually what doctors test for when they check your thyroid hormones.) There is a purpose behind having so many of them bound to a taxi. As long as they are bound, they can't do anything, which is good because you don't want too much of these hormones in your blood. If the body suddenly needs more of either T3 or T4, they can be taken off the taxis and used. So the globulins act as a storage mechanism for these hormones, keeping a steady supply handy in a safe way.

One of these protein taxis, TTR, or **transthyretin** is worth mentioning because it might play a role in Alzheimer's Disease. The "retin" part is for "retinol" which is basically vitamin A. This taxi also appears to be able to carry waste proteins called amyloids. If the TTR protein gets folded wrong when it is made, it won't be able to carry away these waste proteins and they will build up in body tissues. In the brain they may contribute to the symptoms of Alzheimer's. We don't completely understand this disease yet, and there may be other causes, but studies have shown that people with Alzheimer's have reduced levels of TTR in both their blood and their cerebrospinal fluid.

Many body cells use T3 and T4. The list of what these hormones help with is quite impressive, and includes helping children to grow, regulating our heart rate and breathing rate, making our appetite function properly, helping our intestines to absorb nutrients, increasing the action of the mitochondria (which helps to generate body heat), regulating the sleep cycle, regulating the amount of cholesterol in the blood, promoting normal function of the reproductive organs, and telling our bodies whether to store fat or burn it for energy.

The **parathyroid** is actually a collection of "dots" on either side of the thyroid, two on each side. The function of the parathyroid is to make the antagonist to calcitonin, which is simply called **PTH (parathyroid hormone)**. This hormone stimulates those osteoclast cells and tells them to be more active in their destruction of bone tissue. This will release more calcium into the blood.

Just like muscles work in pairs, with an agonist and an antagonist, hormones also work in pairs. For most body process, you need both a stimulator and an inhibitor. Another agonist/antagonist pair that we know already is insulin/glucagon.

Calcitonin and parathyroid hormone work opposite each other, to keep the level of calcium in the blood steady. This process is classified as a negative feedback loop. This means that when the level of something drops, this low level triggers a process that will bring it back up. A high level of calcium will be detected by the thyroid and will trigger the release of calcitonin which will cause the bones to absorb more calcium from the blood. When the level of calcium falls, this triggers the release of PTH from the parathyroid, and osteoclasts begin to break down bone and release calcium into the blood. PTH not only causes osteoclasts to work harder, it also causes the kidneys to reabsorb calcium that might otherwise have gone out as waste, and causes the intestines to absorb more calcium from your food. Vitamin D plays a role in calcium absorption so the kidneys also activate vitamin D.

As an interesting side note, the thyroid begins to develop in an embryo when it is only 3-4 weeks old, before the face has really developed, and before the fingers and toes. By the time the fetus is 5 months old, its hypothalamus, pituitary and thyroid are completely functional and levels of all these hormones can be measured. T3 and T4 play an important role in fetal brain development, and also continuing brain development during infancy.

The thyroid and parathyroid control the level of calcium ions in the blood.
The thyroid also makes hormones that affect the health of all body cells.

The thyroid makes:

1)_____
which inhibits osteoclasts, stopping
them from dissolving bone

2) _____

3) _____

Both of these affect many body cells
and help with normal functioning of:

The parathyroid makes: _____

This hormone acts opposite calcitonin
and raises blood calcium levels by:

1)

2)

Over 99% of T3 and T4 ride around in globulin taxis. While bound to a taxi, they are inactive. This provides safe storage.

NEGATIVE FEEDBACK LOOPS are the body's way of maintaining homeostatis

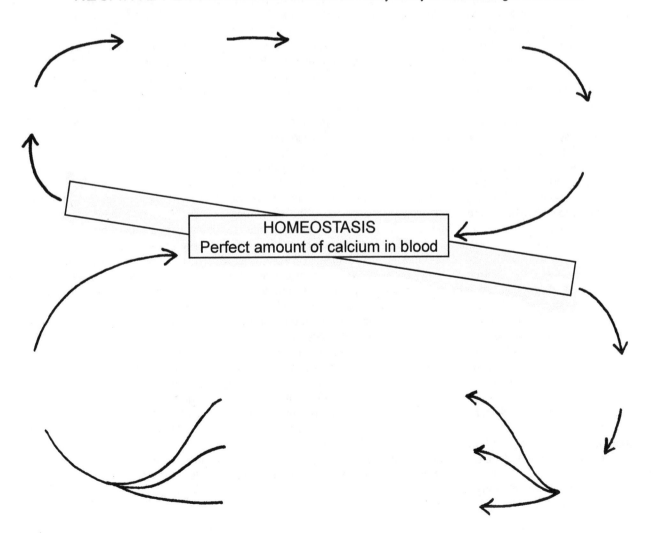

HOMEOSTASIS
Perfect amount of calcium in blood

83: ADRENAL GLANDS

The adrenal glands are all about dealing with stress. The two sections of the adrenals, the medulla (inside) and the cortex (outside) deal with immediate stress (medulla) and long-term stress. (cortex)

The **adrenal medulla** responds to immediate stress, sometimes so immediate that the response has to happen within seconds. Think of emergency situations where you need to act FAST. It could be something simple and not really all that dangerous, or it could be a situation that is life-threatening. Let's think of something very unlikely, such as being face to face with a tiger. When facing a tiger, what does your body need to do? Be fast enough to run away? Be strong enough to kill the tiger? Your options probably all center on thinking fast, running fast, and being extra strong. In a Marvel comic world, we'd be able to tap into our super powers. Since that's not actually an option, our body does the best that it can at being super. Our heart starts beating faster and stronger, our muscles stiffen, our jaw tightens, our pupils get larger, and our brain begins thinking quickly. We are ready to either fight or take flight. The adrenal medulla (along with some special nerve cells) gives us this **"fight or flight"** response.

The medulla is wired directly to the hypothalamus in order to achieve a very fast response. In only a second or two, our frontal lobe processes what we are seeing and hearing, figures out there is danger, sends a signal to the amygdala, which then sends a signal to the hypothalamus. The hypothalamus tells the medulla to release a hormone you've certainly heard of: **adrenaline**. What you may not have heard of is the other name for adrenaline: **epinephrine** *(ep-in-EFF-rin)*. (Unfortunately, you have to learn both words because they get used equally.) The adrenal glands have a large capillary network inside of them, and they are located in an area where the blood flows along pretty quickly, so only seconds after the adrenaline is released into the blood, it is reaching body cells and causing its effects. Adrenaline speeds up the heart, constricts blood vessels in order to increase blood pressure, dilates the pupils, relaxes the bronchioles in the lungs for better oxygen intake, prepares blood for fast clotting, and causes the liver to break down glycogen, putting lots of glucose into the blood. As the seconds tick by, adrenaline begins to work on other organs: Your salivary glands stop making saliva, your digestive system stops doing peristalsis, and your urinary system stops producing urine.

Along with adrenaline (epinephrine), another "sister" hormone is produced, **noradrenaline (norepinephrine)**. "Nor" means "alongside." So these two chemicals are produced alongside each other and work together doing many of the same things. The difference between these two hormones lies in the details of what kind of membrane receptors they bind to. This level of detail is outside the scope of this curriculum. All you need to know is that adrenaline and noradrenaline (epinephrine and norepinephrine) act to keep the body on red alert, ready for action.

The **adrenal cortex** also deals with stress, but not immediate stress. The endocrine system can take its time getting the signals to the cortex. This kind of stress would include things like having a week of final exams, starting a new job, being sick, moving to a new city, coping with divorce, cleaning up after a hurricane, and so on. In times of long-term stress, the hypothalamus secretes a hormone called **ACTH** (adrenalcorticotropin), which tells the adrenal cortex to secrete **glucocorticoid** hormones. This slower, indirect method of cooperation between the brain and the adrenals is often called the **HPA axis** (**H**ypothalamus, **P**ituitary, **A**drenal).

Cortisol (also called hydrocortisone) is the most important glucocorticoid. Just about every cell in the body has receptors for cortisol. Since it is a steroid hormone, it goes through the plasma membrane and enters the cell. There it finds its receptor and binds to it. What happens will depend on what kind of cell it is. This gets into a lot of extremely complicated biochemistry. In the end, all you need to know is that cortisol's main actions are to reduce inflammation and to raise the level of glucose in the blood. The cortisol molecule is very easy to make, so it is an anti-inflammatory drug used very widely by doctors in a variety of diseases. It can even be injected into a joint to reduce inflammation. The downside to using cortisol is that if you take it for a long time, the adrenals stop producing their own supply, and it can be difficult to get them to function again once you stop taking the cortisol.

But wait, there's more! The adrenal cortex does some other things, too. There are actually three separate "zones" in the cortex. The middle zone makes the cortisol we just discussed. The outer zone makes **mineralocorticoids**. As the name suggests, these hormones have something to do with minerals (meaning sodium, potassium, etc.) The most important hormone is this category is aldosterone.

Aldosterone is essential for sodium balance in the kidneys, salivary glands, sweat glands and colon. These are all places where water can be released or recycled. As we learned in the lesson on the kidneys, water molecules follow salt molecules. By the rules of osmosis, water will always go to places where there is a higher concentration of salts, as if it is trying to even things out and make the salty areas less salty. Control where the salt goes, and you control where the water molecules go. Aldosterone's action is to stimulate the sodium/potassium pumps embedded in the plasma membranes of the cells in the distal tubes and collecting tubes of the nephrons in the kidneys. This means that water will be recycled instead of being turned into urine.

Aldosterone is part of a negative feedback loop that starts with the kidneys sensing low blood pressure. The kidneys then produce **Renin**, a hormone whose job is to activate **Angiotensin**. When you see "angio" think "blood vessels." And "tensin" sounds like "tense." Angiotensin tells the tiny muscles around the vessels to contract and make the vessels smaller. This helps to raise blood pressure. Angiotensin also tells the adrenal cortex to make aldosterone, which we have just discussed. Aldosterone makes the kidneys reabsorb salt and water, and produce less urine. Less urine means more blood which means higher pressure.

When blood pressure gets too high, this is sensed by cells in the atria of the heart, and they secrete **ANH, Atrial Natriuretic Hormone**. ("Natri" refers to sodium, and "uretic" can be thought of as "out with the urine.") ANH inhibits the secretion of Renin and also increases the excretion of salts (sodium) in the urine. Water molecules follow the salts, and thus more urine is produced. Producing more urine helps to lower blood volume, which lowers blood pressure.

Lastly, the inner part of the cortex secretes steroid hormones that can be turned into other hormones. Primarily, it makes **DHEA**, which can be turned into either **testosterone** or **estrogen**. The reproductive organs also secrete these hormones, so we will wait to discuss these hormones until we get to those lessons.

ADRENAL GLANDS

The adrenal glands are under the control of the hypothalamus, both directly and indirectly.

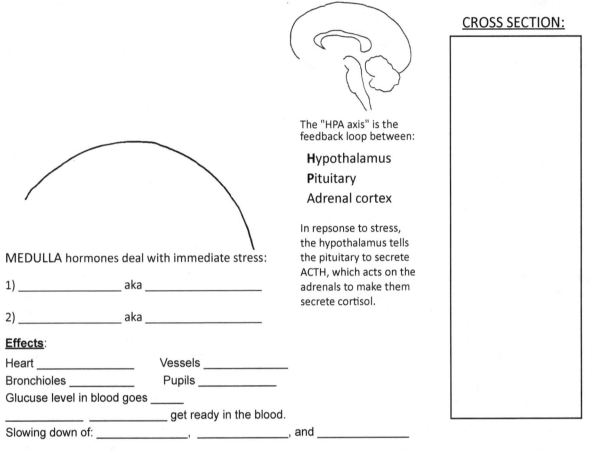

CROSS SECTION:

The "HPA axis" is the feedback loop between:

Hypothalamus
Pituitary
Adrenal cortex

In repsonse to stress, the hypothalamus tells the pituitary to secrete ACTH, which acts on the adrenals to make them secrete cortisol.

MEDULLA hormones deal with immediate stress:

1) _____ aka _____

2) _____ aka _____

Effects:

Heart _____   Vessels _____
Bronchioles _____   Pupils _____
Glucuse level in blood goes _____
_____ _____ get ready in the blood.
Slowing down of: _____, _____, and _____

RECOVERY: In 30- 60 minutes the body will have gotten rid of all the adrenaline and noradrenaline molecules.

ANOTHER NEGATIVE FEEDBACK LOOP:

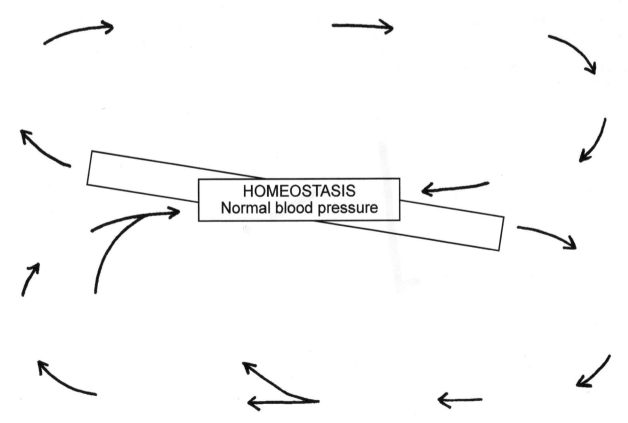

HOMEOSTASIS
Normal blood pressure

84: AUTONOMIC NERVOUS SYSTEM

The autonomic nervous system is part of the peripheral nervous system, meaning it lies outside of the central nervous system (brain and spinal cord). The word "autonomic" might remind you of the word "automatic," and this is a correct association to make. The autonomic nervous system (ANS) does function automatically, without us giving any conscious thought to it. The main controller of the ANS is a part we just looked at just recently: the hypothalamus. The ANS includes all the millions of tiny nerves that go into smooth muscles, cardiac muscles, and glands. The ANS is the controller of bodily functions such as heart rate, digestion, breathing, sweating, shivering, salivation, urination, and reflexes such as sneezing, coughing and vomiting. Although this system is automatic, we saw in the last lesson that it does still have a connection to the conscious brain and the voluntary systems.

The autonomic nervous system has basically two parts, although some people like to add a third. Traditionally, the ANS was always divided into the **sympathetic system**, and the **parasympathetic system**. Recently, some like to classify the nerves going to the digestive system as the **enteric system**. We'll stick to tradition and just talk about the sympathetic and parasympathetic. Also, this will not be an exhaustive list of everything each system does, just a mention of the more obvious functions.

The sympathetic system is the part that does the "fight or flight" response we learned about in the last lesson. The sympathetic system gears you up and makes you ready for action. Your heart rate and breathing rate increase, your pupils dilate, your liver releases glucose, and all your digestive processes from salivation to urination are inhibited. We discussed this in lesson 83. All the nerves of the sympathetic system go out from the thoracic and lumbar areas of the spine (the part inside the thoracic and lumbar vertebrae).

The parasympathetic system does just the opposite. The parasympathetic calms you down. It slows your heart rate, calms your breathing, constricts your pupils, and stimulates your digestive system. The parasympathetic is your "rest and digest" system. All the nerves of the parasympathetic system go out either from the top of the spine, or from the very bottom (sacral area). The ones at the top are part of a group of nerves we call the **cranial nerves** (because they come out of the bottom of the cranium). A very special cranial nerve called the **vagus nerve** drops down and branches off into all the organs. We saw the vagus nerve going to the stomach in lesson 70. When a nerve goes into an organ we say that it "innervates" it. (The "nerve" goes "in.")

Each impulse sent out by the ANS requires only two neurons. The place where they meet is called the **ganglion**. (In our drawing we keep track of which way the nerve is pointing by drawing the cell body as a dot and the axon terminal as a tiny "C" curve.) In the sympathetic system, the nerves go from the spinal cord into a nerve ganglion cord that runs parallel to the spine. Nerves that go from the CNS to a ganglion are called the "pre-synaptic" nerves. Nerves that run from the ganglion to the organ are called the "post-synaptic" nerves. The post-synaptic nerves in the parasympathetic system tend to be very short, and can even be entirely inside the organ. In the sympathetic system the post-synaptic nerves are much longer. (NOTE: None of distances between the neurons in our drawing are very accurate. This is just a rough idea of how they are arranged.)

Although we think of these systems as turning on and off as we respond to strong stimuli in our environment, the truth is that both systems are constantly active at a low level. The digestive organs need to be stimulated. Dilation and constriction of the pupils goes on all the time as our eyes adjust to the level of light around us. Our heart may need to be stronger and faster just because we get up out of a chair and begin to walk, not because we are facing danger. Sometimes the balance changes dramatically, such as when we face danger, or when we are trying to fall asleep, but both are always functioning to some degree.

The autonomic nervous system (ANS) is part of the peripheral nervous system (PNS) and functions automatically.
Each stimulus travels a route that is made of only 2 neurons.

SYMPATHETIC   **PARASYMPATHETIC**

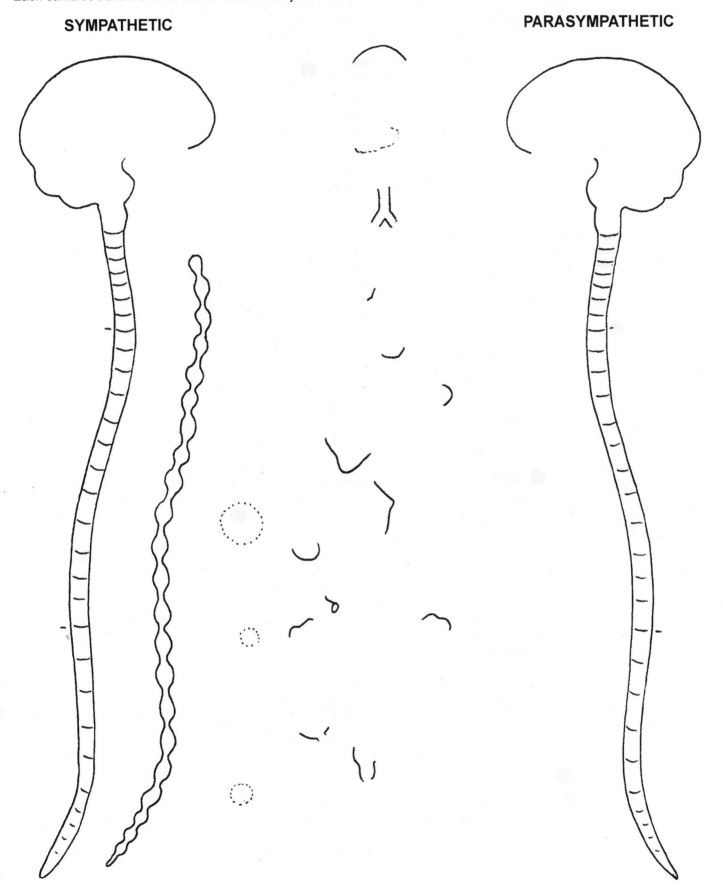

85: MALE REPRODUCTIVE SYSTEM

Mammalian reproductive systems are marvels of engineering. In this lesson we will see how each part is located in exactly the right place, maximizing its efficiency. In other words, if these parts were in a different order, the system would not work. Speaking of efficiency, the male reproductive system "borrows" some parts from the urinary system. Why have an entirely separate system when you can reuse the same parts for a different purpose? Here are the parts classified by which system(s) they belong to:

Urinary only: bladder, ureter   Both urinary and reproductive: urethra, penis
Reproductive only: testes, epididymis, vas deferens, prostate, seminal vesicle, bulbourethral gland (a.k.a. Cowper's gland)

Let's start with the **testes**. (Singular is "testis.") The testis is the oval-shaped thing inside the **testicle**. The word testicle generally refers to the testis and also the things that surround it, like the epididymis and the outer covering of skin. (The outer covering is called the **scrotum**. This is a Latin word, possibly meaning "a leather bag." The word scrotum usually refers to both testicles.) The testis is where the sperm are produced. Remember that we started out saying that everything is the right location? Turns out there is a reason for the testis being outside of the body, instead of hidden inside like the female ovaries. Sperm need to be kept a little cooler than body temperature. At a certain point in fetal development, the testes begin to move to the outside of the body. If this process does not happen, babies will be born with "undescended testicles" and will need to have surgery to bring them outside. Without this surgery, the boy will likely not be able to father children when he grows up because the testes will be too warm all the time.

Inside the testis we find lobes that are filled with coiled tubes. These tubes are called **seminiferous tubules.** If we look at a cross section of these tubes, we find an organized arrangement of cells. The circle around the outside of the tube is the basal lamina, a tissue we've seen in other places, associated with epithelial cells. Just inside this circle, around the outer edge, we find stem cells called **spermatogonia.** Amazingly, these cells came into being when the fetus was just 4 weeks old! They migrated to these tubes once the tubes were formed. The spermatogonia divide constantly, starting at puberty and continuing on throughout the man's lifetime. One "daughter" cell will become the new stem cell, and the other will become a **primary spermatocyte**. The primary spermatocyte then goes through meiosis, meaning that its two "daughter" cells will have only half the original amount of DNA. Normal cells have two sets of chromosomes and are called 2N. After the first step of meiosis, each cell will have only one set of chromosomes, N. (A cell with N chromosomes can also be called "haploid.") These N cells are called **secondary spermatocytes**. Each of these N cells splits in half to produce a **spermatid**. This division process is more like mitosis because the chromosome count stays the same, and the spermatids are still N. The spermatids will then turn into **sperm** (also known as **spermatozoa**), as they develop acrosomes and flagella. (Thousands are produced every second!) This maturation process takes over two months from start to finish. And, even more amazing, the sperm do not become fully **motile** (able to move) until they are already inside a female reproductive system. The female system provides the right environment (the right chemistry) for them to achieve maximum **motility** (motion).

Between all the seminiferous tubules, we find **Leydig cells**, which release the hormone **testosterone**. Testosterone controls masculine body features (large muscles, broad shoulders, hair on face, deep voice) as well as the production of sperm cells. Leydig cells are stimulated into action by the hormone **LH**, secreted by pituitary. The pituitary, in turn, is stimulated by the hypothalamus.

There are supporting cells (**Sertoli cells**) in the seminiferous tubules that nourish the spermatocytes as they develop. They pick up food and oxygen from the circulatory system and pass it along to the spermatogonia and spermatocytes. (This might remind us of the astrocytes that nourish the neurons in the brain, or the pigmented epithelial cells in the retina that nourish the rod and cone cells.) Sertoli cells are controlled by the hormone **FSH**, secreted by the pituitary. (And the pituitary is under the control of the hypothalamus.)

The sperm don't swim in these tubes. A tiny amount of liquid flushes them out into the next structure, the **epididymis**. This is a very long tube that is coiled up around, and on top of, the testis. The epididymis is where the sperm are stored. They don't swim around here, either. They are gradually swept along by ciliated cells that line the tubes. (We saw similar ciliated cells in the trachea sweeping mucus along.) If the sperm are not used, they can be reabsorbed and all their proteins broken down and recycled, just like your body recycles all other dead cells.

From the epididymis the sperm then move into a very long tube called the **vas deferens**. If you follow this tube up and over the bladder you will see that it eventually merges with the **urethra**. This merger happens inside the **prostate gland**. The prostate is a combination of glands and muscles, and has several jobs. The glands make about 30% of the transport fluid, or **semen**, that will carry all the sperm (millions of them). The glands add some interesting things to the semen fluid, like zinc, citric acid, and a protein called Prostate Specific Antigen (PSA) which will help keep the liquid from becoming too thick when other glands add their secretions. The muscles in the prostate do two things: they squeeze shut the entrance to the bladder so that no urine can leak out while sperm are passing through the urethra, and they help to propel the sperm through the urethra and into the penis.

Also at this merger point (where the vas deferens meets the urethra) we see a gland called the **seminal gland**. This gland produces about 60 percent of the semen fluid, and it also adds to the mixture some fructose sugar (energy for the sperm cells), several enzymes, vitamin C, and a hormone called prostaglandin that will help protect the sperm from the female's immune system cells.

Below the prostate gland we see the **bulbourethral gland** (a.k.a Cowper's gland). Its job is to produce a fluid that will go through the urethra before the sperm-containing semen gets there. This fluid will neutralize any acidity left behind by urine, and will also provide mucus to help the sperm have a smooth ride through the urethra. This gland is just where it needs to be to do its job.

For the semen to be able to be released right near the female uterus, the penis needs to change its shape. This is accomplished by two areas of tissue (best seen in the cross section view) called the **corpus cavernosum**. The autonomic nervous system can cause a release of nitric oxide (NO) in this tissue, which tells the capillaries to expand and allow extra blood to come in. As this area swells, the drainage veins become blocked, so the blood can't get out. After the semen leaves the body through a process called **ejaculation**, the drainage veins open again and the penis goes back to its normal shape. Anywhere from 50 million to 1 billion sperm can be released in one ejaculation.

INTERESTING FACT: The testes produce *thousands* of sperm *per second*.

Hypothalamus has ultimate control over the development and functioning of this system. It tells the pituitary to secrete:

1) _____
2) _____

SEMEN is complicated!
_____ from prostate
_____ from sem. vesicle
_____ from bulbourethral

Contains:

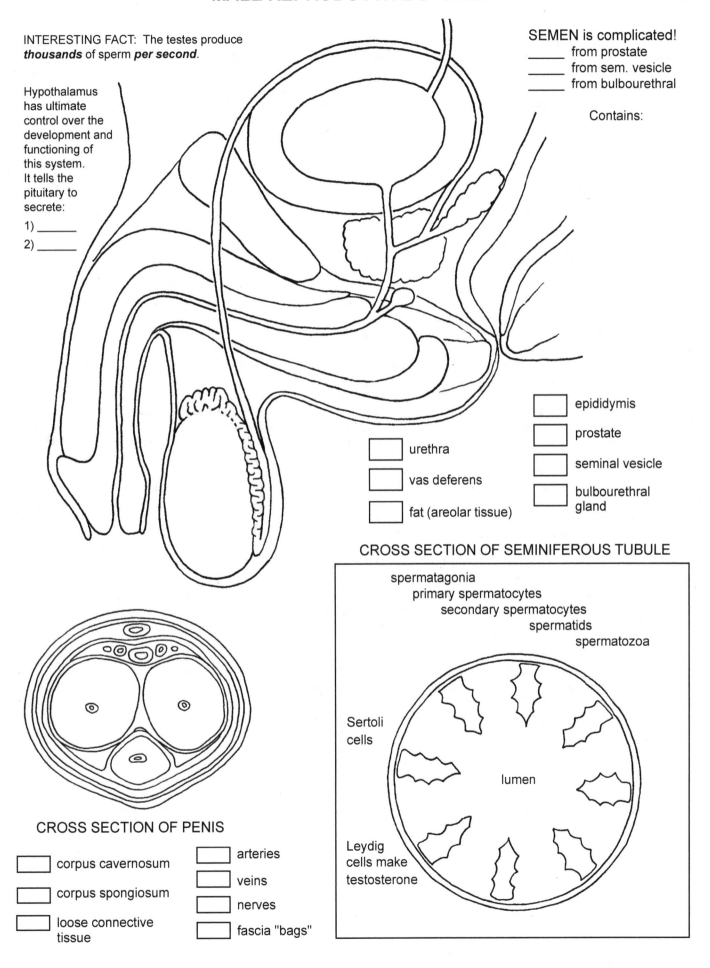

_____ urethra

_____ vas deferens

_____ fat (areolar tissue)

_____ epididymis

_____ prostate

_____ seminal vesicle

_____ bulbourethral gland

CROSS SECTION OF SEMINIFEROUS TUBULE

spermatagonia
primary spermatocytes
secondary spermatocytes
spermatids
spermatozoa

Sertoli cells

Leydig cells make testosterone

lumen

CROSS SECTION OF PENIS

_____ corpus cavernosum

_____ corpus spongiosum

_____ loose connective tissue

_____ arteries

_____ veins

_____ nerves

_____ fascia "bags"

86: FEMALE REPRODUCTIVE SYSTEM

Though in many ways the female reproductive system is quite different from the male system, we will find a few ways in which they are similar. We will see some of the same hormones operating, and we will see a location where we find similar tissue types.

In the female system, the urinary tract and the reproductive tract are completely separate. The female urethra is very short, only a few centimeters. The bladder is right under the uterus, so when the bladder is full, it pushes the uterus upward. This is handy for ultrasounds when you want to image the uterus. Before an ultrasound, the patient must drink a lot of fluid so that the bladder is as full as possible and will raise the uterus to a more vertical position, giving the technician a better view in the ultrasound image.

The **uterus** (where the baby grows) is a very muscular organ. The muscles must be strong enough to push a baby out during childbirth. The uterus has a delicate lining called the **endometrium**. It is made of tissue that is full of capillaries. The bottom of the uterus is called the **cervix**. The **ovary,** where the eggs (**oocytes**) develop, is outside of the uterus but is anchored to it by a ligament. When eggs are released from the ovaries, they are picked up by the **fimbriae** (finger-like things) at the end of the **oviducts** (which are also called **fallopian tubes**). We saw in module 2 what happens to the egg cell as it travels down the oviducts and into the uterus. If there is sperm present and the egg is fertilized, the tiny embryo will try to implant itself in the wall of the uterus. The tube that leads from the uterus to the outside world is called the **vagina.** (The vagina is often called the **birth canal.**)

The lower parts of the female system are the **labia minora** and **labia majora** (basically "flaps" that cover the openings to the urethra and vagina), and the **clitoris**. The clitoris has an interesting embryonic history, in that the same type of embryonic tissue can produce either a penis or a clitoris. Because of this, we find very similar tissue types in both structures (tissue that can enlarge and fill with blood). The clitoris is the primary organ of sexual pleasure for women, although the vagina does play a minor role, as well.

When females are born they have about a million oocytes ready and waiting in the ovaries. Nothing happens until puberty, when the girl's body starts to produce more estrogen and she begins to develop into a woman. By this time the number of oocytes has dropped to about half a million. (Unused eggs just disintegrate and their proteins can be recycled.)

The oocytes are each wrapped in a layer of protective cells called a **follicle**. Once the young woman begins to have her monthly cycle, each month one of these **primary follicles** will begin to grow and to secrete estrogen and progesterone. First it becomes a **secondary follicle**, then it turns into a **vesicular follicle** that contains a vesicle filled with fluid. This fluid will help the next process take place. At about day 14 of the 28-day cycle, the vesicular follicle touches the wall of the ovary and it bursts out, expelling the oocyte. This is called **ovulation**. The follicle then heals itself and turns into a **corpus luteum**. ("Luteum" means "yellow.") The corpus luteum will continue to produce hormones for a few days, but then will quickly start to get smaller and smaller, and eventually disappear and disintegrate.

If pregnancy were to occur, a hormone produced by the embryo (**HCG**) would prolong the life of the corpus luteum so it would go on producing its hormones, especially progesterone. ("Pro" means "for," and "gest" means "pregnancy.") As long as the corpus luteum is active and producing progesterone, the uterus will not shed its lining. This is important because the embryo needs to connect to the capillaries in the thick endometrial lining. (Eventually, the placenta will take over this role of producing hormones and the corpus luteum will disappear.)

If there is not an embryo present, the corpus luteum will do its normal routine and begin to disintegrate. As it does so, the hormone levels will drop and this will initiate the breakdown of the endometrium. The dying endometrial cells produce an enzyme that will prevent the blood from clotting, so that it will flow out smoothly and easily. Occasionally, if the bleeding is very fast and heavy, the cells will not be able to keep up with the need for the anti-clotting chemicals, and the woman will notice clots in the flow.

There is a feedback system between the ovaries and the pituitary. When the pituitary senses the rising levels of estrogen and progesterone, it will stop making so much FSH and LH. Then, when the levels become low, it will start producing more FSH and LH.

Women's bodies also produce a small amount of testosterone. Male bodies produce a small amount of estrogen. So everyone produces everything, but with gender-specific quantities. When women take a lot of testosterone, or when men take a lot of estrogen, their bodies will slowly begin to change and become more like the opposite gender, but the anatomical parts will not change (although they might be reduced in size a bit, or stop working quite so well). However, there will also be long term damage done to other body systems such as the cardiovascular system (heart and blood vessels), as hormones do affect other body tissues, not just reproductive tissues. After a number of years, these people will start to see many negative effects on their overall health.

When women get into their 40s or 50s, a "clock" in their brain decides that it time to stop ovulating and having menstrual periods. Childbirth is a rigorous experience and older women will experience far more problems than younger women. Also, the eggs have aged for decades and chances go up for having a defective egg be fertilized and produce a baby with severe, or fatal, problems. The pituitary and hypothalamus stop producing the hormones that cause the monthly cycle. Ovulation stops and the woman can no longer become pregnant. This is called menopause.

FEMALE REPRODUCTIVE SYSTEM

CROSS SECTION SHOWING ORGANS

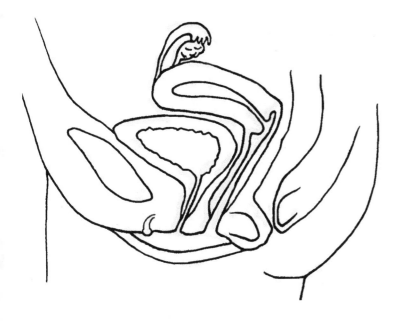

FRONT VIEW: UTERUS, OVARIES

(1) **Primary follicles**: contain an oocyte, and they produce estrogen and progesterone

(2) Primary follicles get larger and are called **secondary follicles.**

(3) A secondary follicle turns into a **vesicular follicle** when it becomes filled with fluid and touches the ovary wall.

(4) The follicle bursts and the oocyte (egg) is released from the ovary. (**ovulation**)

(5) The folllicle turns into a **corpus luteum.** which makes estrogen and progesterone ("pro" means "for," and "gest" means "pregnancy.")

(6) The corpus luteum disintegrates.

☐ = FSH   ☐ = LH   (pituitary)   ☐ = estrogen   ☐ = progesterone (ovaries)

I N O V A R Y	
I N U T E R U S	

Samples of final drawings

(Full size samples are provided on each lesson page on the website. These are just for quick reference.)

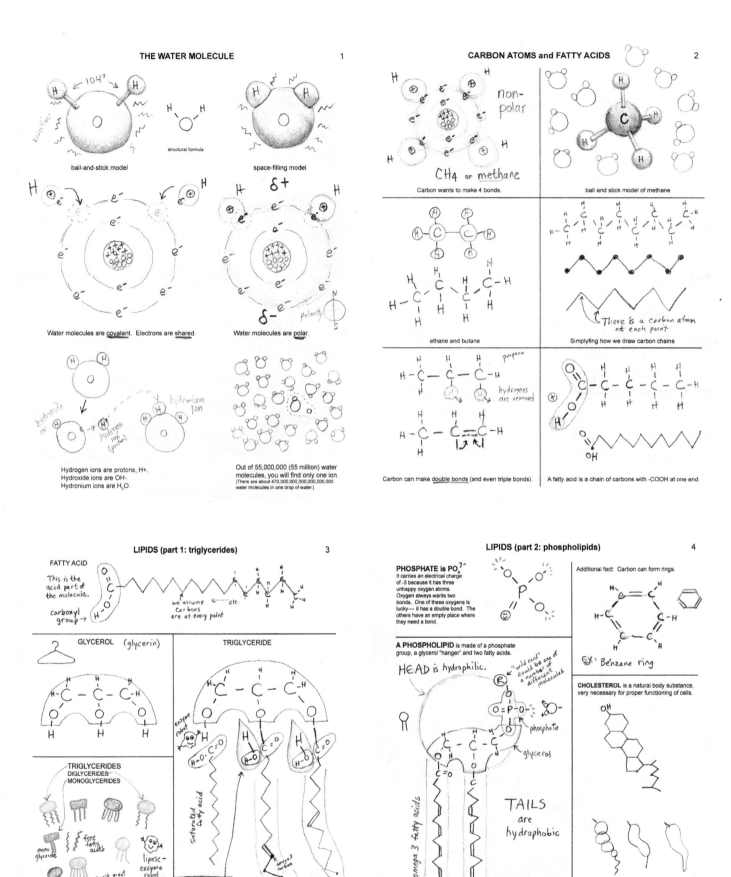

THE WATER MOLECULE

← 104°

structural formula

ball-and-stick model

space-filling model

Water molecules are covalent. Electrons are shared.

δ+

δ- polarity

Water molecules are polar.

hydroxide ion

hydronium ion

hydrogen ion (proton)

Hydrogen ions are protons, H+.
Hydroxide ions are OH-.
Hydronium ions are H₃O.

Out of 55,000,000 (55 million) water molecules, you will find only one ion.
(There are about 470,000,000,000,000,000,000,000 water molecules in one drop of water.)

CARBON ATOMS and FATTY ACIDS   2

non-polar

CH4 or methane

Carbon wants to make 4 bonds.

ball and stick model of methane

ethane and butane

There is a carbon atom at each point

Simplifying how we draw carbon chains

propane

hydrogens are removed

Carbon can make double bonds (and even triple bonds).

A fatty acid is a chain of carbons with -COOH at one end.

LIPIDS (part 1: triglycerides)   3

FATTY ACID

This is the acid part of the molecule.

carboxyl group →

we assume carbons are at every point

etc.

GLYCEROL (glycerin)

TRIGLYCERIDE

enzyme robot

Saturated fatty acid

omega 3 carbon

dehydration synthesis

omega 3 fatty acid

omega 6 fatty acid

TRIGLYCERIDES
DIGLYCERIDES
MONOGLYCERIDES

mono glyceride

free fatty acids

You don't want too many of these in your blood - only in fat cells

lipase - enzyme robot for taking fatty acids off glycerol

What kind of fatty acids might be attached to glycerol?

Lauric acid has only 12 carbon atoms. It is found in coconuts and palms. Palmitic acid has 16 carbon atoms.
Commonly found in our body fats

These fatty acids all have 18 carbons:
1) Stearic acid has 0 double bonds and is found in abundance in animal fats (esp. marrow).
2) Oleic acid has 1 double bond and is found in abundance in olive oil.
3) Linoleic acid has 2 double bonds and is found in abundance in flax and olive oil.
4) α-linolenic (ALA) has 3 double bonds and is found in seeds and nuts.

Our bodies can make some fatty acids. Others must come from our diet and are called essential fatty acids.

LIPIDS (part 2: phospholipids)   4

PHOSPHATE is PO₄³⁻
It carries an electrical charge of -3 because it has three unhappy oxygen atoms. Oxygen always wants two bonds. One of these oxygens is lucky— it has a double bond. The others have an empty place where they need a bond.

A PHOSPHOLIPID is made of a phosphate group, a glycerol "hanger" and two fatty acids.

HEAD is hydrophilic.

"wild card" could be one of a number of different molecules

O=P-O

phosphate

glycerol

omega 3 fatty acids

TAILS are hydrophobic

EPA   DHA
20 carbons   22 carbons
5 double bonds 6 double bonds

Additional fact: Carbon can form rings.

EX: Benzene ring

CHOLESTEROL is a natural body substance, very necessary for proper functioning of cells.

EPA   DHA

In a polar substance (like water) phospholipids form a bilayer with the hydrophobic tails pointing to the inside.

Phospholipids will form a sphere. This is the basic structure of all membranes.

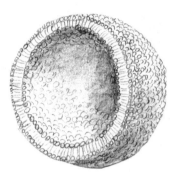

An empty phospholipid sphere is called a liposome. "Lipo" is Greek for "fat," and "soma" is Greek for "body."

Liposomes form basic cell parts such as:

VACUOLES: empty "bubbles" filled with water or air

VESICLES: "storage bags" used to transport things around the cell

minuhes chemical for use by ner cells

LYSOSOMES: vesicles that are acidid inside and contain digestive enyzmes

proton:
protein
enzyl

"Lys" means "to dissolve or break up"

The phospholipid bilayer membrane separates the inside from the outside. However, cells need to bring things in and send things out. There are many methods for getting things in and out, depending on the size and the chemical properties of those things. Some ways require energy and some don't.

PASSIVE TRANSPORT Does not require energy.	**ACTIVE TRANSPORT** Uses energy (often from ATP or NADH)

1) SIMPLE DIFFUSION

Very small, non-polar molecules, such as oxygen and carbon dioxide can go right through the membrane. Small lipids can also diffuse because they get along so well with the fatty acid tails in the middle layer.

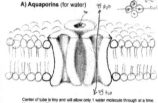

1) PUMPS

Use energy to push molecules across the membrane against the "concentration gradient."
Types of energy: ATP, NADH carrying high-energy electrons

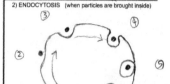

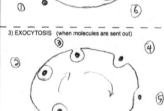

2) FACILITATED DIFFUSION (facil = to make easier)

Molecules that are polar or electrically charged can't use simple diffusion; they must use channel proteins.

A) Aquaporins (for water)

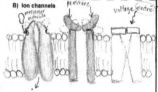

Center of tube is tiny and will allow only 1 water molecule through at a time. However, 1 million water molecules get through every second!

2) ENDOCYTOSIS (when particles are brought inside)

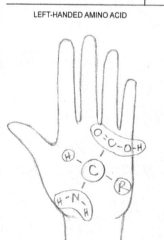

B) Ion channels

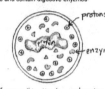

Types of gates: light, temperature, pressure, voltage, binding of messenger molecules

3) EXOCYTOSIS (when molecules are sent out)

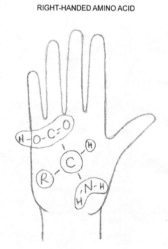

The phospholipid bilayer membrane is a busy place. Not only does it have many channels and pumps, it also has lots of gadgets for sending and receiving messages. Also present are "ID tags" called MHC 1. Every cell in the body has these MHC 1 tags on them so that immune system cells know they belong to the body.

This model is called "fluid" because everything can move around. The word "mosaic" is an art term used to describe a picture made from many small colored tiles. Recent research has revealed that some things in the membrane stay in place more than expected, but this model still seems to be valid, nonetheless.

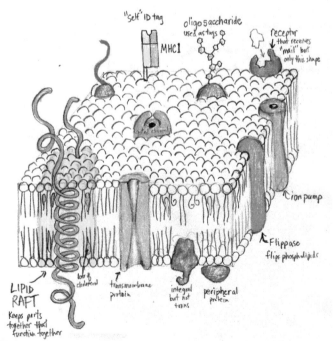

Vocab to know:
1) **integral membrane proteins** are attached to the membrane. They can be on one side, or all the way through.
2) **Transmembrane proteins** go all the way through from one side to the other.
3) There are two words that both means "sugar." **Glyco** is from Greek, and **sacchar** is from Latin.
4) **Oligo** means "few."

The biggest difference between proteins and lipids is the presence of nitrogen, N. Nitrogen is number 7 on the Periodic Table, so it has 5 electrons in its outer shell. This means it wants to make 3 bonds (5+3=8).

·The nitrogen atom:

Nitrogen = N

The basic unit of all proteins is the AMINO ACID:

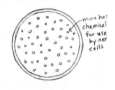

Rest of the molecule

LEFT-HANDED AMINO ACID	RIGHT-HANDED AMINO ACID

"Handedness" is called **chirality.** "Chiro" is Greek for "hand."
Living things are made of left-handed amino acids only!

A protein is a chain of amino acids. Amino acids are attached by a *peptide bond*. When you see the root word "pep," think "protein." (Ex: the enzyme "pepsin" digests protein)

To form a peptide bond, we can use the same attachment method that we used for making triglycerides. We will use *dehydration synthesis*.

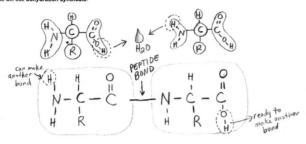

amino acid	dipeptide	tripeptide	polypeptide (poly=many)
Gly	Val—Lys	Pro—Cys—Tyr	P—S—E—K—L—V etc.

They also have 1 letter abbreviations.

There are only 20 amino acids (all left-handed, of course). The R group determines the chemical characteristics. R groups can be hydrophilic or hydrophobic (or neutral), and they can carry an electrical charge (or not).

GLYCINE
R group is just an H. Glycine is smallest amino.

VALINE is hydrophobic
Good for altering shape of polypeptides. methyl group

LYSINE is hydrophilic and ionized
Lysine is good for molecular switches.

GLUTAMIC ACID (glutamate)
Na+ adding a sodium will make MSG
or glutamate
Works to activate ion channels in nerve cells.
• MSG is D form - oops! also it is not connected to other aminos.

CYSTEINE has a sulfur
Sulfur is good for building cross links. For ex: vulcanized rubber
Example in body - insulin
Sulfurs make crosslinks which make molecule more stable.

The PRIMARY structure of a protein is the sequence of amino acids in the chain.

Ala—Gly—Lys—Val—Pro—Glu—Ser—Arg—A—D—W—Q—E—L

The SECONDARY structure of a protein is either an alpha helix or a beta sheet.

HYDROGEN BONDING

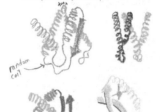
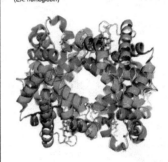

α helix   β sheet

The TERTIARY structure of a protein is the way it folds up into a unique 3D shape.

random coil

zinc finger

Hydrophobic aminos try to hide in the center. Sometimes atoms of other elements (like zinc) are incorporated into the shape.

The QUATERNARY structure is when two or more proteins bond together to form a final protein shape. (EX: hemoglobin)

Proteins are the building blocks of numerous "gadgets" in and around cells.

EMBEDDED GADGETS in the plasma membrane

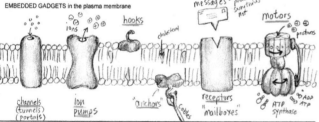

messages - sometimes protein, sometimes not

hooks

motors
protons
ATP ADP
ATP synthase

channels (tunnels) (portals)   ion pumps   "anchors" cables   receptors "mailboxes"

CABLES

Microtubules
subunits of tubulin dimers   20-25 nm nanometers
13!

Cytoskeleton
• structure
• "roadways"
Flagella - whip-like tails
Cilia - fine hairs that move

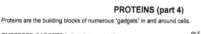

Intermediate filaments   10 nm

Cytoskeleton
Keratin
• hard - hair + nails
• soft - skin

Microfilaments   6-7 nm

Actin
• motion
• abundant in muscles

"SCISSORS" and "STAPLERS" (enzymes)

SCISSORS
substrate

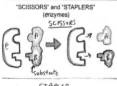

STAPLER
Coenzyme
A and B are called "substrates."

"VEHICLES" transportation around the body

vesicle
drugs
hormones
tails
Kinesin
heads
ions
fatty acids

MOTOR PROTEINS:
--Kinesins
--Dyneins

ALBUMIN
Acts like boat (or taxi) for hydrophobic substances

TAGS mark invaders for destruction by immune cells
this end sticks to invaders
this end sometimes sticks to body cells

They go by three names:
--gamma globulins
--immunoglobulins
--antibodies
All three are correct-- you choose

MONOSACCHARIDES are simple sugars

CH_2OH

GLUCOSE $C_6H_{12}O_6$

sweetest FRUCTOSE $C_6H_{12}O_6$

least sweet GALACTOSE $C_6H_{12}O_6$

DISACCHARIDES are "double" sugars made of two simple sugars

dehydration synthesis!

GLUCOSE   FRUCTOSE

SUCROSE "table sugar"

Don't forget the invisible ENZYME STAPLER ROBOT

GALACTOSE   GLUCOSE

LACTOSE
found in milk

POLYSACCHARIDES are long strings of simple sugars (up to 4,000 units long!)

1) Starch
glucoses
DIGESTIVE ENZYME fits in

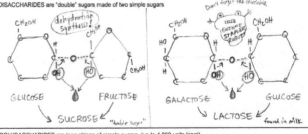

Plants make starches and store them in seeds and roots: wheat, rice, corn, beans, potatoes, beets, carrots, etc)

2) Cellulose
glucoses
Notice alternating flags
gets in the way
DIGESTIVE ENZYME can't fit!

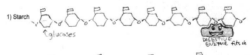

Plants make cellulose and use it for their tough cell walls. We eat it as fiber (leaves and stems).

3) Glycogen is branched.

Our bodies make glycogen as a way to store glucose in the liver and muscles.

The hormone insulin signals the body to turn glucose into glycogen. Adrenaline does the opposite.

RIBOSOMES are the "factories" that make proteins. They are the smallest organelle in a cell.
(NOTE: They are NOT protein gadgets! They are made primarily of RNA with only a few bits of protein mixed in.)

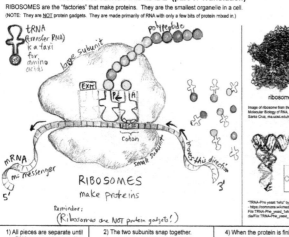

tRNA (transfer RNA) is a taxi for amino acids.

large subunit

polypeptide

Exit · P · A

mRNA
mi messenger

5'

codon

moves this direction

3'

small subunit

RIBOSOMES make proteins

Reminder: (Ribosomes are NOT protein gadgets!)

ribosome
Image of ribosome from the Center for Molecular Biology of RNA, Univ. of CA Santa Cruz, rna.ucsc.edu/rnacenter

tRNA

"tRNA-Phe yeast 1ehz" by Yikrazuul - https://commons.wikimedia.org/wiki/File:TRNA-Phe_yeast_1ehz.png#/media/File:TRNA-Phe_yeast_1ehz.png

1) All pieces are separate until a mRNA snaps onto a small subunit. (There are special molecules that cause the mRNA to stick.)

mRNA

2) The two subunits snap together.

3) tRNAs start bringing amino acids to the ribosome, and they are added one by one to the growing protein chain.

4) When the protein is finished, all parts separate.

5) After the polypeptide is done it gets folded, often with help from chaperone proteins.

shredder

Floating nearby are millions of transfer RNAs (tRNAs) holding amino acids.

RNA is ribonucleic acid. The individual units of RNA are called **nucleotides**.
A nucleotide is made of three pieces: a ribose, a phosphate, and a "base."

Ribose is a simple sugar but has only 5 carbons.

Phosphate is PO4. This is the "acid."

Bases contain nitrogen. There are 5 types.

A NUCLEOTIDE is a ribose, a phosphate and a base.

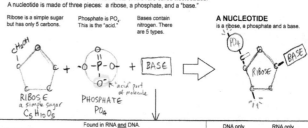

RIBOSE a simple sugar $C_5H_{10}O_5$

PHOSPHATE PO_4

"acid" part of molecule

PO4 · RIBOSE · BASE

Found in RNA and DNA.			DNA only	RNA only

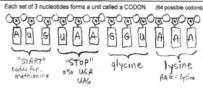   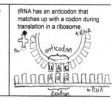

There are 5 kinds of bases. Three of them are in both RNA and DNA.

Adenine   Cytosine   Guanine   Thymine   Uracil

RNA is a long chain of nucleotides. The bases in RNA are A, C, G and U.

We could draw RNA showing the riboses, the phosphates and the bases.

U C G A U C G A C U U G A C

nucleotide

Or we can short-cut and draw it like this:

ect, →

To make it even simpler we can draw it like this:

mRNA   or

Each set of 3 nucleotides forms a unit called a CODON. (64 possible codons)

A U G   U A A   G G U   A A

"START" codes for methionine   "STOP" also UGA UAG   glycine   lysine AAG = lysine

tRNA has an anticodon that matches up with a codon during translation in a ribosome.

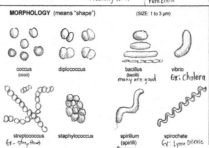

anticodon

U A C
A U G
codon · mRNA

RNA is found in more than one place in a cell.

1) mRNA (messenger RNA) is a copy of a section of DNA and is used to make **proteins**.

2) tRNA (transfer RNA) is used as a taxi for **amino acids**.

3) rRNA (ribosomal RNA) is folded up to make **ribosomes**.

4) miRNA (microRNA) are very short pieces of RNA used to regulate gene expression.

DNA contains all the information a cell will ever need. DNA is so important that it can never leave the protection of the nucleus. Therefore, when information is needed outside the nucleus, a mRNA copy of that section of DNA is created and sent out.

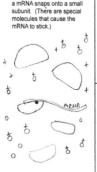

RNA
nucleus
nucleolus is where ribosomes are made
cell

The nucleus is made of TWO bilayers of phospholipids. It has a double-thick membrane.

DNA

DNA is similar to RNA. It is made of nucleotides. These are the DIFFERENCES:

1) DNA is a double helix with 2 strands. RNA is a single helix with 1 strands.

2) DNA has Thymine. RNA has Uracil.

3) DNA has deoxyribose. RNA has ribose.

4) DNA is found only in nucleus. RNA is found in all parts of cell.

A protein gadget called **RNA polymerase** acts as both a scissor (or maybe zipper) and a stapler. It unzips the DNA and then staples nucleotides together to match one of the strands of the DNA. (Notice which strand is being copied.)

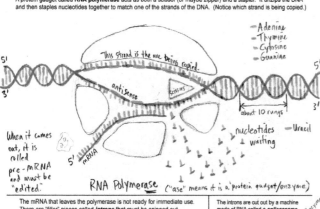

This strand is the one being copied.

— Adenine
— Thymine
— Cytosine
— Guanine

5' 3'
antisense
scissors
about 10 rungs
5' 3'

When it comes out, it is called pre-mRNA and must be "edited."

mRNA 5'

nucleotides waiting
— Uracil

RNA Polymerase ("ase" means it is a protein gadget/enzyme)

The mRNA that leaves the polymerase is not ready for immediate use. There are "filler" pieces called **Introns** that must be snipped out. Also, a special cap must be put on the 5' end, and a poly(A) tail on the 3'.

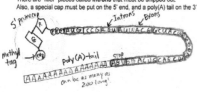

Introns   Exons
methyl tag
poly(A)-tail   STOP
AAAAAA can be as many as 200 long!

The introns are cut out by a machine made of RNA called a *spliceosome*.

ribozyme

After the introns are gone, the mRNA is ready to leave the nucleus.

- For every 1 human cell in your body, you have 10 bacteria cells.
- There are two types of prokaryotes: BACTERIA and ARCHAEA. (Archaea used to be classified as bacteria. They tend to be the "extremophiles" who survive in harsh conditions.)
- Prokaryotes do not have a true nucleus. They have a clump of DNA but it does not have an envelope around it.

ANATOMY of an "average" BACTERIA

full set of DNA
GENOMIC DNA is circular (no true nucleus!)

CYTOSKELETON

VESICLES

PLASMA MEMBRANE

INCLUSIONS (crystals, fat droplets)

RIBOSOMES

CELL WALL made of peptidoglycan (protein, sugar)

PLASMIDS are extra bits of DNA often having information for survival, such as antibiotic resistance.

The info on the plasmids is passed to another bacteria by conjugation.

copying machine
DNA
Conjugating bacteria

CYTOSOL is watery fluid (water, minerals, salts, sugars, enzymes)

CYTOPLASM is the cytosol plus the organelles floating in it

Adenine
methyl · glycines

Penicillin

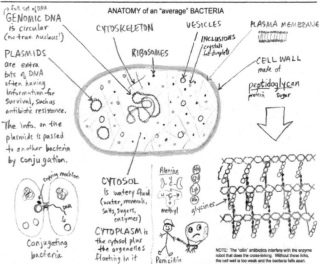

NOTE: The "cillin" antibiotics interfere with the enzyme robot that does the cross-linking. Without these links, the cell wall is too weak and the bacteria falls apart.

MORPHOLOGY (means "shape")   (SIZE: 1 to 3 μm)

coccus (cocci)   diplococcus   bacillus (bacilli) many are good   vibrio Ex: Cholera

streptococcus Ex: strep throat   staphylococcus   spirillum (spirilli) Ex: sewer water   spirochete Ex: Lyme Disease

FIMBRIAE (fim-bree-eye) are little "hairs" that allow the bacteria to stick to surfaces. They are more common in Gram negatives.

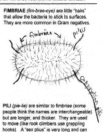

fimbriae   pilus

PILI (pie-lie) are similar to fimbriae (some people think the names are interchangeable) but are longer, and thicker. They are used to move (like rock climbers use grappling hooks). A "sex plus" is very long and can grab another bacteria and "reel it in" until they touch. Then DNA can be shared.

MOTILITY (means "movement")

Bacteria sometimes have "tails" called flagella that are made of microtubules. ("Flagella" means "whip.")

Some bacteria can glide using a sugary slime.

Pili - work like grappling hooks
sugar slime

Spirochetes have an inner flagellum.

CLASSIFICATION by pathologists ("Patho" means "disease.")
The Gram stain is used to find out what kind of cell wall the bacteria has so right kind of antibiotic can be prescribed.

GRAM positive (+)   GRAM negative (-)

Stain is crystal violet

thick cell wall made of peptidoglycan.

toxic sugars (endo-toxin)
very thin peptido-glycan layer
2 plasma membranes

(Gram staining is named after Danish scientist Hans Christian Gram.)

CAPSULES (the "slime layer")

Gram +   Gram −

sugar slime layer

Capsules keep them from drying out and from being eaten by immune cells.

ARCHAEA

- Used to be classified as bacteria (They have their own kingdom now.)
- Most of them live in extreme environments like the bottom of the ocean or in hot mineral springs. However, a few species live in our intestines where they produce methane gas.
- None of them are pathogens. (They won't make you sick.)
- Coccus, bacillus and spirillum shapes
- Some have flagella

Archaea look and act very much like bacteria.

Differences between bacteria and archaea:
1) RNA polymerase structure.
2) Ribosome structure.
3) Archaea phospholipids look like this
4) Archaea cell walls are not peptidoglycan.
5) Archaea DNA is wound on histone spools

... does it mean to be alive? We just looked at bacteria, which are definitely alive. But wha... ...ses? Are they alive? Viruses are also part of our microbiome.

To be alive, an organism must:
1) Grow in some way.
2) Reproduce.
3) Move in some way.
4) Respond to the environment.
5) Use energy and have a metabolism

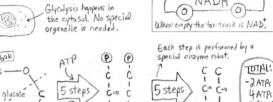

No ribosomes!
protein hooks, lipid bilayer, protein shell

A virus is basically a piece of DNA or RNA inside a protein shell sometimes with a lipid membrane.

HOW DO LIVING CELLS USE ENERGY?
Most cellular processes are powered by **ATP** (adenosine triphosphate). ATP acts like a rechargeable battery.

Nucleotides
G C A U
Adenosine
phosphates mono di tri

Negative charges don't like to be next to each other. They act like a compressed spring.

ATP → ADP → ATP
putting it back on requires energy

GLYCOLYSIS means "breaking glucose"

Glucose is a like a stick of dynamite, full of potential energy. It must be disassembled little by little so the energy is released slowly, not all at once.

Glycolysis happens in the cytosol. No special organelle is needed.

What is NADH? A taxi for electrons!
When empty the taxi truck is NAD⁺.

Each step is performed by a special enzyme robot.

glucose → 5 steps (2 ATP) → 5 steps (4 ATPs, 2 NADHs) → 2 pyruvates

TOTAL:
− 2 ATPs
+ 4 ATPs
+ 2 NADH
● 2 ATPs
● 2 NADH

The sperm cell is the smallest human cell. It has half the normal amount of DNA and has very few organelles.

HEAD 5 μm — MIDPIECE — TAIL (flagellum)

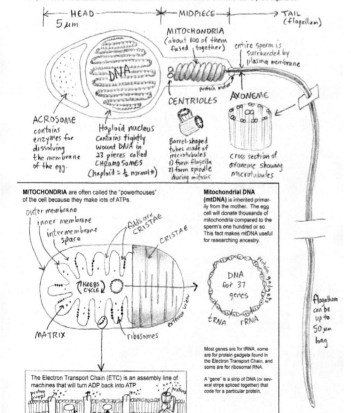

MITOCHONDRIA (about 100 of them fused together)
entire sperm is surrounded by plasma membrane
DNA
protein motor
CENTRIOLES
AXONEME

ACROSOME contains enzymes for dissolving the membrane of the egg.

Haploid nucleus contains tightly wound DNA in 23 pieces called CHROMOSOMES (haploid = ½ normal #)

Barrel-shaped tubes made of microtubules 1) form flagella 2) form spindle during mitosis

cross section of axoneme showing microtubules

MITOCHONDRIA are often called the "powerhouses" of the cell because they make lots of ATPs.

outer membrane
inner membrane
intermembrane space
folds are CRISTAE
CRISTAE
KREBS CYCLE
MATRIX
ribosomes

Mitochondrial DNA (mtDNA) is inherited primarily from the mother. The egg cell will donate thousands of mitochondria compared to the sperm's one hundred or so. This fact makes mtDNA useful for researching ancestry.

DNA for 37 genes
tRNA   rRNA

Most genes are for tRNA, some are for protein gadgets found in the Electron Transport Chain, and some are for ribosomal RNA.

A "gene" is a strip of DNA (or several strips spliced together) that code for a particular protein.

flagellum can be up to 50 μm long

The Electron Transport Chain (ETC) is an assembly line of machines that will turn ADP back into ATP

proton pumps
FADH
NADH
shuttles
ATP synthase

In the mitochondrial matrix, pyruvates are cut apart to make acetyl-CoA's, which are then sent to the Krebs Cycle.

The Krebs Cycle breaks apart the remaining carbon bonds and uses the energy to recharge "taxi" molecules NADH and FADH2. These taxis then go over to the E.T.C.

The end result of all these processes is that one molecule of glucose can yield up to 36 ATPs. CO_2 and H_2O are given off as wastes.

pyruvates enter through partial protein
KREBS   matrix
Glycolysis happens out here.
1) The PRE-KREBS step happens here. This is where CoA gets attached.
2) KREBS CYCLE burns acetyl CoA's
3) Electron Transport Chain   ETC

1) The PRE-KREBS step happens as the pyruvate crosses the inner membrane (into the matrix)

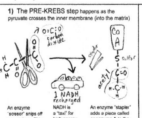

carbon dioxide
CoA
sulfur
acetyl
1 NADH recharged

An enzyme "scissor" snips off one carbon. This carbon has two oxygens attached to it, so the carbon goes sailing off as CO_2.

NADH is a "taxi" for high-energy electrons. One taxi gains two electrons in this step.

An enzyme "stapler" adds a piece called coenzyme A to the remaining 2-carbon molecule (which is called acetyl). CoA is actually a very large molecule.

2) The KREBS CYCLE finishes the "burning" of glucose

citrate (citric acid)
CO_2 NADH
CO_2 NADH
NADH
FADH2
GTP → ATP

This cycle is also called the Citric Acid Cycle

3) The Electron Transport Chain (ETC)
The ETC is an elaborate system of protein gadgets. There are 3 pumps, several shuttles, and a motor at the end. The goal? ATPs!

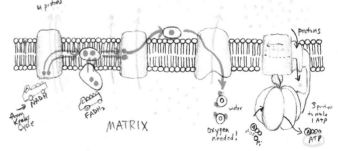

4 protons
protons
from Krebs NADH
FADH2
MATRIX
water
oxygen needed!
3 protons to make 1 ATP
ATP

Since this process requires O_2 and it results in phosphates being put back on, we can call it **oxidative phosphorylation**.

The ovum (egg) is the largest human cell at about 200 microns (the length of a paramecium).

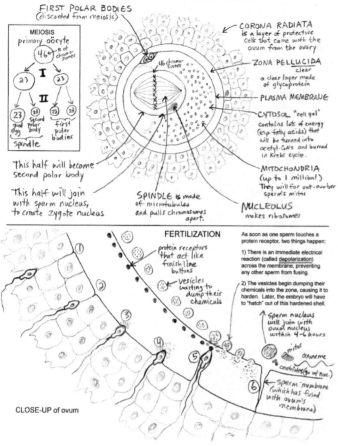

FIRST POLAR BODIES (discarded from meiosis)

MEIOSIS
primary oocyte
46 (# of chromosomes)
I
23 → 23
II
23 good egg → 23 second polar body → 23 first polar bodies
spindle

This half will become second polar body

This half will join with sperm nucleus, to create zygote nucleus

CORONA RADIATA is a layer of protective cells that came with the ovum from the ovary

ZONA PELLUCIDA clear — a clear layer made of glycoprotein

PLASMA MEMBRANE

46 chromosomes

CYTOSOL "cell gel" Contains lots of energy (esp. fatty acids) that will be turned into acetyl-CoA's and burned in Krebs cycle.

MITOCHONDRIA (up to 1 million!) They will out-number sperm's mitos.

NUCLEOLUS makes ribosomes

SPINDLE is made of microtubules and pulls chromosomes apart.

FERTILIZATION

CLOSE-UP of ovum

protein receptors that act like finish line buttons

vesicles waiting to dump their chemicals

As soon as one sperm touches a protein receptor, two things happen:

1) There is an immediate electrical reaction (called depolarization) across the membrane, preventing any other sperm from fusing.

2) The vesicles begin dumping their chemicals into the zona, causing it to harden. Later, the embryo will have to "hatch" out of this hardened shell.

Sperm nucleus will join with ovum nucleus within 4-6 hours.

mitos — axoneme
centrioles (go w/ nuc.)
Sperm membrane (which has fused with ovum's membrane)

When the sperm nucleus fuses with the ovum nucleus, a zygote is formed. This cell is **totipotent** and is capable of turning into any type of cell, including not only body cells, but also placenta cells or umbilical cord cells.

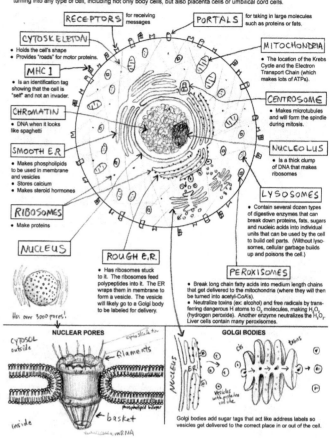

RECEPTORS for receiving messages

PORTALS for letting in large molecules such as proteins or fats.

CYTOSKELETON
• Holds the cell's shape
• Provides "roads" for motor proteins.

MHC 1
• Is an identification tag showing that the cell is "self" and not an invader.

CHROMATIN
• DNA when it looks like spaghetti

SMOOTH E.R.
• Makes phospholipids to be used in membrane and vesicles
• Stores calcium
• Makes steroid hormones

RIBOSOMES
• Make proteins

NUCLEUS

ROUGH E.R.
• Has ribosomes stuck to it. The ribosomes feed polypeptides into it. The ER wraps them in membrane to form a vesicle. The vesicle will likely go to a Golgi body to be labeled for delivery.

MITOCHONDRIA
• The location of the Krebs Cycle and the Electron Transport Chain (which makes lots of ATPs).

CENTROSOME
• Makes microtubules and will form the spindle during mitosis.

NUCLEOLUS
• Is a thick clump of DNA that makes ribosomes

LYSOSOMES
• Contain several dozen types of digestive enzymes that can break down proteins, fats, sugars and nucleic acids into individual units that can be used by the cell to build new cell parts. (Without lysosomes, cellular garbage builds up and poisons the cell.)

PEROXISOMES
• Break long chain fatty acids into medium length chains that get delivered to the mitochondria (where they will then be turned into acetyl-CoA's).
• Neutralize toxins (ex: alcohol) and free radicals by transferring dangerous H atoms to O_2 molecules, making H_2O_2 (hydrogen peroxide). Another enzyme neutralizes the H_2O_2. Liver cells contain many peroxisomes.

Has over 3000 pores!

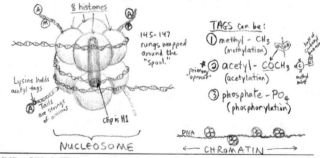

NUCLEAR PORES
CYTOSOL outside
cytosol side
filaments
basket
inside
phospholipid bilayer
mRNA

GOLGI BODIES
Golgi bodies add sugar tags that act like address labels so vesicles get delivered to the correct place in or out of the cell.

Mitosis is the process where body cells (somatic cells) duplicate themselves in order to make new cells. Don't confuse mitosis with meiosis (the process that produces gametes) or with binary fission (how prokaryotes split in half).

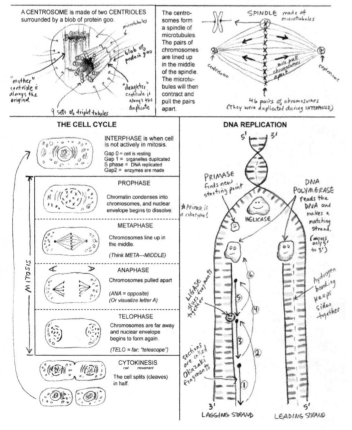

A CENTROSOME is made of two CENTRIOLES surrounded by a blob of protein goo.

"mother" centriole is always the original
"daughter" centriole is always the duplicate
9 sets of triplet tubules
blob of protein goo

The centrosomes form a spindle of microtubules. The pairs of chromosomes are lined up in the middle of the spindle. The microtubules will then contract and pull the pairs apart.

SPINDLE made of microtubules
46 pairs of chromosomes (They were duplicated during INTERPHASE)

THE CELL CYCLE

INTERPHASE is when cell is not actively in mitosis.
Gap 0 = cell is resting
Gap 1 = organelles duplicated
S phase = DNA replicated
Gap2 = enzymes are made

PROPHASE
Chromatin condenses into chromosomes, and nuclear envelope begins to dissolve.

METAPHASE
Chromosomes line up in the middle.
(Think META—MIDDLE)

ANAPHASE
Chromosomes pulled apart.
(ANA = opposite)
(Or visualize letter A)

TELOPHASE
Chromosomes are far away and nuclear envelope begins to form again.
(TELO = far: "telescope")

CYTOKINESIS
cleavage: movement
The cell splits (cleaves) in half.

DNA REPLICATION

PRIMASE finds new starting point
(A Primase is a ribozyme!)

DNA POLYMERASE reads the DNA and makes a matching strand. (moves only 5' to 3')

HELICASE

LIGASE glues sections together
sections are called Okazaki fragments

hydrogen bonding keeps sides together

LAGGING STRAND   LEADING STRAND

The zygote is a human cell but it is not any particular cell. To become a specific type of cell, such as a skin cell or a muscle cell, all the non-skin or non-muscle DNA must be permanently zippered shut. There are three main ways that DNA can be silenced.

1) DNA methylation

This is the most permanent form of locking away information.
Enzymes put methyl tags (CH3) on cytosines in the areas that are to be locked.

Methyl: $= CH_3$

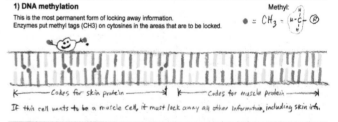

Codes for skin protein   Codes for muscle protein

If this cell wants to be a muscle cell, it must lock away all other information, including skin info.

2) Histone modification

The histone spools on which DNA is wound can control whether a gene is expressed or not.
("Expressed" means that the information is being used and proteins are being made.)

8 histones
145-147 rungs wrapped around the "spool."
Lysine holds acetyl tags
Tails are strings of amines
Clip is H1

TAGS can be:
① methyl - CH_3 (methylation)
② acetyl - $COCH_3$ (acetylation)
③ phosphate - PO_4 (phosphorylation)

NUCLEOSOME
CHROMATIN

3) Micro RNAs (miRNA)

Micro RNAs are non-coding RNAs whose sole purpose is to mess up RNA. When a miRNA attaches, that portion of the RNA becomes unusable. Thus, gene expression is blocked.

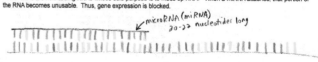

microRNA (miRNA) 20-22 nucleotides long

The zygote is a TOTIPOTENT cell. ("Toti" means "totally" and "potent" means "powerful or capable.") In what sense is this cell totally powerful? It can turn into ANY type of human cell, even supporting cells such as the placenta and amniotic sac. All the DNA in this cell is open and accessible. None of it is methylated or closed in any way. As the embryo develops, the cells will become less "potent" and will have much of their DNA closed.

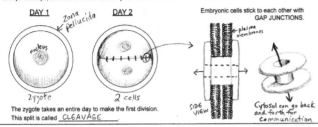

DAY 1 — zona pellucida, nucleus — zygote
DAY 2 — 2 cells
Embryonic cells stick to each other with GAP JUNCTIONS.
plasma membranes — SIDE VIEW — Cytosol can go back and forth for communication

The zygote takes an entire day to make the first division.
This split is called CLEAVAGE

DAY 2.5 — 4 cells
Cells are getting smaller while overall size is staying the same.

DAY 3 — 8 cells
This is a critical stage for unknown reasons. Some embryos don't make it past this stage.

DAY 4 — O_2 CO_2 — 16-32 cells — MORULA ("mulberry") — Inner cells will start to have trouble getting O_2

DAY 5 — cut away view — 32-64 cells — BLASTULA (blastocyst) ("Blast" means "bud.") There is a cavity filled with fluid.

DAY 6 — INNER CELL MASS
These cells are PLURIPOTENT and are often the ones harvested for use in embryonic stem cell research.

DAY 6 or 7 — z.p., blastula
Pre-embryo hatches!
Blastocyst secretes enzymes that soften the zona pellucida, then it enlarges suddenly and breaks free.

Map of where this is happening:
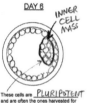
FALLOPIAN TUBE — DAY 2, DAY 3, DAY 1, DAY 4, cilia, anchoring ligament, DAY 5, OVARY, DAY 6, endometrium, DAY 7, UTERUS

AMAZING FACT: The first week is the same for ALL placental mammals, regardless of how long the gestation period is. (mice: 3 wks, elephants: 2 yrs)
SECOND AMAZING FACT: Some mammals can pause pregnancy at this stage and hold the embryo for several months, waiting for the right season.

Week 2 begins with IMPLANTATION. The embryo attaches itself to the wall of the mother's uterus.

DAY 7 - 8
By the end of week 1, the embryonic cells have used most of the energy that the original egg cell had stored up. To get more energy, the embryo will have to tap into the mother's blood supply and use food she has digested.
villi ("fingers") — glands — endometrium (lining of uterus)

DAY 8-9
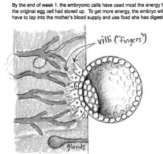
chorionic villi, CHORION, placenta
The chorionic villi begin to secrete HCG, a hormone messenger molecule that will tell the mother's body that an embryo is present.

DAYS 10 - 14
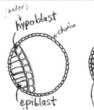
(under) hypoblast, chorion — epiblast (on top) — Bilaminar disc
hypoblast — epiblast — amniotic cavity
notochord starts, future yolk sac — amniotic cavity
connecting stalk, chorion, chorionic cavity, yolk sac, notochord — 3-layer GERM DISC ("germ" means "seed")
GASTRULATION

By the end of week 2, there are 3 GERM LAYERS: ("germ" means "seed")
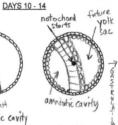
1) ECTODERM will become skin (epidermis), hair, nails, brain, nervous system
2) MESODERM will become muscles, bones, connective tissues, blood, kidneys, heart, bladder, gonads, lymph nodes, spleen, dermis of skin
3) ENDODERM will become lining of gut and lungs, liver, pancreas and lots of glands

SIDE NOTE: After all the body parts are fully developed, the cells are programmed to stop dividing so fast. "Rogue" cells that continue to divide can cause childhood cancers.
Cancers that arise from the *mesoderm* are called sarcomas.
Children and teenagers are more likely to get this kind.
Cancers that arise from *ectoderm and endoderm* are called carcinomas.
This type of cancer is more common in adults.

NOTE: In this drawing we will focus on just the 3-layer germ disc. The disc is still inside the chorion, but we won't see the chorion in this drawing. We will see it again in drawing 28.

TOP VIEW OF 3-LAYER DISC — future mouth area, primitive node, primitive pit, our cross section, primitive streak

SHH Sonic Hedgehog protein — ectoderm, mesoderm, endoderm, notochord
(1) Notochord and primitive streak. Notochord secretes SHH protein.

neural crest cells, neural groove
(2) Neural groove forms

(3) Neural tube begins "Neurulation"

neural crest cells, neural tube
(4) Neural tube will become brain and spinal cord. Neural crests will be peripheral nervous system and more.

heart tubes
(5) Mesoderm begins to differentiate and heart tubes appear

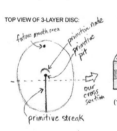 

Here we see the 3-layer disc with amnion and yolk sac.
amnion — Neural crest cells will become peripheral nerves and more. — somites, epidermis, dermis, muscle, bone, urogenital system (kidneys, bladder and repro.), sacs around organs, plus arms and legs, sacs around organs (inner) — endoderm becomes inside of gut tube — yolk sac — This part of yolk sac will make blood cells and blood vessels.

DORSAL SIDE (back) — neural tube, notochord, heart tubes, future kidneys, gut tube, VENTRAL SIDE (stomach), blood islands, yolk sac

(6) "Somites" appear. (Future spine and associated muscles) Mesoderm continues to differentiate and the ends split.

(7) The embryo curls, forming the gut tube and the amniotic sac. Blood islands appear in the yolk sac and will become complete little sections of vessel, with red blood cells already inside.

---

21 DAYS SAGGITAL CROSS SECTION
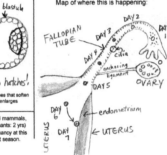

21 DAYS TRANSVERSE CROSS SECTION

ACTUAL SIZE: 1 mm
Embryo would fit into this letter o.

1) endometrium (lining of uterus)
2) maternal capillaries
3) maternal blood
4) chorion
5) chorionic villi (with capillaries)
6) chorionic cavity
7) allantois
8) yolk sac
9) gut tube
10) mesoderm
11) ectoderm/epidermis
12) amniotic cavity
13) amnion
14) umbilical stalk
15) neural tube (future spinal cord)
16) future vertebra
17) notochord
18) nerves
19) muscle
20) dermis
21) future kidneys
22) heart tubes
23) visceral peritoneum
24) parietal peritoneum

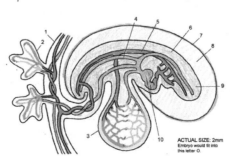

26-28 DAYS
SAGGITAL CROSS SECTION SHOWING VASCULARIZATION

1) chorion
2) chorionic villi
3) yolk sac
4) cardinal vein
5) dorsal aorta
6) embryo
7) amnion
8) amniotic cavity
9) future head/brain
10) future heart — has begun beating

ACTUAL SIZE: 2mm
Embryo would fit into this letter O.

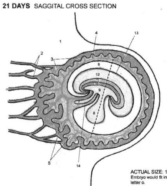

END OF WEEK 4 (28 DAYS)
SAGGITAL SECTION

1) forebrain
2) midbrain
3) hindbrain
4) spinal cord
5) future eye
6) future ear
7) mouth opening
8) heart (has begun beating)
9) throat with glands
10) lungs
11) liver
12) stomach and intestines
13) kidney
14) dorsal aorta
15) allantois
16) yolk sac
17) cloacal opening
18) umbilical stalk

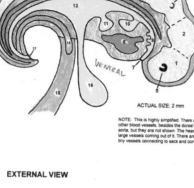

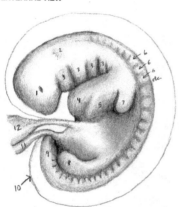

DORSAL   mesoderm   epidermis

VENTRAL

ACTUAL SIZE: 2 mm

NOTE: This is highly simplified. There are other blood vessels, besides the dorsal aorta, but they are not shown. The heart has large vessels coming out of it. There are also tiny vessels connecting to sacs and cord.

END OF WEEK 5 (33-35 DAYS)   EXTERNAL VIEW

1) eye
2) ear
3) pharyngeal arches (jaw + throat)
4) heart
5) liver
6) somites
7) arm bud
8) leg bud
9) "tail" bud
10) amnion
11) umbilical cord
12) yolk sac

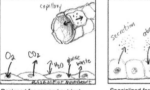
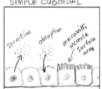
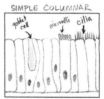

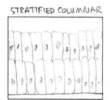
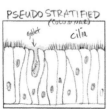
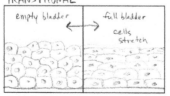

There are many different types of cells in the body, but all of them can be classified into one of the following categories:

1) EPITHELIAL - covers surfaces and lines body cavities. Ex- epidermis of skin, lining of lungs and digestive tract, walls of blood vessels
2) CONNECTIVE - binds and supports parts. Occurs in matrix of solid, liquid or gel. Ex- ligaments, tendons, cartilage, bone, blood and fat, and dermis of skin.
3) MUSCLE - designed for movement. Ex- voluntary, involuntary, cardiac (heart)
4) NERVOUS - designed for communication. Ex- brain, spinal cord, peripheral nerves

EPITHELIAL TISSUE (part 1)

① SQUAMOUS
skin, blood vessels

② CUBOIDAL
glands, kidneys

③ COLUMNAR

GOBLET CELL

All epithelial tissue is built on BASEMENT MEMBRANE (not made of cells)
Cells are held together by ADHESION JUNCTIONS, using DESMOSOMES

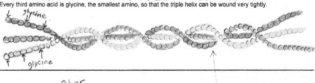

DESMOSOME
attachment plaque
HEMIDESMOSOME
SKIN CELLS (epidermis)
cytoskeleton
basal lamina
reticular lamina (reticulum)
BASEMENT MEMBRANE
CONNECTIVE (dermis) skin

Collagen IV
Laminins (protein)

Collagens I and III
Collagen VII

SIMPLE SQUAMOUS

O_2 CO_2 H_2O glucose   waste

Designed for gas and nutrient exchange because it is very thin. EX: lining of lungs, inside of blood vessels and capillaries

SIMPLE CUBOIDAL

secretion   absorption   microvilli increase surface area

Specialized for secretion and absorption. Some have microvilli. EX: glands such as thyroid, pancreas. Also found in kidney tubules.

SIMPLE COLUMNAR

goblet cell   microvilli   cilia

Is specialized for secretion and absorption, and also for pushing things along. Can have cilia or microvilli. EX: lining of stomach and intestines, Faliopian tubes

STRATIFIED SQUAMOUS

2 Types:
1) KERATINIZED: top layers hard and dead. EX: skin
2) NON-KERATINIZED: top layers soft and alive EX: inside of mouth

STRATIFIED CUBOIDAL

secretion

Designed for secretion. EX: salivary glands, mammary glands NOTE: usually just 2-3 layers

STRATIFIED COLUMNAR

Secretes and protects. Not very common EX: eye, throat, uterus, urethra, salivary glands

PSEUDO STRATIFIED (COLUMNAR)

cilia   goblet

Nuclei are at different levels. EX: lining of trachea

TRANSITIONAL

empty bladder   full bladder   cells stretch

Designed to stretch. EX: Only found in one place: urinary bladder.

CONNECTIVE tissue is made of 3 things:
1) Specialized cells   2) ground substance   and 3) protein fibers
The ground substance and the protein fibers make the MATRIX
The protein fibers can be made of 1) Collagen   2) elastin   or 3) reticular fibers (tiny collagen)

There are three types of connective tissues, and several categories under each:

FIBROUS	CARTILAGENOUS	OTHER
1) Loose (areolar)	1) Hyaline	1) Bone
2) Dense	2) Elastic	2) Blood
3) Adipose (fat)	3) Fibrocartilage	3) Lymph

COLLAGEN is a protein cable made of three separate polypeptide chains (alpha helices). Every third amino acid is glycine, the smallest amino, so that the triple helix can be wound very tightly.

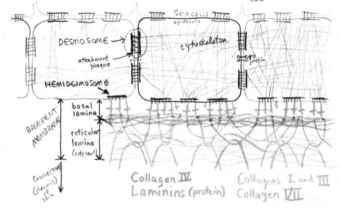

glycine   glycine

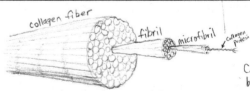

collagen fiber   fibril   microfibril   collagen protein

Collagen fibers are bundled for strength.

One type of specialized cell is the **FIBROBLAST**, which makes collagen proteins and exports them (using exocytosis) outside the cell, where they then join together and make collagen fibers.

Fibroblasts also made the ground substance which is a mixture of water (90%) and glycoproteins (10%).

Fibroblasts live 2 to 3 months. They multiply rapidly after an injury. Scar tissue is a result of very active fibroblasts.

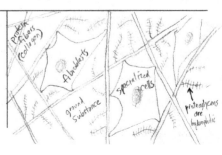

protein fibers (collagen)   fibroblasts   ground substance   specialized cells   proteoglycans are hydrophilic

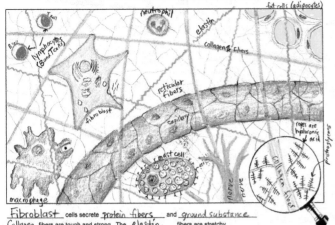

fat cells (adipocytes)

Fibroblast cells secrete protein fibers and ground substance.
Collagen fibers are tough and strong. The elastin fibers are stretchy.
Reticular fibers serve as an anchoring network for capillaries, fibroblasts, and adipocytes.
Macrophages are the "big eaters," swallowing bacteria, viruses, fungus, dirt, debris, dead cells, etc.
Neutrophils are also eaters and gobble up pathogens. Lymphocytes are white cells (T and B cells)
that recognize and tag foreign invaders. Mast cells start the inflammatory process.
The ground substance is made of "ropes" of hyaluronic acid to which are attached "bottle brushes" called
proteoglycans which act like sponges and soak up water.

HOW MAST CELLS START INFLAMMATION:

Mast cells are covered with receptors of all kinds that can detect just about any kind of irritation (pathogens, injury, allergens).

Inside, there are thousands of vesicles (often called granules) that are filled with histamine and cytokines. If the receptors are activated, the vesicles merge with the membrane (exocytosis) and dump their contents outside the cell.

HOW INFLAMMATION HAPPENS:

1) Mast cells are triggered by pathogens or allergens or injury, which causes them to release their chemicals

2) The histamine causes capillaries to dilate and become leaky. This causes swelling (edema).
3) Histamine also sends signals to nerve endings, causing pain and/or itching.
4) The granules also contain cytokines which are chemical messengers that go and recruit more white blood to come to this area.

Fibrous connective tissues have specialized cells called fibroblasts.

DENSE

regular
- Almost all collagen and very little elastin
- Strong in 1 direction ↕
- EX's TENDONS- connects bone to muscle
  LIGAMENTS- bone to bone
  DURA MATER - covering tough mother around brain

irregular
- Not quite as strong but still pretty strong in all directions
- EX's DERMIS of skin (bottom layer)
  PERIOSTEUM around bones
  FASCIA - bags around muscles
  SUBMUCOSA of gut (under mucosa layer)

ADIPOSE (fat)

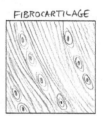

a few collagen or reticular fibers
Cells are called ADIPOCYTES. The vacuoles are filled with fat molecules - triglycerides. Stores energy, provides padding, insulation
EX's - under the skin, around eyes, heart, kidneys

CARTILAGINOUS CONNECTIVE TISSUE

Cartilaginous tissues have specialized cells called chondrocytes
Cartilaginous tissues have no capillaries and no nerves.

HYALINE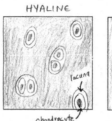
lacuna / chondrocyte
Mostly collagen.
EX's NOSE (below bone)
TRACHEA (windpipe)
- ENDS of LONG BONES
- ENDS of RIBS
- EMBRYONIC BONES (until 4th month)

ELASTIC
Fibers are mostly elastin, very little collagen
EX's EAR (outer ear)
EPIGLOTTIS (covers trachea when you swallow)

FIBROCARTILAGE
Very tough - mostly collagen
EX's
- PADS between VERTEBRAE in spine
- PUBIC BONE in PELVIS
Pubic bone has separation with cartilage

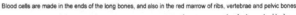

Bone is classified as a connective tissue, so it has:
1) Specialized cells (called osteoblasts and osteocytes
2) Protein fibers (called collagen
3) Ground substance, which is a solid made of minerals such as calcium (Ca) and phosphorus (P).

OSTEOBLASTS secrete collagen, then fill the empty spaces with minerals. Cells are cuboidal when active, flat when not.

Osteoblasts can become OSTEOCYTES.

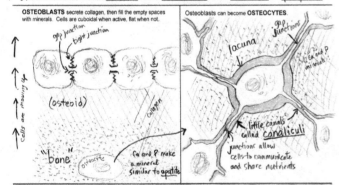

little canals called canaliculi
junctions allow cells to communicate and share nutrients
Ca and P make a mineral similar to apatite

When osteoblasts are done, they have created an OSTEON.

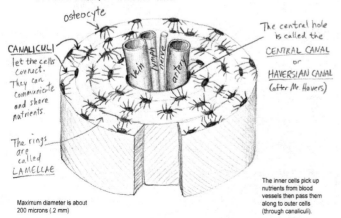

osteocyte

CANALICULI let the cells connect. They can communicate and share nutrients.

The rings are called LAMELLAE

The central hole is called the CENTRAL CANAL or HAVERSIAN CANAL (after Mr. Havers)

The inner cells pick up nutrients from blood vessels then pass them along to outer cells (through canaliculi).

Maximum diameter is about 200 microns (.2 mm)

Blood cells are made in the ends of the long bones, and also in the red marrow of ribs, vertebrae and pelvic bones.

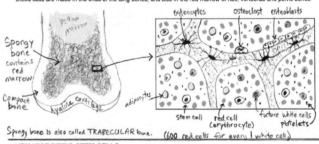

Spongy bone contains red marrow
Compact bone
Spongy bone is also called TRABECULAR bone.
osteocytes osteoclast osteoblasts
adipocytes
stem cell / red cell (erythrocyte) / future white cells / platelets
(600 red cells for every 1 white cell)

HEMATOPOIETIC STEM CELLS make all kinds of blood cells.
(He-MAT-o-po-ee-ET-ic)
"Hema" is Greek for "blood," and "poiein" is Greek for "makes."

("HEMATOPOIESIS" is the process of making blood cells.)

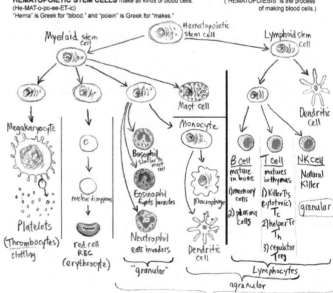

Myeloid stem cell / Hematopoietic stem cell / Lymphoid stem cell
Mast cell / Monocyte / Dendritic Cell
Megakaryocyte
nucleus disappears
Basophil (similar to mast cell)
Eosinophil fights parasites
Macrophage
B Cell matures in bone 1) memory cells 2) plasma cells
T Cell matures in thymus 1) Killer Ts (cytotoxic) Tc 2) helper Ts Th 3) regulator Treg
NK cell Natural Killer (granular)
Platelets (Thrombocytes) clotting
red cell RBC (erythrocyte)
Neutrophil eats invaders
Dendritic cell
"granular"
Lymphocytes
agranular
Leukocytes

Blood is classified as a connective tissue because it has:

- 1) specialized cells : many types of blood cells
- 2) protein fibers : FIBRINOGEN and other non-fibrous proteins
- 3) ground substance : liquid called PLASMA (mostly water)

(matrix)

A centrifuge can separate blood into its 3 parts: plasma, "Buffy coat," and red cells.

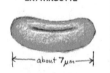

55%

NOTE: SERUM is plasma minus fibrinogen

BUFFY COAT white cells and platelets

<1%

45%

PLASMA:

| SIDE NOTE: Blood has a pH of 7.4. |

1 Water (91%)

2 Proteins (7%) Produced by the liver (except for gamma globulins)

  ① Fibrinogen: fibers that make FIBRIN which forms clots

  ② Clotting factors • Some act on fibrinogen
      • Some act on platelets

  ③ Albumins: (Greek "albumen" meaning egg white.)
     "Taxis" for hydrophobic substance such as: fats, cholesterol, hormones, ions (calcium + copper) and bilirubin
     *also maintains blood pressure

  ④ Globulins:

     (1) α alpha-- transports hormones, cholesterol, ions
         HDL

     (2) β beta-- transports hormones, cholesterol, ions, helps to dissolve clots
         LDL

     (3) γ gamma-- a.k.a "antibodies" or immunoglobulins (Ig's) produced by B cells

3 Solutes (2%) nutrients (glucose), O_2, CO_2, wastes, vitamins, urea ammonium minerals

---

FIBRINOGEN -- the "fibers" in blood

Fibrinogen is a protein molecule that looks like this:

We can simplify the shape like this:

Clotting factors can act on fibrinogen to form **FIBRIN** but only after a special message is sent.

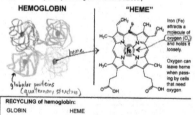

fibrinogen

clotting factors (THROMBIN)

Both float in blood, but don't interact because fibrinogen has a "safety cap" on its active site.

Message goes out: "Help! Blood is leaking out!"

Thrombin gets a message.

safety caps

THROMBIN is told to take the safety caps off the central active site of fibrinogen

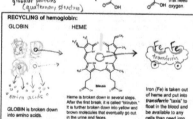

The fibrinogens bond with each other to form strings of FIBRIN.

FIBRIN

The body's way of dealing with damage to blood vessels is called **HEMOSTASIS** (blood stay in)

The PRE-STEP is: Vasoconstriction (blood vessels shrink)

STEP 1: PLATELET PLUG ("primary hemostasis")

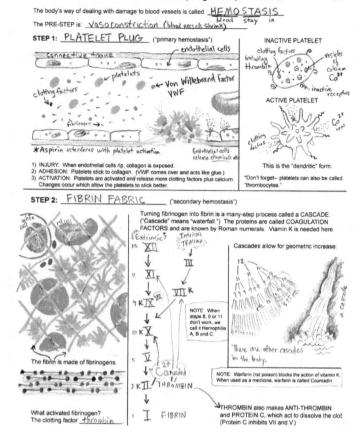

connective tissue

endothelial cells

platelets

clotting factors

Von Willebrand factor VWF

fibrinogen

Endothelial cells release chemicals at...

*Aspirin interferes with platelet activation.

1) INJURY: When endothelial cells rip, collagen is exposed.
2) ADHESION: Platelets stick to collagen. (VWF comes over and acts like glue.)
3) ACTIVATION: Platelets are activated and release more clotting factors plus calcium. Changes occur which allow the platelets to stick better.

INACTIVE PLATELET

clotting factors including thrombin

vesicles of calcium Ca^{2+}

inactive receptors

ACTIVE PLATELET

clotting factors

Ca^{2+} released

This is the "dendritic" form.

*Don't forget-- platelets can also be called "thrombocytes."

STEP 2: FIBRIN FABRIC ("secondary hemostasis")

white cell

red cell

Turning fibrinogen into fibrin is a many-step process called a CASCADE. ("Cascade" means "waterfall.") The proteins are called COAGULATION FACTORS and are known by Roman numerals. Vitamin K is needed here.

Cascades allow for geometric increase:

a Cascade

{Extrinsic} {Intrinsic TRAUMA}

12 XII
III
11 XI F
VII K
9 K IX
NOTE: When steps 8, 9 or 11 don't work, we call it Hemophilia A, B and C.
10 X
5 V
Ca^{2+} Ca needed
3 K II THROMBIN
1 I FIBRIN

"There are other cascades in the body."

NOTE: Warfarin (rat poison) blocks the action of vitamin K. When used as a medicine, warfarin is called Coumadin.

The fibrin is made of fibrinogens.

What activated fibrinogen? The clotting factor **thrombin**

THROMBIN also makes ANTI-THROMBIN and PROTEIN C, which act to dissolve the clot. (Protein C inhibits VII and V.)

Erythrocytes are produced by the myeloid stem cells in bone marrow. The **KIDNEYS** control how many are produced. Low oxygen levels in the blood cause the kidneys to make a substance called **erythropoietin** which acts as a signal to the myeloid cells to differentiate into more red cells. This process takes a few days. (This is what happens when you adjust to higher or lower altitudes.)

ERYTHROCYTE

about 7 μm

Does **NOT** have: nucleus, DNA, ER, Golgi bodies, mitochondria

DOES have: 250 million hemoglobin molecules

Bone marrow can make 2 million red cells per second !

The body has 20 trillion red cells at any given time.

Red cells live for 3 months. Old cells are eaten by macrophages in the liver and spleen.

HEMOGLOBIN

heme

globular proteins (quaternary structure)

"HEME"

Iron (Fe) attracts a molecule of oxygen (O_2) and holds it loosely.

Oxygen can leave heme when passing by cells that need oxygen.

RECYCLING of hemoglobin

GLOBIN

HEME

Fe Fe transferrin

GLOBIN is broken down into amino acids.

amino acids

Heme is broken down in several steps. After the first break, it is called "bilirubin." It is further broken down into yellow and brown molecules that eventually go out in the urine and feces.

Iron (Fe) is taken out of heme and put into transferrin "taxis" to float in the blood and be available to any cells that need iron.

BLOOD TYPES

An erythrocyte has hundreds of proteins on its surface. The most critical ones are A, B and Rh.

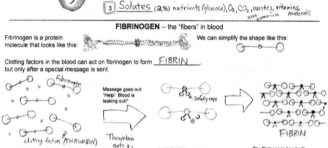

| TYPE A | TYPE B | TYPE AB | TYPE O |

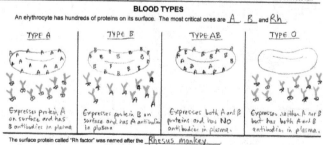

Expresses protein A on surface and has B antibodies in plasma.

Expresses protein B on surface and has A antibodies in plasma.

Expresses both A and B proteins and has NO antibodies in plasma.

Expresses neither A nor B but has both A and B antibodies in plasma.

The surface protein called "Rh factor" was named after the **Rhesus monkey**.

Rh+

Rh-

Normally, no antibodies but if sensitized, then blood will have Y Rh

---

BASOPHILS (.5% of leukocytes in blood)

10-12 μm

antigen (not self)

IgE

Blue dots are vesicles (granules)
Blue color is from basic stain.

1) Normally, they float around in the blood, but they can be recruited into tissues if other cells "call" for them to come.

2) Are extremely similar to mast cells.

3) Vesicles are filled with histamine (and other chemicals). Histamine dilates blood vessels and makes them leak. (If this happens too fast and too strong, you get a sudden drop in blood pressure and you faint.)

4) Basophils have many IgE antibodies (from B cells) attached to them. IgE's trigger release of histamine when antigens bind to them.

5) Basophils also have the ability to "call" other white cells to come and help, including eosinophils, neutrophils and basophils.

MAST CELLS (not in blood)

10-12 μm

when touching signal is sent

IgE

1) Found only in tissues, not in blood. We met these in loose connective tissue, sitting next to capillaries.

2) Are extremely similar to basophils.

3) They start the inflammatory process when endothelial cells are damaged.

4) Mast cells are covered with IgE antibodies (from B cells). When allergens bind to the IgE's, histamine is released.

EOSINOPHILS (2% of leukocytes in blood)

10-12 μm

reminds us of worms

NOT TO SCALE!

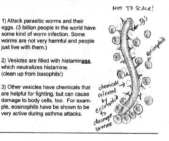

Ig's

eosinophils

Red color is from eosin stain (acidic).

1) Attack parasitic worms and their eggs. (3 billion people in the world have some kind of worm infection. Some worms are not very harmful and people just live with them.)

2) Vesicles are filled with histaminase, which neutralizes histamine. (clean up from basophils!)

3) Other vesicles have chemicals that are helpful for fighting, but can cause damage to body cells, too. For example, eosinophils have been shown to be very active during asthma attacks.

chemicals released by eosinophils to damage worms

NEUTROPHILS (60-65% of leukocytes in blood)

These guys are so amazing that they need their own page... (They make several kinds of chemical weapons!)

NEUTROPHILS (60-65% of leukocytes in blood)

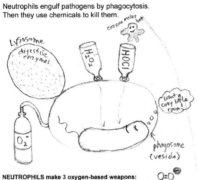

multi-lobed nucleus; vesicles (granules) stain neutral; receptors; First responder

in females the inactive X chromosome looks like a clump (Barr body)

CD3 "hook" for the handshake

1) Our body makes about 100 billion per day.

2) We have 5 times as many neutrophils in reserve (in marrow mostly) as we do in circulation.

3) They float in blood until needed in tissues. When they get chemical signals that they are needed, they leave the vessels by squeezing through the cracks between the endothelial cells.

4) Lifespan: a few days

5) "Pus" is mostly dead neutrophils.

HOW THEY GET FROM BLOOD INTO TISSUES:

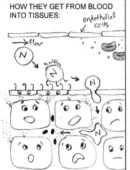

endothelial cells

flow, N

* Pathogens are often already coated with chemical tags called opsonins.

Chemical messages are sent out by cells in distress. The endothelial cells put out "hooks" to slow down and catch the neutrophils that are floating past.

The neutrophils then squeeze through the cracks and get into the *interstitial* space. (the "empty" space between cells)

The neutrophils have chemicals that can dissolve the junctions between the endothelial cells, in order to make the crack larger. The endothelial cells then quickly repair the damage.

Neutrophils can sense the bacteria, but are also helped out by "yummy" tags placed on the invaders by other parts of the immune system.

Neutrophils engulf pathogens by phagocytosis. Then they use chemicals to kill them.

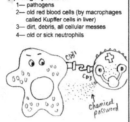

Lysosome; digestive enzymes; enzyme maker; H_2O_2; HOCl; O_2; phagosome (vesicle); What a cozy little room!

O_2^-

NEUTROPHILS make 3 oxygen-based weapons:

1) "Super-oxide" is an oxygen molecule, O_2, with an extra electron stuck on (by a special enzyme). The electron will go flying off like a bullet, striking the invader. Super-oxide is a very common "free radical" found in your body. It kills pathogens, but it can also damage your cells.

2) Hydrogen peroxide, H_2O_2 (yes, the same stuff that is in your First Aid kit for sterilizing wounds). The "bullet" here is the second O molecule.

3) HOCl, a form of "bleach." (hypochlorous acid) When the neutrophil is making a lot of this, your mucus turns green. (The enzyme that makes it is greenish in color.) Green mucus suggests bacterial infection rather than viral.

Other strategies:

4) Digestive enzymes: Lysosomes filled with enzymes can merge (join) with the phagosome and dump enzymes all over the pathogen.

5) Iron: Neutrophils can hide (Fe) from bacteria. Bacteria die without iron.

Monocytes in the blood can go into tissues and differentiate into either macrophages or dendritic cells.

MONOCYTES (about 5% of leukocytes in blood)

-- 50% of them are in spleen
-- They can do phagocytosis, but are less efficient than neutrophils.

10-15 μm; no visible granules

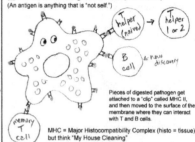

DENDRITIC CELL: Similar to macrophage except that it stays small, and is less involved in secreting cytokines (messenger molecules).

Dendritic cells are "professional" **A**ntigen **P**resenting **C**ells (APCs). They eat pathogens by phagocytosis then put tiny pieces of the antigen on their plasma membranes so they can show them to T cells in the lymph nodes.

Many dendritic cells are found in the skin, the lining of the intestines, and in lymph nodes and spleen.

antigen

MACROPHAGES

40-50 μm

CD31 "hook" (NOT TO SCALE!)

Many receptors for detecting many kinds of pathogens, and for receiving messages.

lots of lysosomes

Macrophages are found in all body tissues. In some tissues they go by other names:

— Skin: Langerhans cells
— Liver: Kupffer cells (Macrophages in the liver clear out bacteria and also eat old red blood cells.)
— Lungs: "dust cells"
— Brain: microglia
— Bone: (osteoclasts?) This is being debated...

MACROPHAGES have basically 3 jobs:

(1) Phagocytosis of:
1— pathogens
2— old red blood cells (by macrophages called Kupffer cells in liver)
3— dirt, debris, all cellular messes
4— old or sick neutrophils

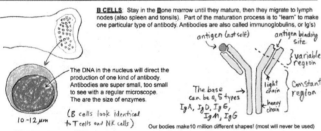

chemical password

The "CD31 handshake" is when a macrophage grabs a neutrophil and won't let it go until it gives a chemical password. Sick or infected neutrophils will not be able do do this. They will get eaten.

(2) Presentation of antigens to T cells (and B cells)
(An antigen is anything that is "not self.")

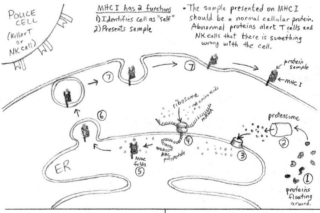

helper (naive) → T helper 1 or 2

B cell ← new discovery

memory T cell

Pieces of digested pathogen get attached to a "clip" called MHC II, and then moved to the surface of the membrane where they can interact with T and B cells.

MHC = Major Histocompatibility Complex (histo = tissue) but think "My House Cleaning"

(3) Release of cytokines (messenger molecules), especially the kind called "interleukins" (IL). Interleukins are numbered, such IL1, IL6, IL10. This is how white cells talk to each other and coordinate their actions.

Lymphoid stem cells differentiate into 3 types of cells: B cells, T cells and NK cells.

B CELLS: Stay in the Bone marrow until they mature, then they migrate to lymph nodes (also spleen and tonsils). Part of the maturation process is to "learn" to make one particular type of antibody. Antibodies are also called immunoglobulins, or Ig's)

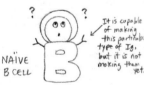

The DNA in the nucleus will direct the production of one kind of antibody. Antibodies are super small, too small to see with a regular microscope. The are the size of enzymes.

10-12 μm

(B cells look identical to T cells and NK cells)

antigen (not self); antigen binding site; variable region; light chain; Constant region; heavy chain

The base can be a 5 types IgA, IgD, IgE, IgM, IgG

Our bodies make 10 million different shapes! (most will never be used)

Mature B cells go to lymph nodes to set and wait until they are needed. These cells are called NAÏVE because they don't know what to do yet.

NAÏVE B CELL

It is capable of making this particular type of Ig, but it is not making them yet.

When activated they can become:

OR

PLASMA CELL making millions of copies of its antibody

MEMORY CELL remembers that at one time this Ig was needed.

I remember...

Q: What do antibodies do?
A: Stick to antigens
Q: What is an antigen?
A: Anything "not self"

AGGLUTINATION:

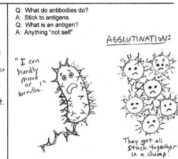

"I can hardly move or breathe!"

They get all stuck together in a clump!

Antibodies function as "flags" to alert other cells to the presence of an intruder. They also act like signs that say "Eat me!"

"Yummy Igs!"

He is opsonized!

lysosome; macrophage

To understand how T cells work, we first need to learn a little more about how body cells work. Cells are like little houses with no windows. How will their neighbors know what is going on inside? What if a thief is inside? (a virus)

The body has a roaming police force that constantly scans for trouble. The police will kill any cell that cannot prove that everything is okay inside. Body cells must "clean house" and cover their outer membranes with samples of the proteins that are floating around inside. If the cellular police detect an intruder or a sickness (virus or cancer) they will kill the cell.

POLICE CELL (Killer T or NK cell)

MHC I has 2 functions
1) Identifies cell as "self"
2) Presents sample

• The sample presented on MHC I should be a normal cellular protein. Abnormal proteins alert T cells and NK cells that there is something wrong with the cell.

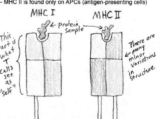

ER; ribosome; amino acids; MHC folds; MHC polypeptide; proteasome; protein sample; MHC I; proteins floating around.

WHAT'S GOING ON IN THE PICTURE ABOVE:

1) There are always proteins floating around in the cytosol.

2) A tiny organelle called a proteasome acts like a shredder and chops these proteins up into tiny pieces. The shredded pieces are 5-10 amino acids long.

3) The tiny protein bits go through a portal and into the ER.

4) Meanwhile a ribosome is making a polypeptide, spitting it into the ER.

5) The poly peptide folds up and forms an MHC clip shape. A protein is attracted to the clip part and sticks to it.

6) The ER puts the "loaded" MHC clip into a vesicle.

7) The vesicle goes to the plasma membrane and joins with it (exocytosis). That's how the MHC gets to the surface.

The little "clips" are called MHC molecules:
Major Histocompatibility Complex. (My House Cleaning!)
– MHC I is found on all body cells
– MHC II is found only on APCs (antigen-presenting cells)

MHC I   MHC II

protein sample

This part is what T cells see as "self"

There are many minor variations in structure.

NOTE: Each person has about 50 types of MHC's. This is what they try to match in transplants.

Now we need to learn about the MHC clip on the APCs (antigen presenting cells such as macrophages).

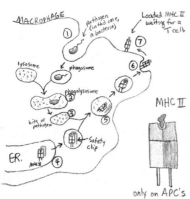

MACROPHAGE

lysosome
phagosome
phagolysosome
bits of pathogen
safety clip
ER.
MHC II

Loaded MHC II waiting for a T cell

MHC II

only on APC's

1) A pathogen gets eaten (phagocytosis). The vesicle containing the pathogen is called a phagosome.

2) The phagosome merges with a lysosome which contains digestive enzymes.

3) This merger is called a phagolysosome. The lysosome digests the pathogen and dissolves it into small pieces. The pieces are 15-20 amino acids in length.

4) Meanwhile an MHC II protein is coming out of the ER. (It was made by a ribosome, of course.) Notice the little safety device on the MHC II, making sure the clips stays open until it meets a pathogen protein.

5) The phagolysosome merges with the vesicle containing the MHC II. The safety device falls off and a piece of pathogen protein sticks to the MHC II.

6) The vesicle merges with the membrane.

7) The MHC II is on the outside of the cell.

T CELL RECEPTORS:

T cells have receptors on their membranes that will lock on to either MHC I or MHC II (not both). This means there are two different types of T cells.

CD4⁺ T cell (helper)

This protein is called CD4. It recognizes MHC II.

TCR (T cell Receptor)

This end is very similar to antibodies and only recognizes one type of pathogen

When CD4 cells are activated, they go and alert B cells and other T's.

CD8⁺ T cell (Killer/cytotoxic)

CD8 recognizes MHC I.

Each T cell receptor has a unique shape. If pathogen shape matches the TCR shape, then the CD8 will kill the body cell.

MACROPHAGE | BODY CELL | T CELL

T cells look just like B cells.

thymus (under ribcage)

Thymus is large in children and small in adults.

T cells are "born" in the bone marrow, like all leukocytes, then they migrate to the Thymus to mature.
In the thymus the T cells will...
1) differentiate into either CD4 or CD8 cells
2) be tested to see if they will not attack body cells, but will attack pathogens (98% fail and are discarded!)

(T cells don't "know" to go to the thymus. They have special receptors that match those found on the epithelial cells in the thymus. The T cells float in the blood stream until they come into contact with the thymus cells, then they stick there.)

Th (CD4 helper)   Tc (CD8 killer/cytotoxic) — perforin gun

When CD4 and CD8 cells "graduate" from the thymus training school, they go sit in lymph nodes and tonsils until they are needed. (Most will never be needed.)

Since they have no work experience, they are called NAIVE T cells.

Now we must talk about the bad guys: PATHOGENS. They can be viruses, bacteria, protozoans, yeasts, or multicellular animals like worms. Some of these critters stay outside your cells and some like to get inside and hide. Don't forget that cells don't have eyes-- they can't see the pathogens in or out of cells! Yet they must decide where to look for the pathogens.

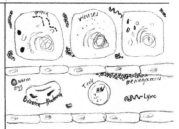

Th — chemical signals (cytokines)

Inside! Will go to alert T cells — 1

Outside! Go to alert B cells — 2

The decision is made by the macrophages who ate the pathogens. They tell the T helper cells whether to activate the B cells (to fight pathogens hiding outside the cells) or to activate CD8 T cells (for pathogens inside cells).

Examples of pathogens that generally stay outside of cells (extracellular):

-- anthrax (affects livestock)
-- most E. coli
-- cholera (in intestines)
-- meningococcus
-- clostridium (in intestines)
-- Borrelia (Lyme)
-- some strep (group A)
-- staph (S. aureus)

Examples of pathogens that like to hide inside of cells (intracellular):

-- all viruses
-- Listeria (found on food)
-- some strep (group B)
-- candida yeast
-- Yersinia (the plague)
-- Salmonella (food poisoning)
-- mycobacterium (TB)
-- malaria (in erythrocytes)

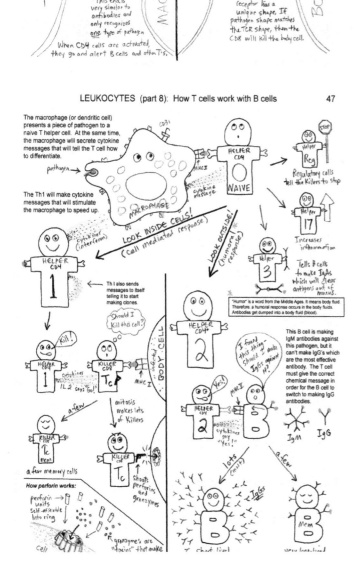

The macrophage (or dendritic cell) presents a piece of pathogen to a naive T helper cell. At the same time, the macrophage will secrete cytokine messages that will tell the T cell how to differentiate.

The Th1 will make cytokine messages that will stimulate the macrophage to speed up.

pathogen
MACROPHAGE
MHC II
cytokine message
HELPER CD4 — NAIVE

HELPER Reg — STOP — Regulatory cells tell the Killers to stop

Helper 17 — Increases inflammation

Helper 3 — Tells B cells to make IgAs which will clear antigens out of mucus.

cytokines (interferon)
LOOK INSIDE CELLS! (Cell mediated response)
LOOK OUTSIDE. (Humoral response)

HELPER CD4 — 1

Th1 also sends messages to itself telling it to start making clones.

Should I kill this cell?

HELPER — 1 — Kill!
cytokines IL2 says "Go!"

KILLER CD8 — Tc
MHC I
BODY CELL

a few — mitosis makes lots of Killers

KILLER Tc

a few memory cells

How perforin works:

perforin units self-assemble into ring

cell

granzymes are "toxins" that make...

HELPER CD4⁺ — 2

"Humor" is a word from the Middle Ages. It means body fluid. Therefore, a humoral response occurs in the body fluids. Antibodies get dumped into a body fluid (blood).

This B cell is making IgM antibodies against this pathogen, but it can't make IgG's which are the most effective antibody. The T cell must give the correct chemical message in order for the B cell to switch to making IgG antibodies.

MHC II

I found this thing... Should I make IgGs against it?

HELPER CD4 — 2 — Yes!

cytokines say "Yes!"

B — Yes!

IgM   IgG

lots (many)

B — IgGs — short lived

a few

B — Mem — very long-lived

THE IMMUNE SYSTEM OVERVIEW   48

The body has several layers of defenses. Scientists have identified three basic layers and given them names. They are: 1) physical barriers, 2) the innate (non-specific) system, and 3) the adaptive (specific) system.

1) **PHYSICAL BARRIERS** that can block pathogens from entering the body.

SKIN | CILIATED COLUMNAR (trachea) | MUCOSA OF GUT (mucosal layer acts as barrier) | STOMACH ACID (Acid kills most pathogens) | TEARS (Tears also have antimicrobial properties) | MUCUS IN NOSE

2) **THE INNATE SYSTEM** (also known as the NON-SPECIFIC system) ("Nat-" is Latin for "born," so innate means you are born with it.)

BASOPHIL | MAST CELL | EOSINOPHIL | NEUTROPHIL | NATURAL KILLER (NK) | LIVER

LINKS BETWEEN INNATE and ADAPTIVE

APC's — Macrophage — Dendritic cell

3) **ADAPTIVE SYSTEM** (also known as SPECIFIC or ACQUIRED)

HELPER Th CD4 | KILLER Tc CD8 | B cells

Natural Killer cells are lymphocytes (related to T cells). They mature in bone, thymus, tonsils, spleen and lymph nodes. NK cells have multiple types of receptors and can sense both bad antigens and missing MHC I on body cells.

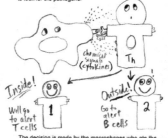

perforin gun — missing MHC I's

receptors that can sense MHC I and also bad proteins

MHC I — BODY CELL

The liver produces tiny proteins called COMPLEMENT (written in the singular, though this sounds strange!). These proteins function as a CASCADE so the response can be fast, hopefully faster than the rate at which the pathogens can multiply! Just like coagulation proteins, the complement proteins float in the blood waiting until they are needed. Once the first protein is activated, then the cascade starts and all the others are activated. Most of the complement proteins are named with the letter C (C1, C2, C3... C9).

Complement proteins accomplish 4 things:

1) Activates __mast cells.__

2) __Membrane Attack Complex__

3) Acts as __opsonins__ ("eat me" tags)

4) __Agglutination__ (sticks pathogens together into clumps) so that they are easier for phagocytes to catch

C3 is a key protein:

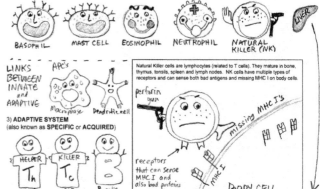

The "b" part sticks to a membrane and attracts C5, C6, C7, C8 and C9.

A scissor enzyme cuts C3 apart.

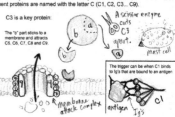

membrane attack complex

The trigger can be when C1 binds to Ig's that are bound to an antigen

antigen — Ig's — C1

This is a **MOTOR NEURON** (often connected to muscle fibers). Only found in the peripheral nervous system (not in brain or spine).

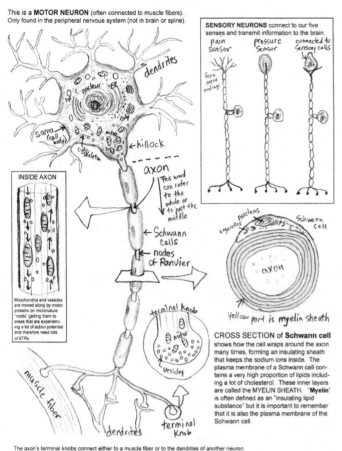

soma (cell body)
nucleus
dendrites
hillock
cytoskeleton
←hillock
axon
This word can refer to the whole or to just the middle
← Schwann Cells
nodes of Ranvier
terminal knob
vesicles

INSIDE AXON

Mitochondria and vesicles are moved along by motor proteins on microtubule "roads" getting them to areas that are experiencing a lot of action potential and therefore need lots of ATPs.

muscle fiber
dendrites
terminal knob

The axon's terminal knobs connect either to a muscle fiber or to the dendrites of another neuron.

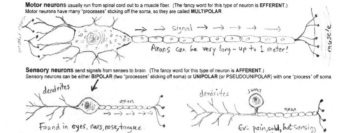

SENSORY NEURONS connect to our five senses and transmit information to the brain.

pain sensor
pressure sensor
connected to sensory cells
free nerve endings

organelles nucleus Schwann cell
axon

Yellow part is myelin sheath

CROSS SECTION of Schwann cell shows how the cell wraps around the axon many times, forming an insulating sheath that keeps the sodium ions inside. The plasma membrane of a Schwann cell contains a very high proportion of lipids including a lot of cholesterol. These inner layers are called the MYELIN SHEATH. "**Myelin**" is often defined as an "insulating lipid substance" but it is important to remember that it is also the plasma membrane of the Schwann cell.

Neurons in the PNS (Peripheral Nervous System) are specialized for the jobs they do. The size and length of the axon and the location of the soma can vary.

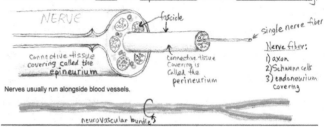

Motor neurons usually run from spinal cord to a muscle fiber. (The fancy word for this type of neuron is EFFERENT.) Motor neurons have many "processes" sticking off the soma, so they are called MULTIPOLAR

signal
Axons can be very long – up to 1 meter!
muscle

Sensory neurons send signals from senses to brain. (The fancy word for this type of neuron is AFFERENT.) Sensory neurons can be either BIPOLAR (two "processes" sticking off soma) or UNIPOLAR (or PSEUDOUNIPOLAR) with one "process" off soma.

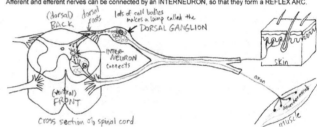

dendrites
axon
Found in eyes, ears, nose, tongue.

dendrites soma
axon
Ex: pain, cold, hot sensing in skin

A NERVE is a bundle of _fascicles_, which is a bundle of _nerve fibers_.
A NERVE FIBER is made of an _axon_ and its _myelin sheath_ and also the _endoneurium covering_.

NERVE fascicle
Connective tissue covering called the epineurium
Connective tissue covering is called the perineurium
single nerve fiber
Nerve fiber:
1) axon
2) Schwann cells
3) endoneurium covering

Nerves usually run alongside blood vessels.

neurovascular bundle

Afferent and efferent nerves can be connected by an INTERNEURON, so that they form a REFLEX ARC.

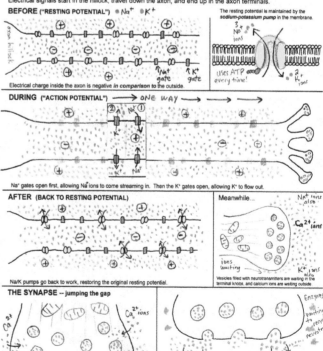

(dorsal) BACK dorsal roots
lots of cell bodies makes a lump called the DORSAL GANGLION
INTER-NEURON connects
skin
axon
(ventral) FRONT
Cross section of spinal cord
muscle

The CNS (Central Nervous System) consists of the brain and the spinal cord. This is a drawing of the cells of the brain. Most of these cells would be in the spine, also, but the arrangment might be different.

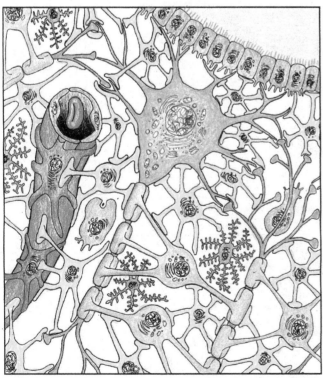

- Neurons (transmit electrical impulses)
- Oligodendrocytes (act like Schwann cells)
- Astrocytes (protect and nourish neurons)
- Microglia (macrophages of the brain)
- Ependymal cells (secrete fluid into ventricles)
- Endothelial cells (form capillary walls)
- Erythrocyte (carries oxygen)
- Pericytes (wrap around vessels, regulate blood flow)

Electrical signals start in the hillock, travel down the axon, and end up in the axon terminals.

BEFORE ("RESTING POTENTIAL") ● Na⁺ ● K⁺

axon hillock
Na⁺ gate K⁺ gate

Electrical charge inside the axon is negative *in comparison* to the outside.

The resting potential is maintained by the *sodium-potassium pump* in the membrane.

3 Na⁺ ions
uses ATP every time!
2 K⁺ ions

DURING ("ACTION POTENTIAL") → ONE WAY →

Na⁺ K⁺

Na⁺ gates open first, allowing Na⁺ ions to come streaming in. Then the K⁺ gates open, allowing K⁺ to flow out.

AFTER (BACK TO RESTING POTENTIAL)

Na/K pumps go back to work, restoring the original resting potential.

Meanwhile...

Na⁺ ions also
Ca²⁺ ions
ions waiting
K⁺ ions also

Vesicles filled with neurotransmitters are waiting in the terminal knobs, and calcium ions are waiting outside

THE SYNAPSE -- jumping the gap

Ca²⁺ ions
Ca²⁺ ions

This space is called the synaptic cleft.

When the action potential reaches the terminal knob, a sudden influx of Ca²⁺ ions causes the vesicles to do exocytosis.

Enzymes wanting to recycle neurotrans.

Na⁺ channel influx of Na⁺
START a new action potential!
K⁺ channel
influx of K⁺
No action potential

The neurotransmitters cross the synaptic cleft and bind to receptor sites on ion channels. Some neurotransmitters are "excitatory" and will start a new electrical signal by opening Na⁺ channels. Other neurotransmitters are "inhibitory" and will prevent a new signal from starting by opening K⁺ channels.

Na⁺ and K⁺ ion channels that have binding sites (receptors)

Muscles are bundles of bundles of bundles. Nerves run through muscles and attach to muscle fibers.

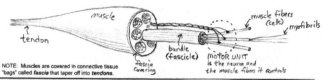

NOTE: Muscles are covered in connective tissue "bags" called *fascia* that taper off into *tendons*.

MUSCLE CELLS are called MUSCLE FIBERS

Muscle cells are called *muscle fibers* because lots of cells join together to make one long fiber.

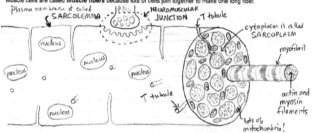

The T tubules carry the action potential down into the fiber so it can reach the myofibrils at the center.

THE NEUROMUSCULAR JUNCTION

This synapse works just like the ones we learned about in the last lesson. A sudden influx of Ca²⁺ ions makes the neurotransmitters flow across the gap and stick to receptors on Na⁺ channels on the other side. The Na⁺ ions begin an action potential.

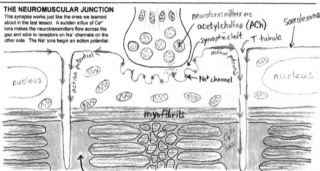

The smooth ER is called the SARCOPLASMIC RETICULUM. It stores calcium ions that will be needed for contraction.

Muscle fibers are made of myofibrils. Each myofibril is made of two types of protein filaments: *actin* and *myosin*. Actin and myosin overlap in such a way that the myofibril appears to have stripes. Dark places are where many fibers overlap and light places are where few overlap. The repeating patterns are called *sarcomeres*.

SARCOMERES

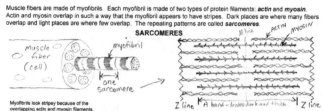

Myofibrils look stripey because of the overlapping actin and myosin filaments.

How Ca²⁺ ions allow myosin to bind to actin:

Tropomyosin blocks myosin from binding.

The action potential causes calcium to be released from the SR. Calcium binds to troponin, which causes tropomyosin to move away.

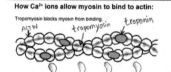
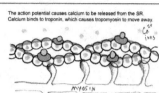

How ATPs are used by actin and myosin:

This is a continuous cycle that can repeat in a split second.

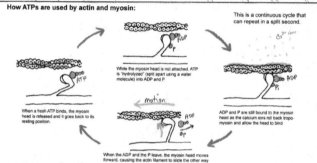

While the myosin head is not attached, ATP is "hydrolyzed" (split apart using a water molecule) into ADP and P.

When a fresh ATP binds, the myosin head is released and it goes back to its resting position.

ADP and P are still bound to the myosin head as the calcium ions roll back tropomyosin and allow the head to bind.

When the ADP and the P leave, the myosin head moves forward, causing the actin filament to slide the other way.

Where do the ATPs come from?

1) CREATINE is a molecule that holds onto a P. An enzyme can take the P off, and then put it onto an ADP, making ATP. No O₂ needed.

2) The Electron Transport Chain (ETC) This takes place in the mitochondria. Oxygen must be available so that it can receive the "tired" electrons at the end of the chain.

3) LACTIC ACID FERMENTATION is a process that enables glycolysis to take place over and over again, generating 2 ATPs each time. No O₂ needed.

The lymph system has two sides: left and right.
– The left side drains both legs, the left arm, the chest and the left side of the head.
– The right side drains only the right arm and the right side of the head.

The lymph system is made of a network of vessels that drain extra fluid from body tissues (the interstitial spaces between cells) into vessels that carry the fluid through lymph nodes and then up (against gravity!) to the top of the rib cage where lymph vessels dump the fluid back into the bloodstream. As the fluid makes its way up through the lymph vessels, it passes through lymph nodes where lymphocytes and macrophages can recognize and/or destroy any pathogens in the fluid. Also, macrophages and dendritic cells from the tissues can intentionally hop into the lymph vessels and drift to the nearest nodes so that they can present their antigens to T cells.

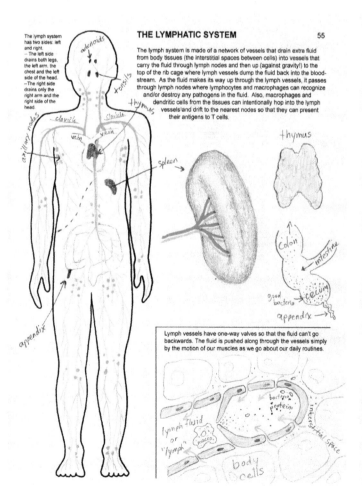

Lymph vessels have one-way valves so that the fluid can't go backwards. The fluid is pushed along through the vessels simply by the motion of our muscles as we go about our daily routines.

LYMPH NODE

Reticular fibers with macrophages and dendritic cells crawling along framework

Naive B cells (can be activated by T cells and turn into plasma cells)

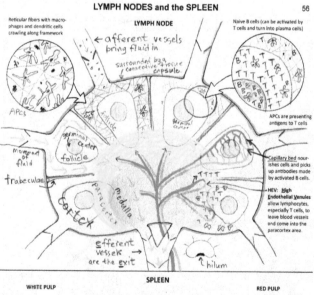

APCs are presenting antigens to T cells.

Capillary bed nourishes cells and picks up antibodies made by activated B cells.

HEV: High Endothelial Venules allow lymphocytes, especially T cells, to leave blood vessels and come into the paracortex area.

SPLEEN

WHITE PULP

RED PULP

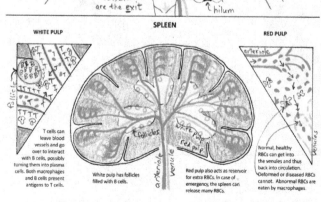

T cells can leave blood vessels and go over to interact with B cells, possibly turning them into plasma cells. Both macrophages and B cells present antigens to T cells.

White pulp has follicles filled with B cells.

Red pulp also acts as reservoir for extra RBCs. In case of emergency, the spleen can release many RBCs.

Normal, healthy RBCs can get into the venules and thus back into circulation. Deformed or diseased RBCs cannot. Abnormal RBCs are eaten by macrophages.

Though we think of skin as primarily being epithelial cells, it is actually a blend of all 4 tissue types. The dermis and epidermis are separated by the basement membrane (of the epidermis). We have 4 kinds of connective tissue: loose, irregular, adipose and blood cells. We have muscle, and both sensory and motor neurons.

1) basement membrane
2) basal layer (keratinocytes)
3) melanocytes
4) dermal papillae
5) fibroblasts
6) collagen and elastin

7) deep touch sensors (Pacinian corpuscle)
8) light touch sensor (Meissner's corpuscle)
9) free nerve endings (pain sensors)
10) cold sensors
11) heat sensors
12) motor neuron

13) sweat gland
14) arrector pili muscle
15) sebaceous glands
16) "bulge" (stem cells)
17) lymph vessels

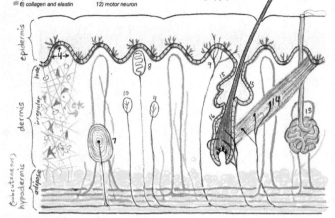

THE EPIDERMIS is made of mostly **keratinocytes**. The keratinocytes in the basal layer are the only ones that go through mitosis. As they divide, the new cells go upwards. As the cells mature, they begin producing a lot of a waxy protein called **keratin** and they also begin to lose their organelles.
By the time they reach the top, they have lost everything, even their nucleus. They are dead cells filled with keratin. We have these dead cells flaking off our skin all the time.

Top layer of dead cells is called the **STRATUM CORNEUM**

KERATINOCYTES: all cells except melanocytes and Langerhans cells

LANGERHANS cells are a type of macrophage and are the only immune cells in the epidermis.

MELANOCYTES produced the pigment **melanin**. Melanin is brown or red (usually brown). Dark skin has more melanin than light skin. Melanocytes release the melanin in little vesicles and these vesicles are taken in by keratinocytes.

FIBROBLASTS
COLLAGEN AND ELASTIN

BASAL LAYER (of keratinocytes)
BASEMENT MEMBRANE

Hair and nails are part of skin and grow in much the same way. However, unlike skin, the keratinocytes in hair and nails don't produce destructive enzymes that cut the bonds between the cells. The cells stay firmly connected.

HAIR GROWTH CYCLE

1) ANAGEN: active growth
2) CATAGEN: growth slows
3) TELOGEN: resting, shrinks and disconnects from papilla
1) back to ANAGEN: stem cells begin a new hair

CROSS SECTIONS showing shapes of hair shaft:
straight   wavy   kinky

NAILS

TOP VIEW:
free edge
plate
lunula
skin around edges is called 'nailfolds'

SIDE VIEW CROSS SECTION:
free edge   plate   lunula   cuticle   nail root
epidermis
matrix
nail bed
bone

Most of a tooth lies below the surface. Teeth have deep roots that go down into the bone. The white enamel is the hardest substance in the body and is non-living. The inner pulp is alive.

The tongue is much larger than it appears. We see the body of the tonuge, but the large "root" is below the surface. The tongue is made of many muscles, and it connects to many others.

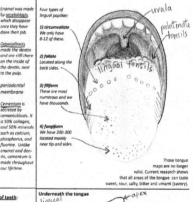

Enamel was made by ameloblasts, which disappear once they have done their job.

Odontoblasts made the dentin and are still there on the inside of the dentin, next to the pulp.

Cementum is secreted by cementoblasts. It is 50% collagen, and 50% minerals such as calcium, phosphorus, and fluorine. Unlike enamel and dentin, cementum is made throughout our lifetime.

Four types of lingual papillae:

1) circumvallate
We only have 8-12 of these.

2) foliate
Located along the back sides.

3) filiform
These are most numerous and we have thousands.

4) fungiform
We have 200-300 located mainly near tip and sides.

Those tongue maps are no longer valid. Current research shows that all areas of the tongue can taste sweet, sour, salty, bitter and umami (savory).

Underneath the tongue

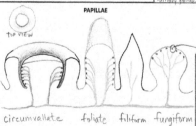

Types of teeth:
Incisors
Cuspids ("canines")
Bicuspids (premolars)
Molars

PAPILLAE

TOP VIEW

circumvallate   foliate   filiform   fungiform

The bumps you see on your tongue are papillae, not taste buds. Taste buds are microscopic. The filiform papillae do not have taste buds. They simply provide friction and sensation.
(Animal tongues (notably cats) often have very large and long papillae, making their tongues feel rough.)

TASTEBUDS

Taste receptors in taste buds are similar to smell receptors in the nose. Both are triggered by chemicals. The cells start an action potential in the nearby neurons, which travels to the brain where it is interpreted as an odor.

basal cells can trigger taste bud

Taste buds can only sense sweet, sour, bitter, salty and umami (savory). Most of taste involves smell.

---

bone
cartilage
fat
muscle
lymph
nervous
mucosa (epithelium)

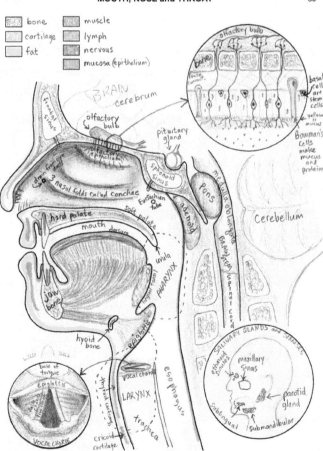

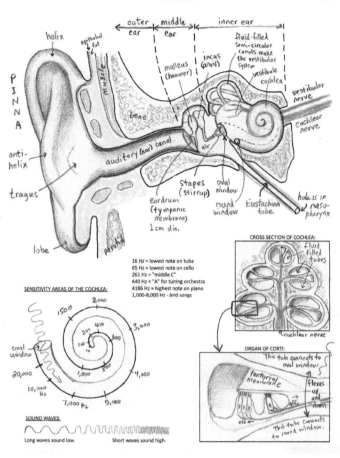

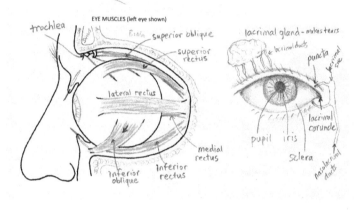

16 Hz = lowest note on tuba
65 Hz = lowest note on cello
261 Hz = "middle C"
440 Hz = "A" for tuning orchestra
4186 Hz = highest note on piano
1,000-8,000 Hz - bird songs

SENSITIVITY AREAS OF THE COCHLEA:

SOUND WAVES:

Long waves sound low.   Short waves sound high.

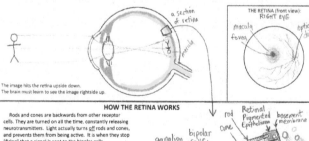

How does the eye focus? The ciliary body controls the shape of the lens.

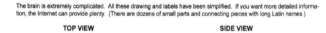

To focus on distant objects, the ciliary body relaxes, causing the zonules to tighten, making the lens become more flat.

To focus on objects that are close, the ciliary body tightens, causing the lens to become more round.

The image hits the retina upside down.
The brain must learn to see the image rightside up.

HOW THE RETINA WORKS

Rods and cones are backwards from other receptor cells. They are turned on all the time, constantly releasing neurotransmitters. Light actually turns off rods and cones, and prevents them from being active. It is when they stop "firing" that a signal is sent to the bipolar cells.

The mechanism that starts the turning-off process is a pigment molecule called **rhodopsin**. Rhodopsin is found in the phospholipid membranes in the "pancakes" (discs) in the ends of the rods and cones. It holds a smaller molecule called **retinal**. When light hits retinal, its shape changes and this starts a chemical cascade that results in sodium ions rushing into the cell. The influx of sodium stops the cell from releasing its inhibitory neuro-transmitters. The bipolar cells are then activated.

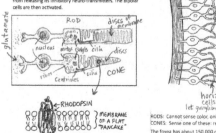

RODS: Cannot sense color, only light/dark. Function in low light conditions.
CONES: Sense one of these: red, green blue. Need lots of light to function.

The fovea has about 150,000 cones per mm². Other parts of the retina might have 10,000 or fewer cones per mm².

The brain is extremely complicated. All these drawing and labels have been simplified. If you want more detailed information, the Internet can provide plenty. (There are dozens of small parts and connecting pieces with long Latin names.)

TOP VIEW   **SIDE VIEW**

The purpose of wrinkles is to provide more surface area. The surface is where all the neuron cell bodies are and where most of our "thinking" takes place.

The LEFT hemisphere controls the right side of the body.   The RIGHT hemisphere controls the left side of the body.

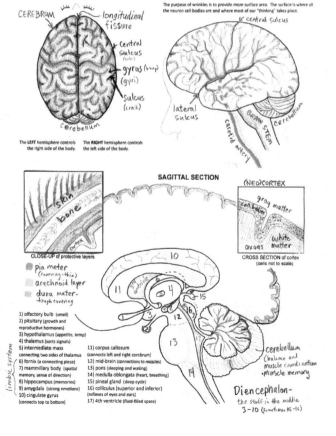

SAGITTAL SECTION   **(NEO)CORTEX**

CLOSE-UP of protective layers   CROSS SECTION of cortex (cells not to scale)

pia mater (covering - thin)
arachnoid layer
dura mater - tough covering

1) olfactory bulb (smell)
2) pituitary (growth and reproductive hormones)
3) hypothalamus (appetite, temp)
4) thalamus (sorts signals)
5) intermediate mass connecting two sides of thalamus
6) fornix (a connecting piece)
7) mammillary body (spatial memory, sense of direction)
8) hippocampus (memories)
9) amygdala (strong emotions)
10) cingulate gyrus (connects top to bottom)
11) corpus callosum (connects left and right cerebrum)
12) mid-brain (connections to muscles)
13) pons (sleeping and waking)
14) medulla oblongata (heart, breathing)
15) pineal gland (sleep cycle)
16) colliculus (superior and inferior) (reflexes of eyes and ears)
17) 4th ventricle (fluid-filled space)

Diencephalon - the stuff in the middle 3-10 (sometimes 15-16)

The LEFT hemisphere is known for:

1) _Speech (90th of time)_
2) _math calculations_
3) _logic and analysis_
4) _Sequencing (parts)_
5) _time_
6) _Symbols (♡ ★ ☺)_

The RIGHT hemisphere is known for:

1) _creativity_
2) _art_
3) _music_
4) _flashes of insight "Eureka!"_
5) _seeing the whole picture_
6) _rhythm of speech_

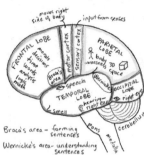

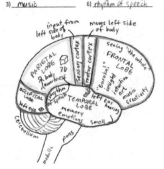

Braca's area - forming sentences

Wernicke's area - understanding sentences

HOW SOME MEMORIES ARE ENCODED:
(This would be declarative/explicit memories.)

Short term involves glutamate crossing synapses.
Long term involves production of new spines on dendrites.

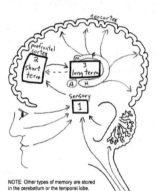

NOTE: Other types of memory are stored in the cerebellum or the temporal lobe.

VENTRICLES:

Ventricles are filled with cerebrospinal fluid (CSF).
Ventricles and CSF help the brain by providing:

1) _buoyancy (brain floats)_
2) _cushioning (protection against bumps)_
3) _nutrition (food for cells)_
4) _communication (hormone messengers)_

CSF is found in the 4 ventricles, as well as under the arachnoid layer around the exterior. The CSF circulates around these spaces, then drains either into veins at the top and middle of the brain, or down into the central canal that goes into the spinal cord.

jugular vein

☐ CHOROID PLEXUS: epithelial cells that produce CSF

There are 3 main types of vessels: **arteries** (away from the heart), **veins** (toward the heart) and **capillaires** (microscopic).

ARTERIES : built for high pressure

The heart is a very strong pump. When blood leaves the heart, it does so under high pressure. Arteries must be able to withstand high pressure. Smooth muscles in the vessels contract with each pump.

VEINS: built for low pressure

Veins experience much less pressure because they are farther away from the heart. In fact, they have one-way valves to ensure that blood does not flow the wrong way.

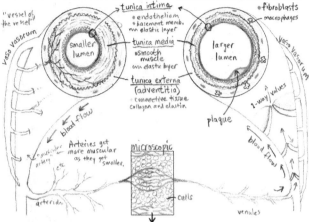

CAPILLARIES form "beds" (networks)

ARTERIOLE (small artery)
The smooth muscles of arterioles control how much blood goes to which parts of the body.

O_2 CO_2 urea Wastes

nutrients

H_2O glucose H_2O creatinine lactic acid

VENULE (small vein)

TYPES of CAPILLARIES:

1) **Continuous**
Where? _most places_

one layer thick epithelial cells (endothelial)

2) **Fenestrated** ("fenestra" = "window")
Where? _intestines, glands, pancreas, kidneys_

3) **Sinusoidal**
Where? _liver, spleen_

Even basement membrane has gaps!

large gaps

ARTERIES go away from the heart.   VEINS go toward the heart.

EXTERIOR ANATOMY

INTERIOR ANATOMY

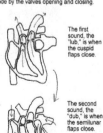

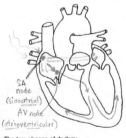

The **pericardium** is a membrane "bag" that goes around (peri) the heart (cardi).

VALVES:
1: Tricuspid
2: Bicuspid (a.k.a. mitral valve)
3: Semilunar valves

"LUB DUB" (the cardiac cycle)

The familiar "lub dub" sound of a beating heart is made by the valves opening and closing.

The first sound, the "lub," is when the cuspid flaps close.

The second sound, the "dub," is when the semilunar flaps close.

Intrinsic Conduction System
(how the heart beats in rhythm)

SA node (sinoatrial)

AV node (atrioventricular)

The two phases of rhythm:
1) _SYSTOLE_ : _contraction_
2) _DIASTOLE_ : _relaxation_

Although the lungs are the main organs of the respiratory system, the action of the lungs is actually called **ventilation**, not respiration. Respiration is what happens in the cells ("cellular respiration" from lessons 18 and 20 in module 1). When the lungs take in air, this is called **inhalation**. When the lungs expel air, this is called **exhalation**.

The **THORACIC CAVITY** contains the lungs, the heart, the trachea and the esophagus.

The **diaphragm** separates the thoracic cavity from the abdominal cavity below.
(The pleura are called serous membranes.)

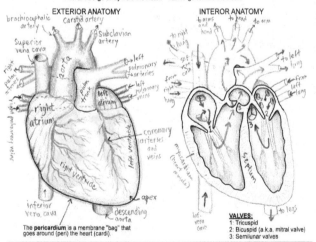

At the end of each bronchiole is a **lobule** made of microscopic **alveoli**. Each alveolus is covered with a bed of capillaries.

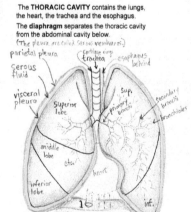

ALVEOLI

ONE ALVEOLUS:

The diaphragm is shaped like an upside down bowl.

Three holes in diaphragm:
1) inferior vena cava, 2) descending aorta, and 3) the esophagus.

✱ Contraction of the diaphragm causes the chest to expand, causing air to rush in.

1) ultra-thin squamous epithelial cells
2) cuboidal epithelial cells that secrete **surfactant**
3) macrophages called "dust cells"
4) layer of water that contains phospholipids that lower the surface tension of water so that O_2 can pass through)
5) endothelial cells of the capillaries
6) red blood cells that pick up the O_2

WHAT CAUSES US TO BREATHE:
The diaphragm is "wired" to the medulla oblongata in the brain. The medulla is very good at sensing small changes in the CO_2 level in the blood and sends signals for the diaphragm and the intercostal muscles (between the ribs) to contract when CO_2 gets too high.

HOW O_2 IS TRANSPORTED IN THE BLOOD:

RBCs contain billions of hemoglobin molecules.

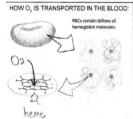

O_2

heme

HOW CO_2 IS TRANSPORTED IN THE BLOOD:

1) A small amount is carried by the globin part of hemoglobin, or is dissolved directly into the plasma.

2) Most CO_2 is combined with water to form carbonic acid (H_2CO_3), then bicarbonate ions (HCO_3^-) and hydrogen ions (H^+). HCO_3^- diffuses out into the plasma. To keep the pH even, Cl^- ions diffuse in to replace HCO_3^-.

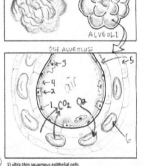

CO_2

$$CO_2 + H_2O \rightarrow H_2CO_3$$
$$H^+ + HCO_3^-$$

Cl^-

HCO_3

The liver is the largest gland in the body. It weighs about 3 lbs (1.5 kg). It is the ultimate "mulit-tasker" and by some counts does as many as 500 jobs! The gall bladder is simply a storage bag for one of the products that the liver makes.

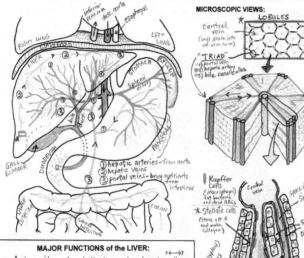

MICROSCOPIC VIEWS:

LOBULES

Central vein (will drain into and vein-ion)

"TRIAD"
1) portal vein
2) hepatic artery
3) bile canaliculus

① hepatic arteries—from aorta
② hepatic veins
③ portal veins—bring nutrients from intestines

Kupffer Cells (macrophages) eat bacteria and dead RBCs

★ Stellate cells (store vit A and make collagen)

Space of Disse

Sinusoid

bile canaliculur

portal vein   hepatic artery

bile canaliculus

smooth ER   mito's
rough ER   lysosomes
space of Disse   Golgi's
peroxisomes for detox   Heme & globin

sometimes double DNA! (polyploidy)

HEPATOCYTES

MAJOR FUNCTIONS of the LIVER:

A: Amino acids : Can make 14 of them, and can transform them. Albumin (blood protein) and Angiotensin (blood pressure)

B: Bile : Emulsifies fats (like dish soap) for digestion. Made of water, bile acids/salts, cholesterol, phospholipids

C: Cholesterol : Liver makes it, also proteins for transport. HDL (high density lipoprotein), LDL (low), VLDL (very low)

C: Clotting factors : fibrinogen, prothrombin, V, VI, VII, IX, X etc. (lesson 38)

C: C-reactive protein : CrP (indicates inflammation). Complexes of immune system C3, C5 etc. (lesson 48)

D: Detox : Breaks down yucky stuff - alcohol, drugs, insulin, NH₃

E: Erythrocytes recycled : Kupffer cells recycle hemoglobin into (lesson 29)

F: Fe (iron) and Fat-soluble vitamins stored (A, D, E, K, B-12)

G: Glucose / glycogen : Liver stores glucose as glycogen

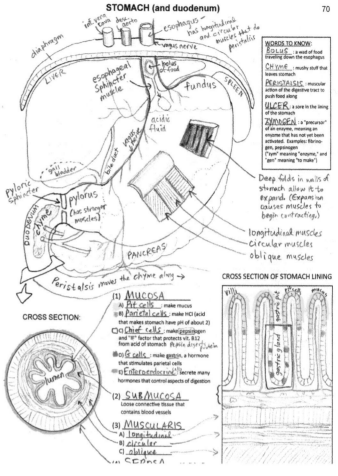

diaphragm

esophagus - has longitudinal and circular muscles that do peristalsis

esophageal sphincter muscle

bolus of food

fundus

LIVER   SPLEEN

vagus nerve

acidic fluid

gall bladder

bile duct

Pyloric Sphincter

pylorus (has stronger muscles)

chyme

DUODENUM

PANCREAS

Peristalsis moves the chyme along →

longitudinal muscles
circular muscles
oblique muscles

WORDS TO KNOW:
Bolus : a wad of food traveling down the esophagus

Chyme : mushy stuff that leaves stomach

Peristalsis : muscular action of the digestive tract to push food along

Ulcer : a sore in the lining of the stomach

Zymogen : a "precursor" of an enzyme, meaning an enzyme that has not yet been activated. Examples: fibrinogen, pepsinogen ("zym" meaning "enzyme," and "gen" meaning "to make")

Deep folds in walls of stomach allow it to expand. (Expansion causes muscles to begin contracting.)

CROSS SECTION:

lumen

CROSS SECTION OF STOMACH LINING

villi   gastric pit   gastric gland

(1) MUCOSA
A) Pit cells : make mucus
B) Parietal cells : make HCl (acid that makes stomach have pH of about 2)
C) Chief cells : make pepsinogen and "IF" factor that protects vit. B12 from acid of stomach. Pepsin digests protein
D) G cells : make gastrin, a hormone that stimulates parietal cells
E) Enteroendocrine cells : secrete many hormones that control aspects of digestion

(2) SUBMUCOSA
Loose connective tissue that contains blood vessels

(3) MUSCULARIS
A) longitudinal
B) circular
C) oblique

(4) SEROSA

The pancreas is a "mixed gland" meaning that it performs both **endocrine** and **exocrine** functions. The exocrine products go into the duodenum. The endocrine products go into the blood.

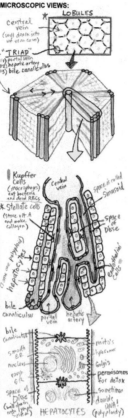

LIVER   STOMACH

gall bladder   PYLORUS

DUODENUM

acidic chyme

sphincters

pancreatic ducts

vesicles filled with zymogens

Golgi   ER   nucleus

ONE ACINUS

The bacteria in our gut like pH 7.

Acid is neutralized by NaHCO₃ from pancreas

Organs communicate using hormones. When the duodenum detects protein, fats and stomach acid coming into it, its cells start making **secretin** and **CCK**. These hormones go into the blood and eventually reach the pancreas and gall bladder, causing them to increase their output.

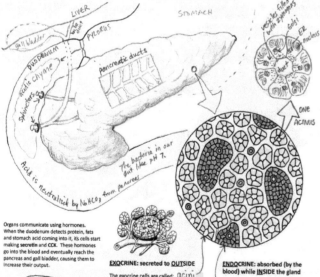

EXOCRINE: secreted to OUTSIDE

The exocrine cells are called: acini

Exocrine products of pancreas:

1) Sodium bicarbonate (NaHCO₃ that neutralizes stomach acid) (In kitchen we call it: baking soda)

2) Proteases break proteins
ex: pepsin (-ogen)
ex: trypsin (-ogen)
ex: chymotrypsin (-ogen)
These are made as zymogens which are activated by an enzyme in the intestines.

3) Lipase breaks apart fats

4) Amylase breaks starch into units of maltose (disaccharide)

5) Nucleases break RNA and DNA

ENDOCRINE: absorbed (by the blood) while INSIDE the gland

The endocrine cells are organized into: islets of Langerhans

Endocrine products of pancreas:

1) α cells make glucagon that tells cells to release glucose into blood (from glycogen)

2) β cells make insulin that tells cells to get glucose out of blood (forming glycogen)

3) δ cells make somatostatin that gives the signal to stop. (stops insulin, glucagon, gastrin)
4) and a few others

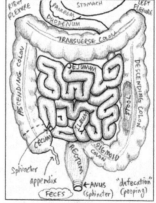

■ = gastrin
■ = secretin
■ = CCK

The intestines consist of two distinct regions: the small intestine (divided into **duodenum**, **jejunum**, and **ileum**), and the **large intestine** (also called the **colon**).

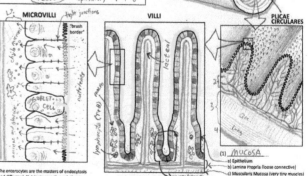

RIGHT FLEXURE   STOMACH   LEFT FLEXURE

PANCREAS   DUODENUM

TRANSVERSE COLON

ASCENDING COLON   DESCENDING COLON

JEJUNUM   ILEUM

CECUM   SIGMOID COLON

RECTUM

Sphincter   appendix   FECES   ANUS (Sphincter)   "defecation" (pooping)

What do the jejunum and ileum have in common?
1) They secrete digestive enzymes and absorb nutrients into blood and lymph vessels
2) They have circular folds that increase surface area.

Differences between jejunum and ileum:

Jejunum ("fasting")	Ileum ("twisting")
1) shorter (2 meters)	1) longer (4-5 meters)
2) thicker walls	2) thinner walls
3) stronger peristalsis	3) bile salts and B12 absorbed
4) more blood vessels	4) Peyer's patches (lymph tissue)

The colon's function is to reabsorb:
1) water   2) minerals   3) vitamins (K) B's

mesentery   visceral peritoneum

MICROVILLI   **VILLI**   **PLICAE CIRCULARES**

"brush border"

GOBLET CELL

lacteal

nutrients

crypts/glands

The enterocytes are the masters of endocytosis and diffusion! Nutrients pass through these cells and out the other side. The cells package triglycerides into chylomicrons that go into the lacteals (lymph vessels). Glucose and amino acids go into the blood capillaries.

■ = enterocytes   ■ = goblet cells
■ = enteroendocrine cells   ■ = stem cells
■ = Paneth cells (antibiotic chemicals)

(1) MUCOSA
a) Epithelium
b) Lamina Propria (loose connective)
c) Muscularis Mucosa (very tiny muscles)

(2) SUBMUCOSA

(3) MUSCULARIS (circular, long.)

(4) SEROSA (visceral peritoneum)

BODY CAVITIES are large sections of the body that are enclosed by membranes.

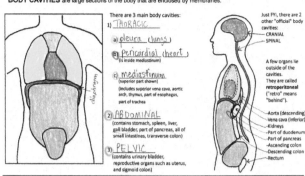

There are 3 main body cavities:

1) THORACIC
   a) pleura (lungs)
   b) pericardial (heart)
      (is inside mediastinum)
   c) mediastinum
      (superior part shown)
      (includes superior vena cava, aortic arch, thymus, part of esophagus, part of trachea)
2) ABDOMINAL
   (contains stomach, spleen, liver, gall bladder, part of pancreas, all of small intestines, transverse colon)
3) PELVIC
   (contains urinary bladder, reproductive organs such as uterus, and sigmoid colon)

Just FYI, there are 2 other "official" body cavities:
- CRANIAL
- SPINAL

A few organs lie outside of the cavities. They are called **retroperitoneal** ("retro" means "behind").

- Aorta (descending)
- Vena cava (inferior)
- Kidneys
- Part of duodenum
- Part of pancreas
- Ascending colon
- Descending colon
- Rectum

THE MESENTERY is a very thin membrane that holds all the organs in place. The mesentery also provides a surface for nerves, blood vessels, and lymph vessels. Mesentery is made of serous membrane (which is made of a layer of simple squamous epithelial cells stuck to a layer of connective tissue).

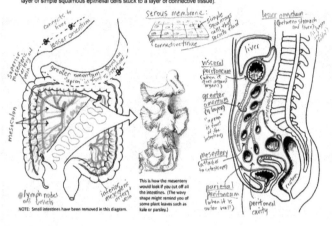

NOTE: Small intestines have been removed in this diagram.

This is how the mesentery would look if you cut off all the intestines. (The wavy shape might remind you of some plant leaves such as kale or parsley.)

The functional unit of the kidney is the **NEPHRON**

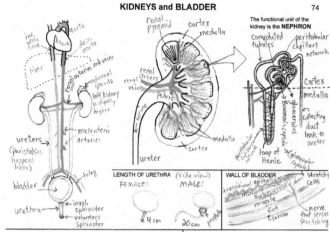

left kidney is slightly higher

ureters (peristalsis happens here)

bladder

urethra — invol. sphincter, voluntary sphincter

LENGTH OF URETHRA (side view)
FEMALE: 4 cm
MALE: 20 cm

WALL OF BLADDER

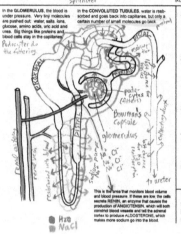

In the **GLOMERULUS**, the blood is under pressure. Very tiny molecules are pushed out: water, salts, ions, glucose, amino acids, uric acid and urea. Big things like proteins and blood cells stay in the capillaries. Podocytes do the filtering.

In the **CONVOLUTED TUBULES**, water is reabsorbed and goes back into capillaries, but only a certain number of small molecules go back.

This is the area that monitors blood volume and blood pressure. If these are low, the cells secrete RENIN, an enzyme that causes the production of ANGIOTENSIN, which will both constrict blood vessels and tell the adrenal cortex to produce ALDOSTERONE, which makes more sodium go into the blood.

SUMMARY OF KIDNEY FUNCTIONS:

1) Excretion of metabolic wastes
   Nitrogenous wastes come from the breakdown of amino acids, which have a nitrogen atom. (The liver turns ammonia into urea.)

2) Maintenance of blood pressure
   This is achieved through a balance of water and salt in the blood. The more salt in the blood, the more water goes into the blood, and that means greater blood volume and greater pressure. To increase blood pressure, the kidneys secrete renin, which activates angiotensin, which tells the adrenals to make aldosterone which causes more sodium to go into the blood.

3) pH balance (blood must be 7.4)
   The kidneys can excrete or reabsorb both H+ ions (which make things acidic) and HCO_3^- ions (alkaline).

4) Secretion of erythropoietin
   If the kidneys sense that there is less oxygen in the blood, they will begin to produce more erythropoietin, which tells the hematopoietic stem cells to make more red blood cells.

5) Converting vit.D to active form
   Vitamin D from the diet must be converted to a more active form that the digestive system can use to absorb calcium ions.

NOTE: The pituitary gland (in the brain) secretes a chemical called ADH (anti-diuretic hormone) at night, which causes more water to be reabsorbed, making less urine.

You will want to go back and review drawings 35 and 36 before doing this drawing.

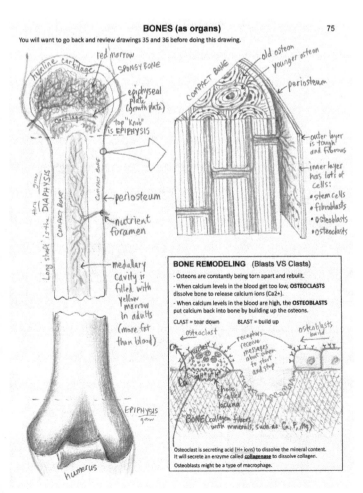

red marrow
SPONGY BONE
epiphyseal plate (growth plate)
top "knob" is EPIPHYSIS
hyaline cartilage
COMPACT BONE
periosteum
nutrient foramen
medullary cavity is filled with yellow marrow in adults (more fat than blood)
Long shaft is the DIAPHYSIS
EPIPHYSIS

old osteon
younger osteon
periosteum
outer layer is tough and fibrous
inner layer has lots of cells:
- stem cells
- fibroblasts
- osteoblasts
- osteoclasts

humerus

BONE REMODELING (Blasts VS Clasts)

- Osteons are constantly being torn apart and rebuilt.
- When calcium levels in the blood get too low, OSTEOCLASTS dissolve bone to release calcium ions (Ca2+).
- When calcium levels in the blood are high, the OSTEOBLASTS put calcium back into bone by building up the osteons.

CLAST = tear down   BLAST = build up

osteoclast   osteoblasts build

receptors receive messages about when to start and stop

hole called lacuna

BONE (collagen fibers with minerals, such as Ca, P, Mg)

Osteoclast is secreting acid (H+ ions) to dissolve the mineral content. It will secrete an enzyme called <u>collagenase</u> to dissolve collagen.

Osteoblasts might be a type of macrophage.

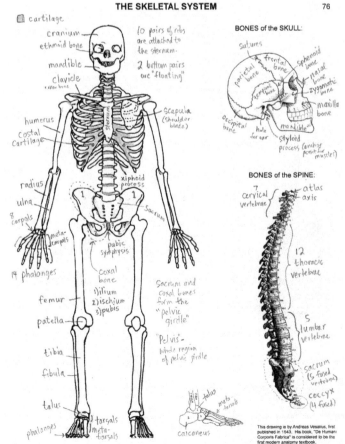

cartilage
cranium
ethmoid bone
mandible
clavicle (collar bone)
humerus
costal cartilage
radius
ulna
8 carpals
metacarpals
14 phalanges
femur
patella
tibia
fibula
talus
tarsals
metatarsals
phalanges

10 pairs of ribs are attached to the sternum.
2 bottom pairs are "floating"

scapula (shoulder blade)
xiphoid process
sacrum
pubic symphysis
coxal bone
1) ilium
2) ischium
3) pubis

Sacrum and coxal bones form the "pelvic girdle"

Pelvis - whole region of pelvic girdle

talus
metatarsals
calcaneus

BONES of the SKULL:
sutures
parietal bone
frontal bone
sphenoid bone
nasal bone
zygomatic bone
maxilla bone
temporal bone
occipital bone
hole for ear
mandible
styloid process (anchor point for muscles)

BONES of the SPINE:
7 cervical vertebrae
atlas
axis
12 thoracic vertebrae
5 lumbar vertebrae
sacrum (5 fused vertebrae)
coccyx (4 fused)

This drawing is by Andreas Vesalius, first published in 1543. His book, "De Humani Corporis Fabrica" is considered to be the first modern anatomy textbook.

There are three kinds of joints: 1) FIBROUS, 2) CARTILAGINOUS, 3) SYNOVIAL

FIBROUS: (don't move at all)

Ex: sutures in skull, teeth in sockets, ends of ulna/radius, tibia/fibula

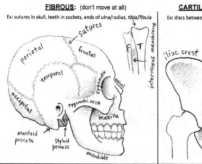

CARTILAGINOUS: (move only slightly)

Ex: discs between vertebrae, pubic symphysis

SYNOVIAL: (very flexible)

Synovial joints have fluid-filled capsules in and around the joint to decrease friction. They also have slippery (white) hyaline pads. There are 6 types of synovial joints: hinge, ball and socket, pivot, saddle, plane and ellipsoidal.

HINGE: the knee (shown) and the elbow

▨ hyaline cartilage	▢ synovial cavity	▨ fat
▨ ligaments, tendons	▨ bursa (fluid filled sac)	fluid

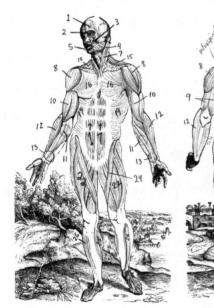

BALL AND SOCKET: hip and shoulder

HIP:

SHOULDER:

There are 3 kinds of muscles: 1) SKELETAL (voluntary), 2) SMOOTH (involuntary), 3) CARDIAC (heart)

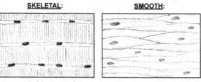

SKELETAL:

1) no individual cells (fibers)
2) stripes (sarcomeres)
3) uses creatine for ATPs
4) When creatine runs out, it burns glucose from glycogen in cellular respiration (ETC)

SMOOTH:

1) single nucleus cells
2) no stripes (spindles)
3) low energy needs (uses ETC)
4) EX's: peristalsis, iris, sphincters, blood vessels, ducts, bladder, uterus, and more.

CARDIAC:

1) single nucleus cells
2) tubular, branched
3) cellular resp. burns fatty acids
4) Intercalated discs have gap junctions that connect cells.

SKELETAL MUSCLES WORK IN PAIRS

Muscles can only do one thing: CONTRACT.
A prime mover and its antagonist work together.

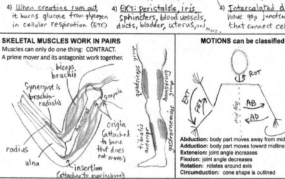

biceps brachii
Synergist is brachio-radialis
radius
ulna
scapula
origin (attached to bone that does not move)
insertion (attaches to moving bone)
quadriceps
hamstring
tibialis anterior

MOTIONS can be classified

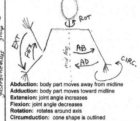
ROT
EXT
FLEX
mid hip
AB
AD
CIRC.

Abduction: body part moves away from midline
Adduction: body part moves toward midline
Extension: joint angle increases
Flexion: joint angle decreases
Rotation: rotates around axis
Circumduction: cone shape is outlined

WHERE DO SKELETAL MUSCLES GET THEIR ENERGY? Here they are, in order of preference.

1) CREATINE PHOSPHATE

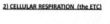

CREATINE is frist choice, but can be sustained for only 8 seconds. Creatine holds onto a phosphate. An enzyme can take the P off, and then put it onto an ADP, making ATP. No oxygen is needed.

2) CELLULAR RESPIRATION (the ETC)

needs O_2
NADH
ATP

After 8 seconds, cellular respiration kicks in. Oxygen is needed for the Electron Transport Chain to turn ADP back into ATP. Glucose from glycogen is the preferred fuel for the ETC in skeletal muscles.

3) FERMENTATION

Glycolysis → 2 pyruvates
NADH
← lactic acid

Lactic acid fermentation is the third and last choice if oxygen is not available. Lactic acid gives that burning sensation in muscles when they are fatigued.

In this lesson, we will be using drawings made by famous anatomist Andreas Vesalius in the year 1555.

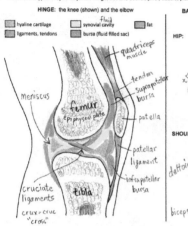
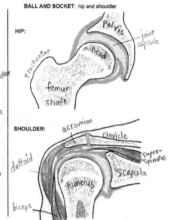

HEAD and NECK

1) **Frontalis:** wrinkles forehead and moves eyebrows.
2) **Orbicularis oculi:** closes eyes
3) **Zygomaticus:** smiling
4) **Masseter:** closes jaw
5) **Orbicularis oris:** closes and protrudes lips (like a kiss)
6) **Occipitalis:** moves scalp backwards
7) **Sternocleidomastoid:** turns and twists head

UPPER LIMBS

8) **Deltoid:** raises arm at shoulder joint ("delts")
9) **Triceps brachii:** straightens arm
10) **Biceps brachii:** bends arm at elbow
11) **Flexor carpi group:** bends hand down at wrist
12) **Extensor carpi:** pulls hand up at wrist
13) **Flexor digitorum:** closes hand
14) **Extensor digitorum:** opens hand

TORSO

15) **Trapezius:** moves head, shrugs shoulders ("traps")
16) **Pectoralis major:** ("pecs") pulls arm across chest
17) **Rectus abdominis:** "abs" "sit-up" muscles
18) **Latissimus dorsi:** ("lats") pulls arm across back and extends shoulders
19) **External oblique:** rotates torso
20) **Teres major and minor:** pulls arm down and back

LOWER LIMBS

21) **Gluteus maximus:** going from sitting to standing
22) **Quadriceps group:** straightens leg
23) **Hamstring group:** bends leg at knee
24) **Sartorius:** rotates thigh (so you can sit cross-legged)
25) **Gastrocnemius:** points toes ("calf")
26) **Tibialis anterior:** pulls toes up, and inverts foot
27) **Achilles tendon**

Endocrine glands secrete hormones into the blood. Hormones are messenger molecules.

PEPTIDE HORMONES are made of amino acids

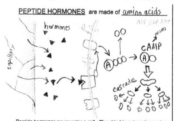

hormones
cAMP
cascade

Peptide hormones never enter a cell. They bind to external receptors. Usually, ATP is turned into cAMP, which starts a cascade reaction. Cascades allow for rapid manufacturing.

STEROID HORMONES are made using cholesterol

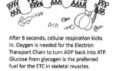

hormones
receptor
coding needed
ribosomes

Steroid hormones enter the cell and bind to a receptor inside. That receptor molecule will attach to DNA and cause a certain part to be copied into mRNA, which will then build a protein.

THE ENDOCRINE GLANDS

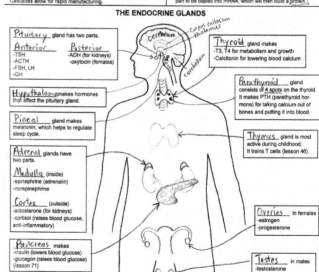
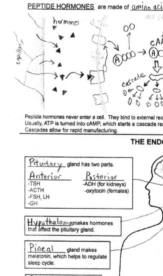

Pituitary gland has two parts.

Anterior	**Posterior**
-TSH	-ADH (for kidneys)
-ACTH	-oxytocin (females)
-FSH, LH	
-GH	

Hypothalamus makes hormones that affect the pituitary gland.

Pineal gland makes melatonin, which helps to regulate sleep cycle.

Adrenal glands have two parts.

Medulla (inside)
-epinephrine (adrenalin)
-norepinephrine

Cortex (outside)
-aldosterone (for kidneys)
-cortisol (raises blood glucose, anti-inflammatory)

Pancreas makes
-insulin (lowers blood glucose)
-glucagon (raises blood glucose) (lesson 71)

Thyroid gland makes
-T3, T4 for metabolism and growth
-Calcitonin for lowering blood calcium

Parathyroid gland consists of 4 spots on the thyroid. It makes PTH (parathyroid hormone) for taking calcium out of bones and putting it into blood.

Thymus gland is most active during childhood. It trains T cells (lesson 46).

Ovaries in females
-estrogen
-progesterone

Testes in males
-testosterone

Cerebrum, Corpus callosum, thalamus, cerebellum

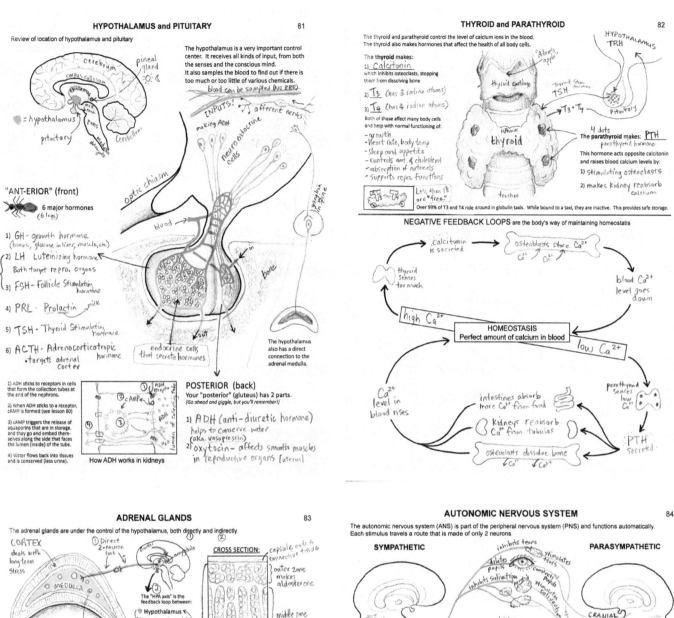

Review of location of hypothalamus and pituitary

cerebrum
pineal gland
corpus callosum
thalamus
= hypothalamus
pons
Cerebellum
pituitary

The hypothalamus is a very important control center. It receives all kinds of input, from both the senses and the conscious mind.
It also samples the blood to find out if there is too much or too little of various chemicals.
blood can be sampled (No BBB)

INPUTS! afferent nerves
making ADH
neuroendocrine cells

optic chiasm

blood

in

bone

"ANT-ERIOR" (front)

6 major hormones
(6 legs)

1) GH - growth hormone
   (bones, glucose in liver, muscle, etc)

2) LH - Luteinizing hormone
3) FSH - Follicle Stimulating hormone
   Both target repro. organs

4) PRL - Prolactin → milk

5) TSH - Thyroid Stimulating hormone

6) ACTH - Adrenocorticotropic hormone
   • targets adrenal cortex

endocrine cells that secrete hormones

The hypothalamus also has a direct connection to the adrenal medulla.

1) ADH sticks to receptors in cells that form the collection tubes at the end of the nephrons.

2) When ADH sticks to a receptor, cAMP is formed (see lesson 80)

3) cAMP triggers the release of aquaporins that are in storage, and they go and embed themselves along the side that faces the lumen (inside) of the tube.

4) Water flows back into tissues and is conserved (less urine).

How ADH works in kidneys

POSTERIOR (back)
Your "posterior" (gluteus) has 2 parts.
(Go ahead and giggle, but you'll remember!)

1) ADH (anti-diuretic hormone)
   helps to conserve water
   (aka. vasopressin)
2) oxytocin - affects smooth muscles in reproductive organs [uterus]

The thyroid and parathyroid control the level of calcium ions in the blood.
The thyroid also makes hormones that affect the health of all body cells.

The **thyroid** makes:

1) Calcitonin
   which inhibits osteoclasts, stopping them from dissolving bone

2) T_3 (has 3 iodine atoms)

3) T_4 (has 4 iodine atoms)

Both of these affect many body cells and help with normal functioning of:
- growth
- heart rate, body temp
- sleep and appetite
- controls amt. of cholesterol
- absorption of nutrients
- supports repro. functions

HYPOTHALAMUS
TRH
TSH
Pituitary
$T_3 + T_4$

Thyroid cartilage
isthmus
thyroid
trachea
Adam's apple

4 dots
The **parathyroid** makes: PTH
parathyroid hormone
This hormone acts opposite calcitonin and raises blood calcium levels by:
1) stimulating osteoclasts
2) makes kidney reabsorb calcium

T_3 T_4 Less than 1% are "free"

Over 99% of T3 and T4 ride around in globulin taxis. While bound to a taxi, they are inactive. This provides safe storage.

NEGATIVE FEEDBACK LOOPS are the body's way of maintaining homeostasis

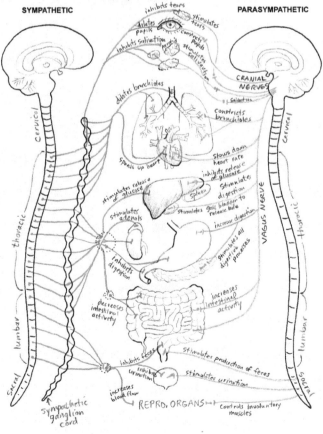

calcitonin is secreted → osteoblasts store Ca^{2+}

Thyroid senses too much
high Ca^{2+}
blood Ca^{2+} level goes down

HOMEOSTASIS
Perfect amount of calcium in blood

low Ca^{2+}

Ca^{2+} level in blood rises
intestines absorb more Ca^{2+} from food
kidneys reabsorb Ca^{2+} from tubules
osteoclasts dissolve bone

parathyroid senses low Ca^{2+}
PTH secreted

The adrenal glands are under the control of the hypothalamus, both directly and indirectly.

CORTEX
deals with long term stress

MEDULLA

KIDNEY

blood vessels (for fast release)

① Direct 2-neuron link
amygdala

The "HPA axis" is the feedback loop between:
▸ Hypothalamus
▸ Pituitary (anterior)
▸ Adrenal cortex

In response to stress, the hypothalamus tells the pituitary to secrete ACTH, which acts on the adrenals to make them secrete cortisol.

CROSS SECTION:
capsule made of connective tissue
outer zone makes aldosterone
middle zone makes cortisol (raises blood glucose, reduces inflammation)
inner zone makes DHEA sex hormones
MEDULLA makes adrenalin
nerve

MEDULLA hormones deal with immediate stress:

1) adrenaline aka epinephrine
2) noradrenaline aka norepinephrine

Effects: "Fight or Flight"
Heart Speeds up   Vessels Constrict
Bronchioles relax   Pupils dilate
Glucose level in blood goes Up
Clotting factors get ready in the blood.
Slowing down of: digestion, salivation, and urination

RECOVERY: In 30-60 minutes the body will have gotten rid of all the adrenaline and noradrenaline molecules.

ANOTHER NEGATIVE FEEDBACK LOOP:

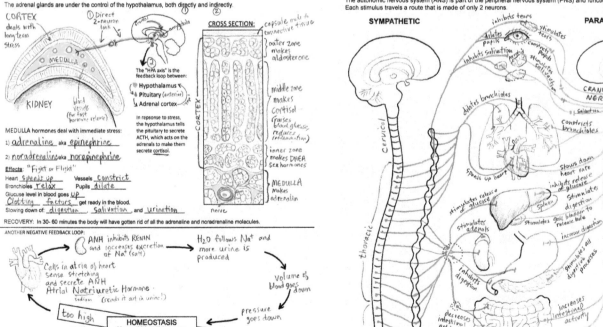

ANH inhibits RENIN and increases excretion of Na^+ (salt)

H_2O follows Na^+ and more urine is produced

Cells in atria of heart sense stretching and secrete ANH
Atrial Natriuretic Hormone
sodium (sends it out in urine!)

Volume of blood goes down

too high
pressure

HOMEOSTASIS
Normal blood pressure

pressure goes down

too low

less urine

Na^+ and H_2O reabsorbed by kidneys

Vessels constrict

Kidneys sense low pressure

Adrenal cortex is told to make ALDOSTERONE

ANGIOTENSIN is activated in blood by RENIN

RENIN is secreted

The autonomic nervous system (ANS) is part of the peripheral nervous system (PNS) and functions automatically.
Each stimulus travels a route that is made of only 2 neurons.

SYMPATHETIC   **PARASYMPATHETIC**

inhibits tears   stimulates tears
dilates pupils   constricts pupils
inhibits salivation   stimulates salivation

CRANIAL NERVES

Cervical
Thoracic
Lumbar
Sacral

dilates bronchioles   constricts bronchioles
Speeds up heart   slows down heart rate
stimulates release of glucose   inhibits release of glucose
stimulates digestion
spleen   stimulates gall bladder to release bile
stimulates adrenals
inhibits digestion   stimulates all digestive processes
increase digestion
decreases intestinal activity   increases intestinal activity

VAGUS NERVE

inhibits feces   stimulates production of feces
inhibits urination   stimulates urination
increases blood flow
REPRO. ORGANS   controls involuntary muscles

sympathetic ganglion cord

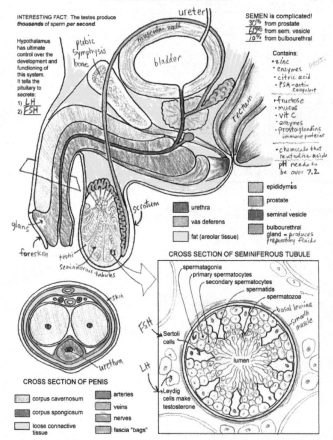

INTERESTING FACT: The testes produce *thousands* of sperm *per second*.

Hypothalamus has ultimate control over the development and functioning of this system. It tells the pituitary to secrete:
1) LH
2) FSH

SEMEN is complicated!
30% from prostate
60% from sem. vesicle
10% from bulbourethral

Contains:
- zinc
- enzymes
- citric acid
- PSA - anti-coagulant
- fructose
- mucus
- vit C
- enzymes
- prostaglandins — immune proteins
- chemicals that neutralize acids
pH needs to be over 7.2

■ epididymis
□ urethra
□ vas deferens
□ fat (areolar tissue)
■ prostate
■ seminal vesicle
■ bulbourethral gland – produces preparatory fluids

CROSS SECTION OF SEMINIFEROUS TUBULE

spermatogonia
primary spermatocytes
secondary spermatocytes
spermatids
spermatozoa
basal lamina
smooth muscle
Sertoli cells
lumen
Leydig cells make testosterone
FSH
LH

CROSS SECTION OF PENIS

□ corpus cavernosum
■ corpus spongiosum
□ loose connective tissue
■ arteries
■ veins
■ nerves
■ fascia "bags"

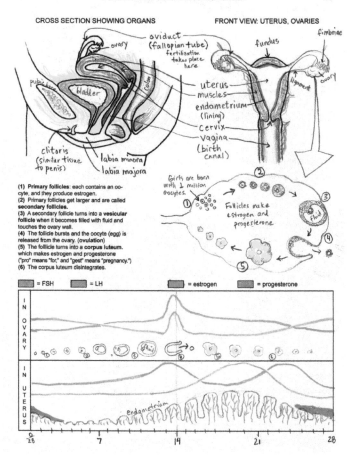

CROSS SECTION SHOWING ORGANS   **FRONT VIEW: UTERUS, OVARIES**

ovary
oviduct (fallopian tube) — fertilization takes place here
fundus
fimbriae
ovary
ligament
pubic bone
bladder
colon
uterus
muscles
endometrium (lining)
cervix
vagina (birth canal)
clitoris (similar tissue to penis)
labia minora
labia majora

(1) **Primary follicles:** each contains an oocyte, and they produce estrogen.
(2) Primary follicles get larger and are called **secondary follicles**.
(3) A secondary follicle turns into a **vesicular follicle** when it becomes filled with fluid and touches the ovary wall.
(4) The follicle bursts and the oocyte (egg) is released from the ovary. (ovulation)
(5) The follicle turns into a **corpus luteum**, which makes estrogen and progesterone ("pro" means "for," and "gest" means "pregnancy.")
(6) The corpus luteum disintegrates.

Girls are born with 1 million oocytes.
Follicles make estrogen and progesterone.
fluid

■ = FSH   ■ = LH   ■ = estrogen   ■ = progesterone

IN OVARY

IN UTERUS
endometrium

0/28   7   14   21   28

Removable activity pages

For your convenience (and to save money on color printing) here are some of the color pages that go with the recommended activities for various lessons.

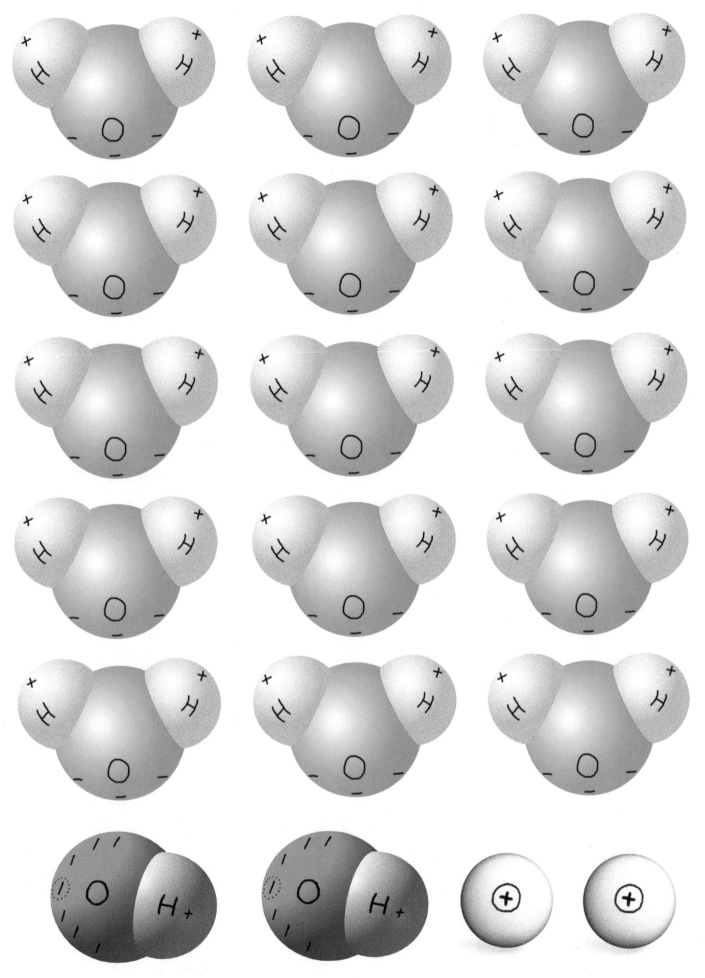

CUT THESE OUT AND SAVE THEM IN A PLASTIC BAG OR ENVELOPE FOR FUTURE USE.

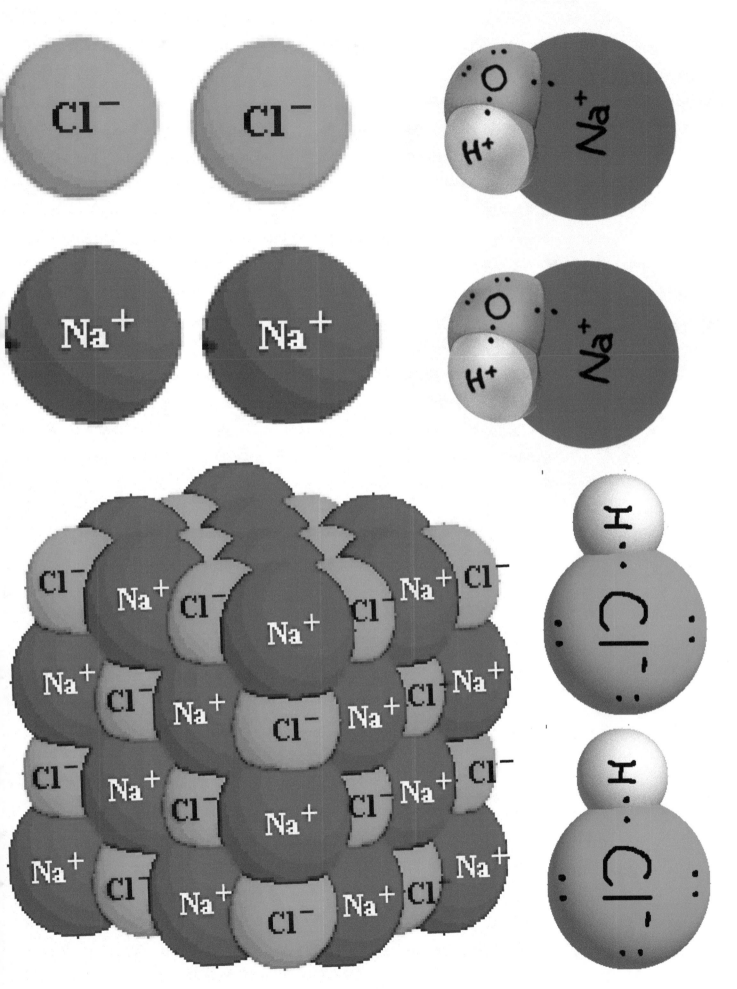

CUT THESE OUT AND SAVE THEM IN A PLASTIC BAG OR ENVELOPE FOR FUTURE USE.

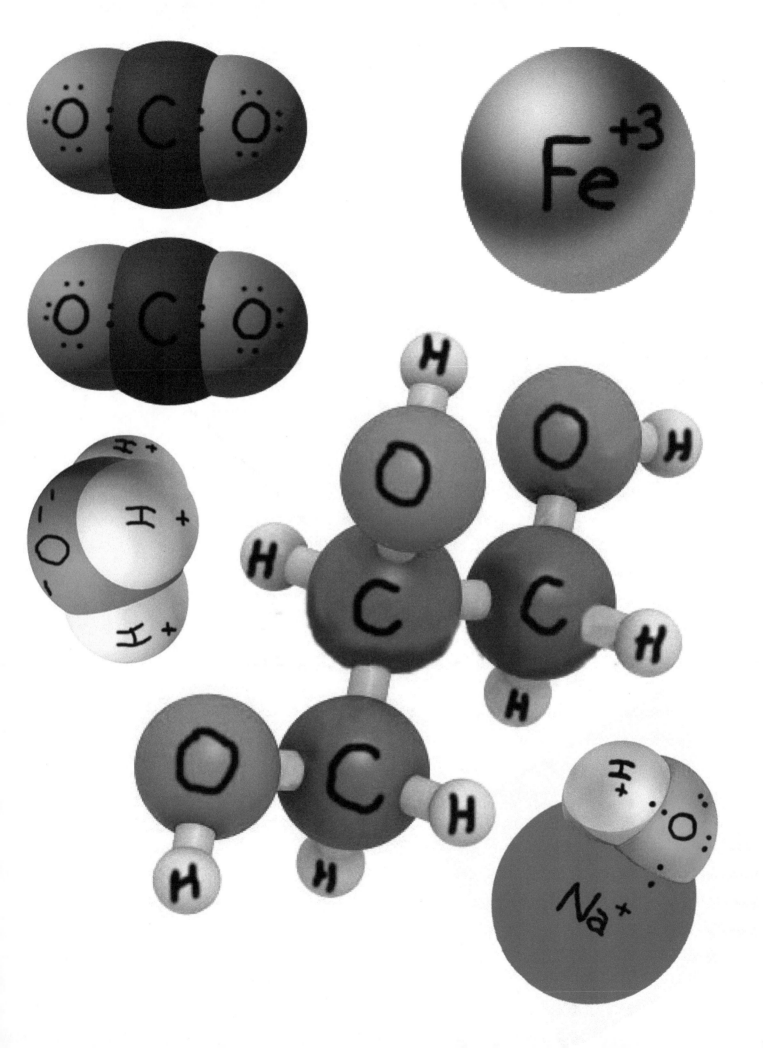

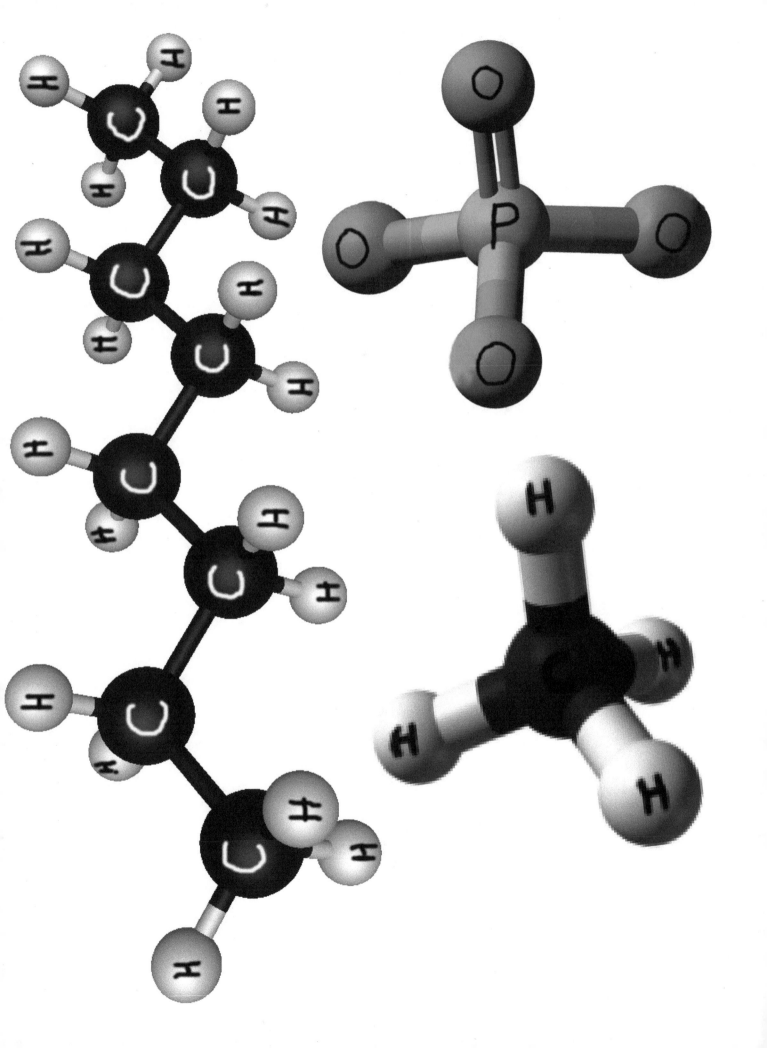

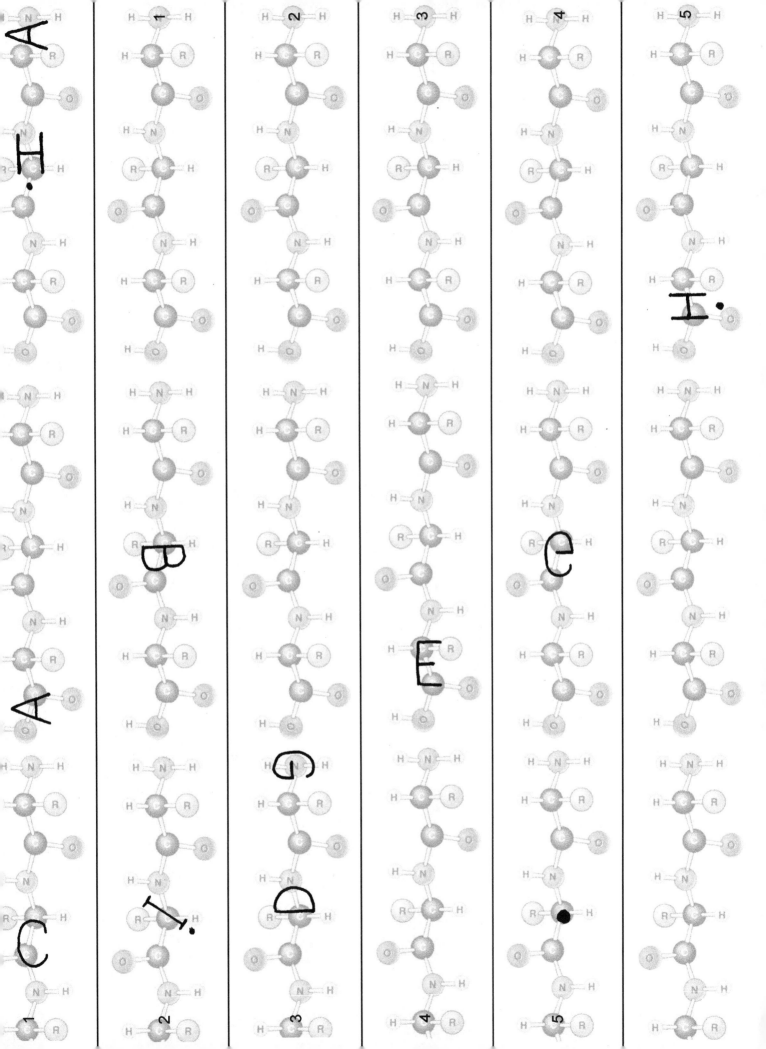

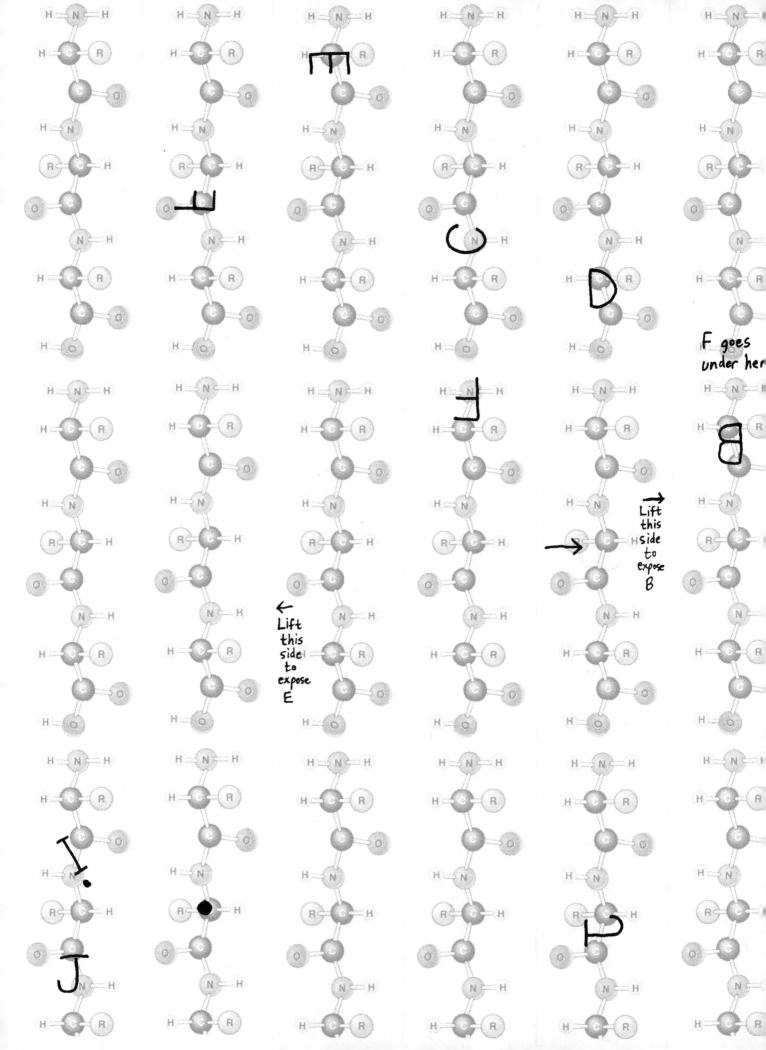

F goes
under her

Lift
this
side
to
expose
B

Lift
this
side
to
expose
E

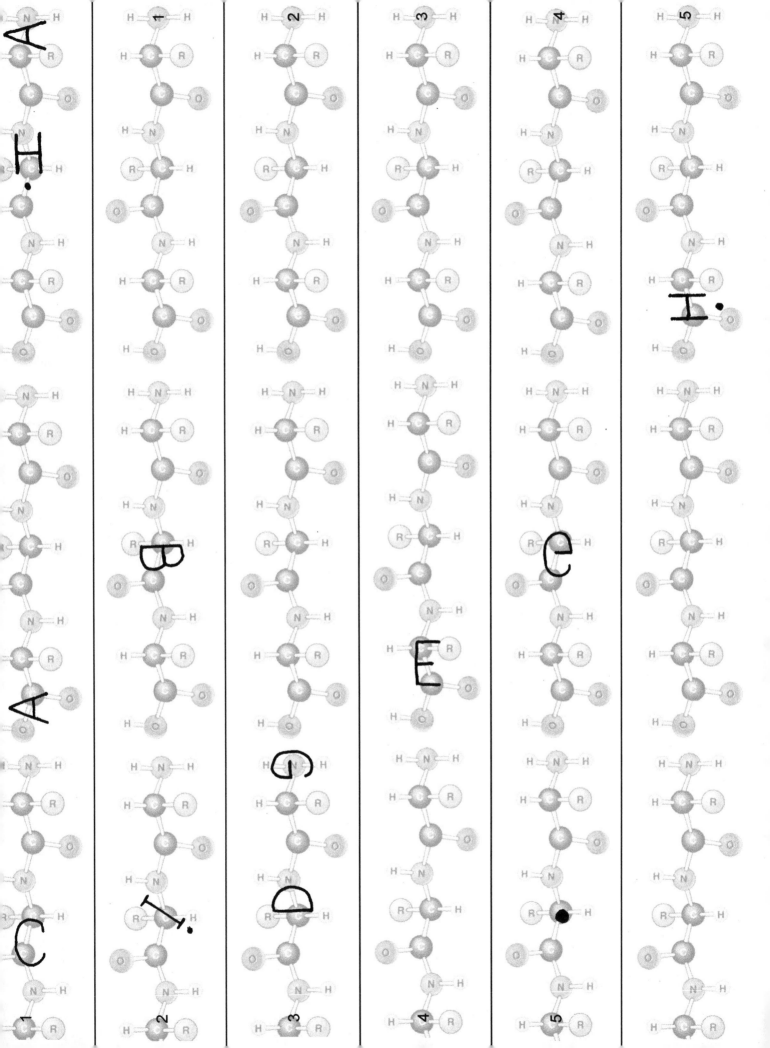

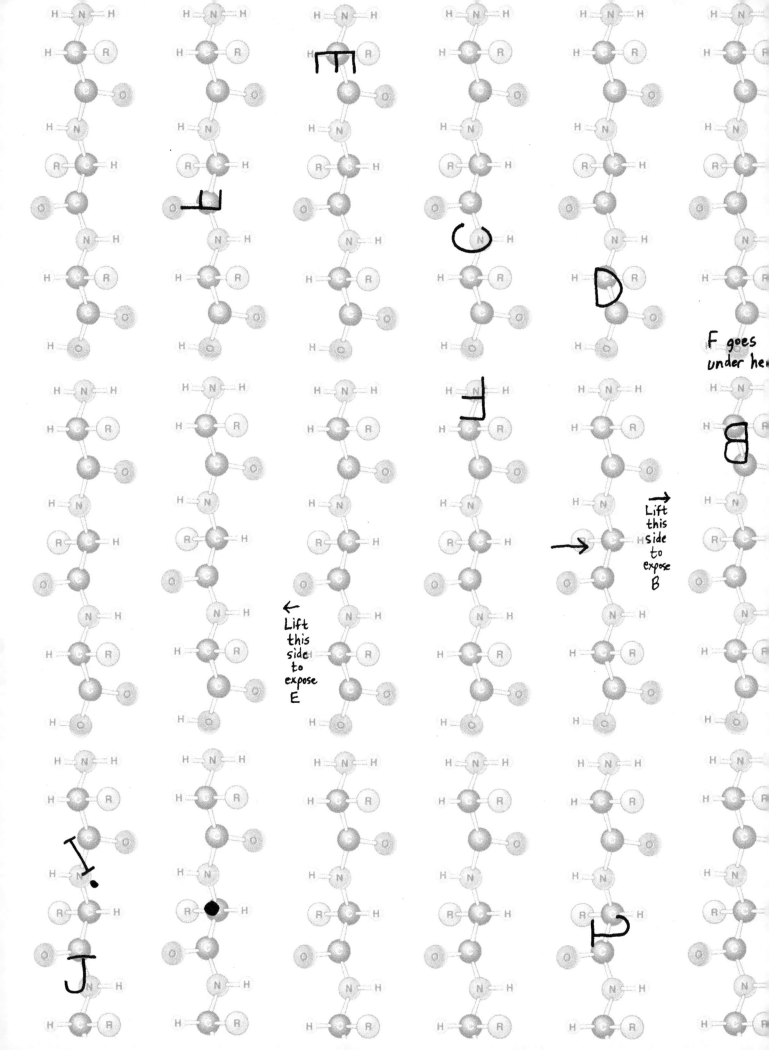

F goes
under he[...]

Lift
this
side
to
expose
B

Lift
this
side
to
expose
E

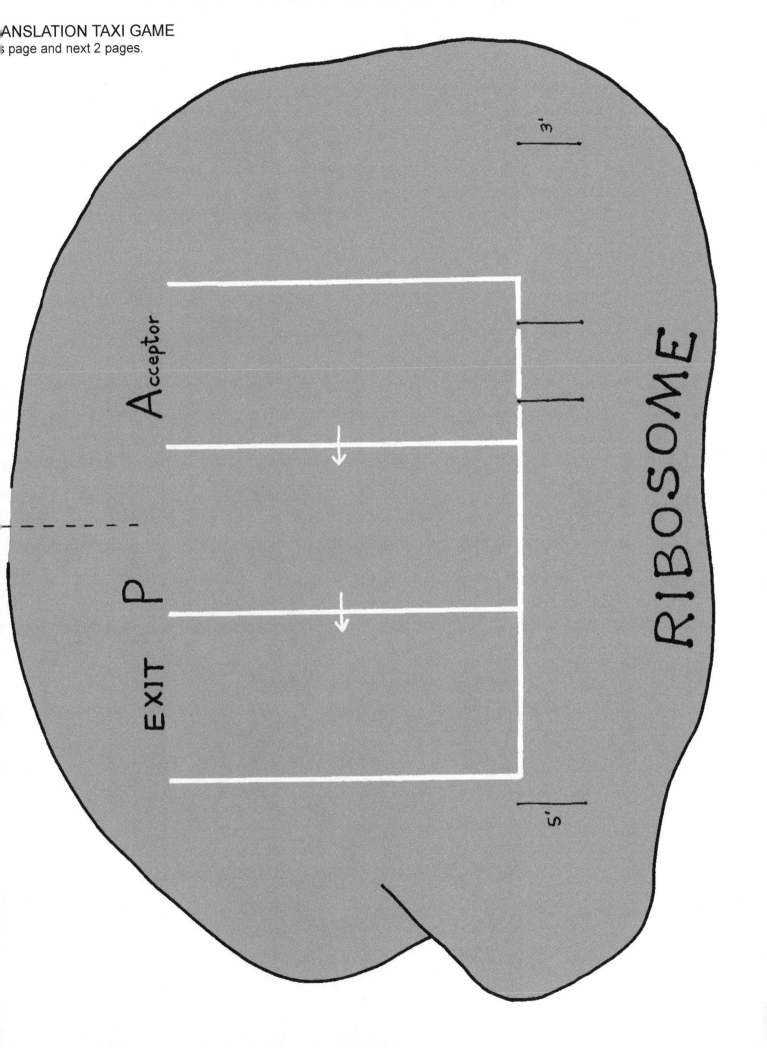

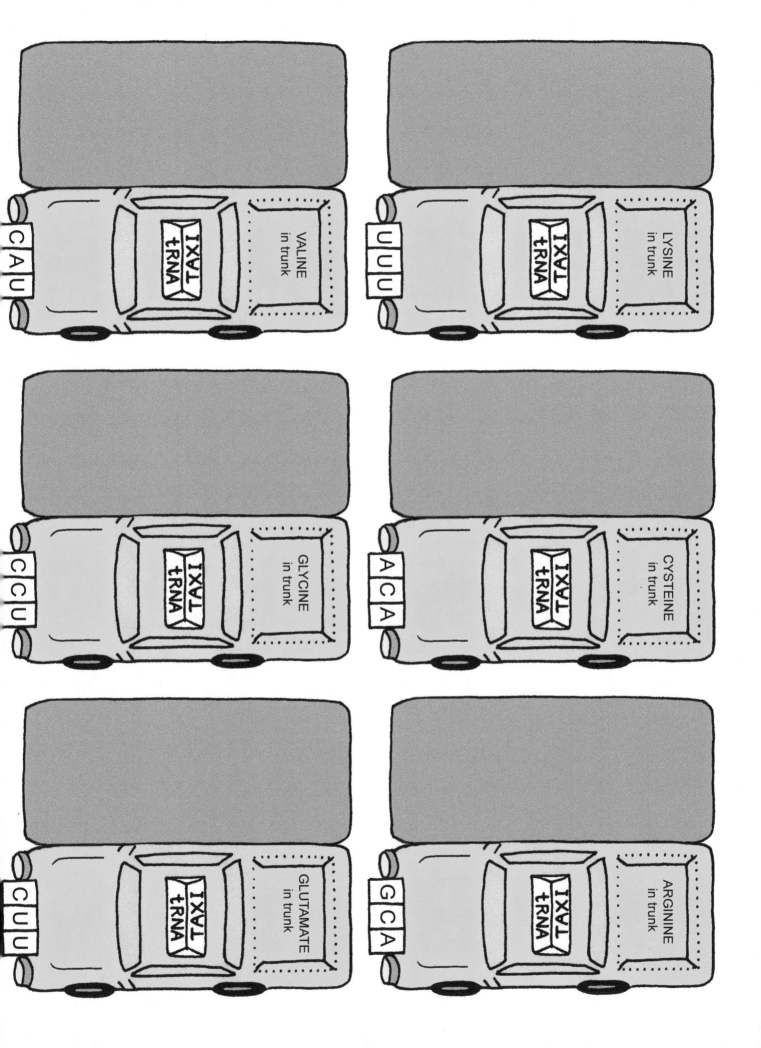

mRNA¹: 5′ start — GAA CGU GUA CGU UGU GAA AAA — AUG GUA AAA GGA UGU end 3′

mRNA²: 5′ start — UGU GGA AAA CGU GAA UGU — AUG AAA GUA GAA GGA end 3′

mRNA³: 5′ start — GUA AAA UGU GAA GGA CGU GGA — AUG GGA UGU GUA AAA end 3′

mRNA⁴: 5′ start — AAA GGA AAA GUA GAA GGA UGU — AUG GGA GAA GUA end 3′

mRNA⁵: 5′ start — GUA GGA GUA UGU UGU GGA GAA AAA — AUG GGA UGU AAA GAA end 3′

CELLULAR RESPIRATION GAME
This page and next 4 pages.

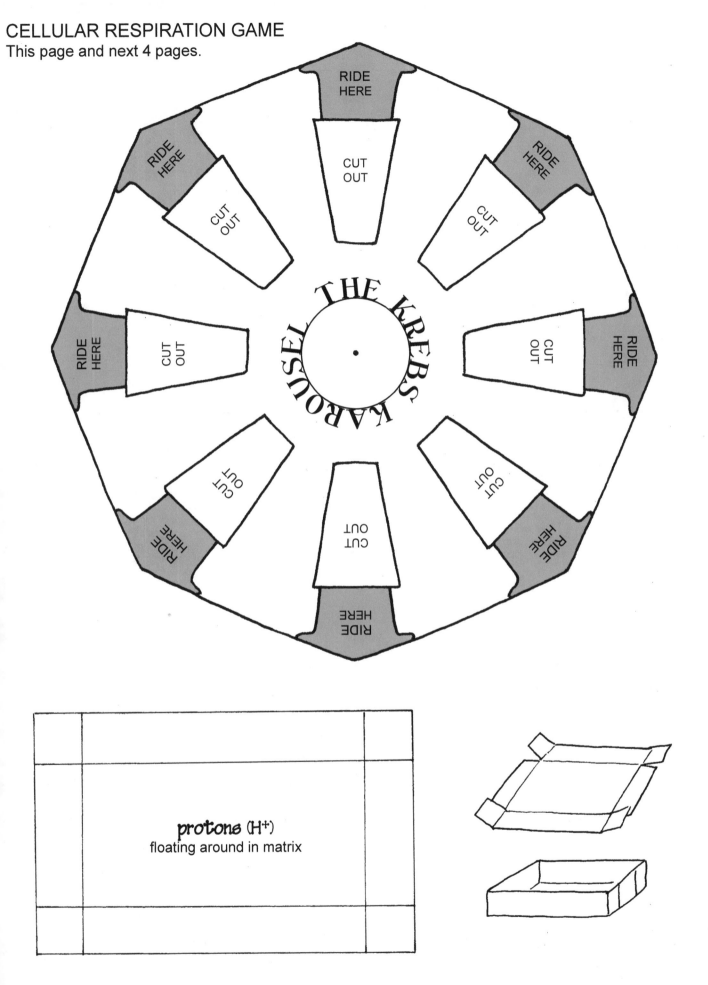

THE KREBS KAROUSEL

RIDE HERE

CUT OUT

protons (H+)
floating around in matrix

COPY ONTO CARD STOCK

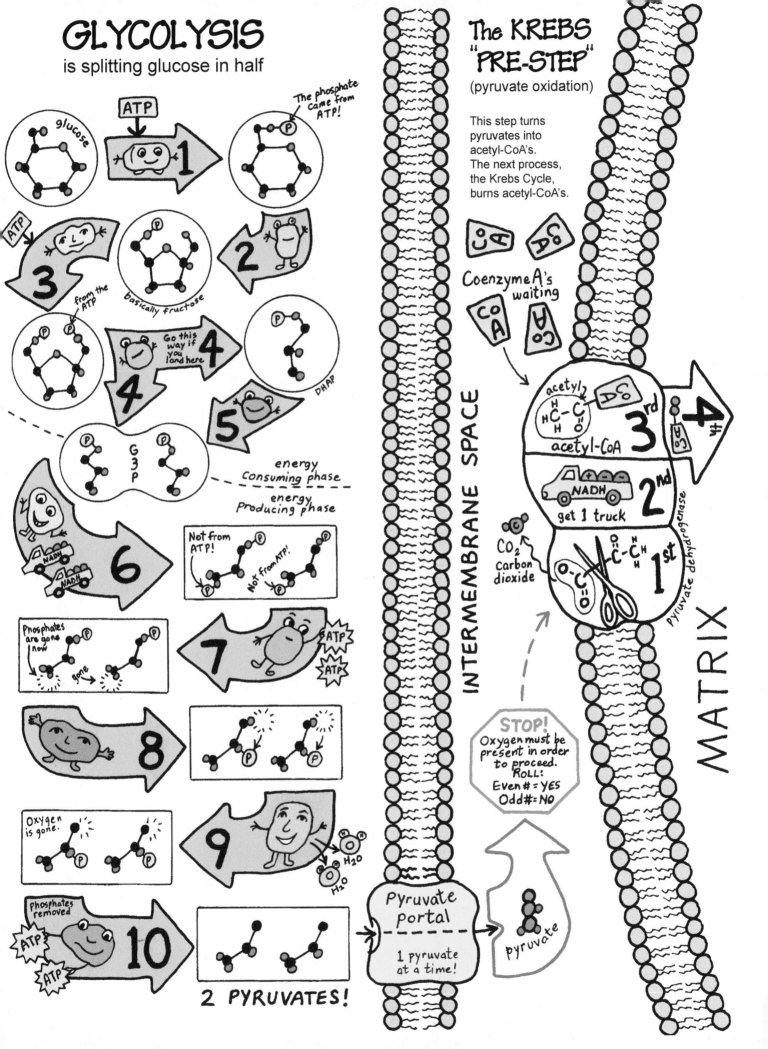

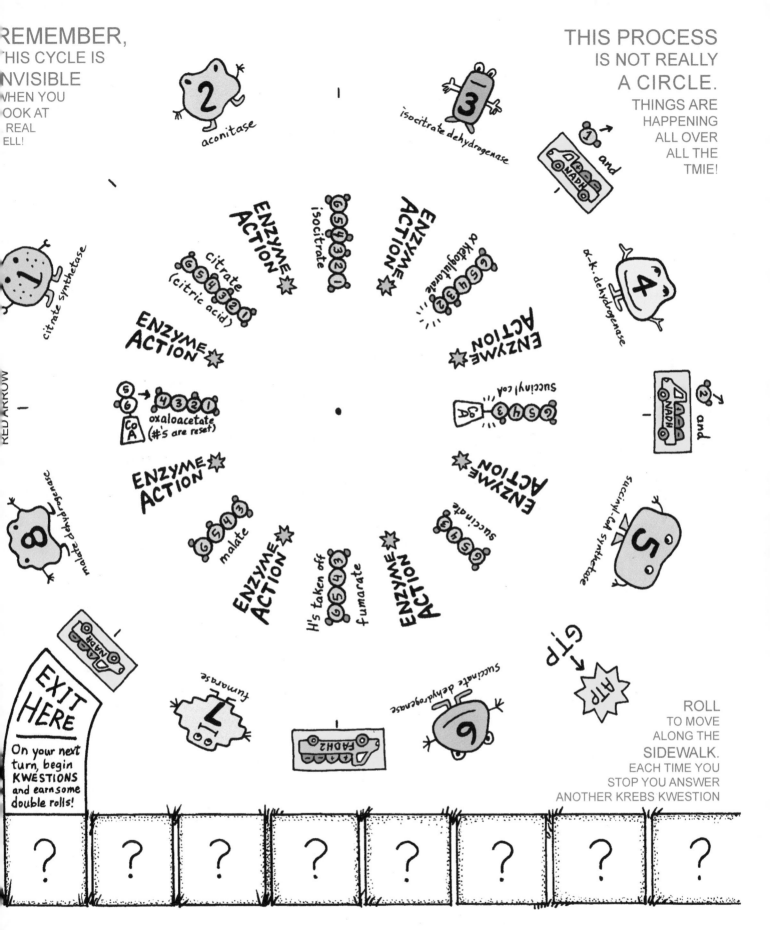

REMEMBER, THIS CYCLE IS INVISIBLE WHEN YOU LOOK AT A REAL CELL!

THIS PROCESS IS NOT REALLY A CIRCLE. THINGS ARE HAPPENING ALL OVER ALL THE TMIE!

2 aconitase

3 isocitrate dehydrogenase

1 and

ENZYME ACTION

isocitrate 6 5 4 3 2 1

ENZYME ACTION

α-ketoglutarate 6 5 4 3 2

α-k. dehydrogenase 4

citrate (citric acid) 6 5 4 3 2 1

ENZYME ACTION

1 citrate synthetase

ENZYME ACTION

succinyl CoA 6 5 4 3 2

2 and

RED ARROW

5 6 CoA → 4 3 2 1 oxaloacetate (#'s are reset)

ENZYME ACTION

ENZYME ACTION

succinyl-CoA synthetase 5

8 malate dehydrogenase

6 5 4 3 malate

ENZYME ACTION

H's taken off 6 5 4 fumarate

ENZYME ACTION

6 5 4 3 succinate

ATP ← GTP

ROLL TO MOVE ALONG THE SIDEWALK. EACH TIME YOU STOP YOU ANSWER ANOTHER KREBS KWESTION

EXIT HERE
On your next turn, begin KWESTIONS and earn some double rolls!

7 fumarase

FADH2

6 succinate dehydrogenase

? ? ? ? ? ? ? ?

KREBS CYCLE
(also known as the "Citric Acid Cycle" or the "Tricarboxylic Acid Cycle")

THE GOAL IS TO FILL TRUCKS!
They will provide the fuel for the last step.

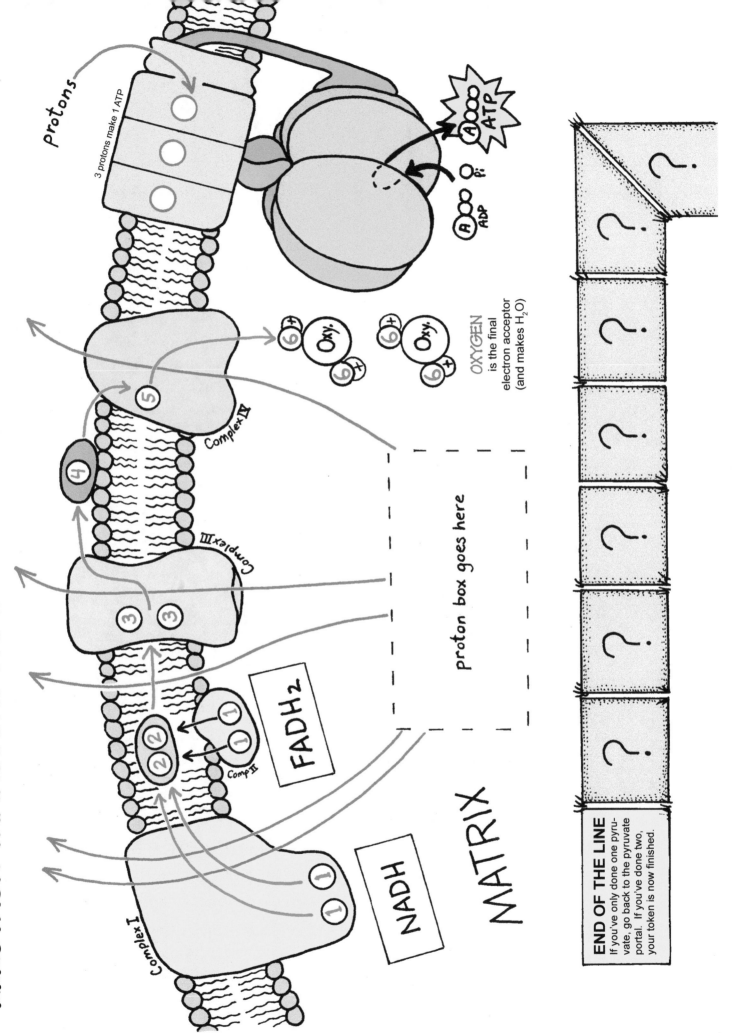

ELECTRON TRANSPORT CHAIN

INTERMEMBRANE SPACE

Protons

3 protons make 1 ATP

ATP

O₂
P₁

ADP

OXYGEN
is the final
electron acceptor
(and makes H₂O)

Complex IV

Complex III

proton box goes here

FADH₂

Comp II

NADH

MATRIX

Complex I

END OF THE LINE
If you've only done one pyruvate, go back to the pyruvate
portal. If you've done two,
your token is now finished.

?. ?. ?. ?. ?. ?. ?.

PRINT ONE COPY PER GAME (card stock if possible)   This gives you enough trucks for 4 players.

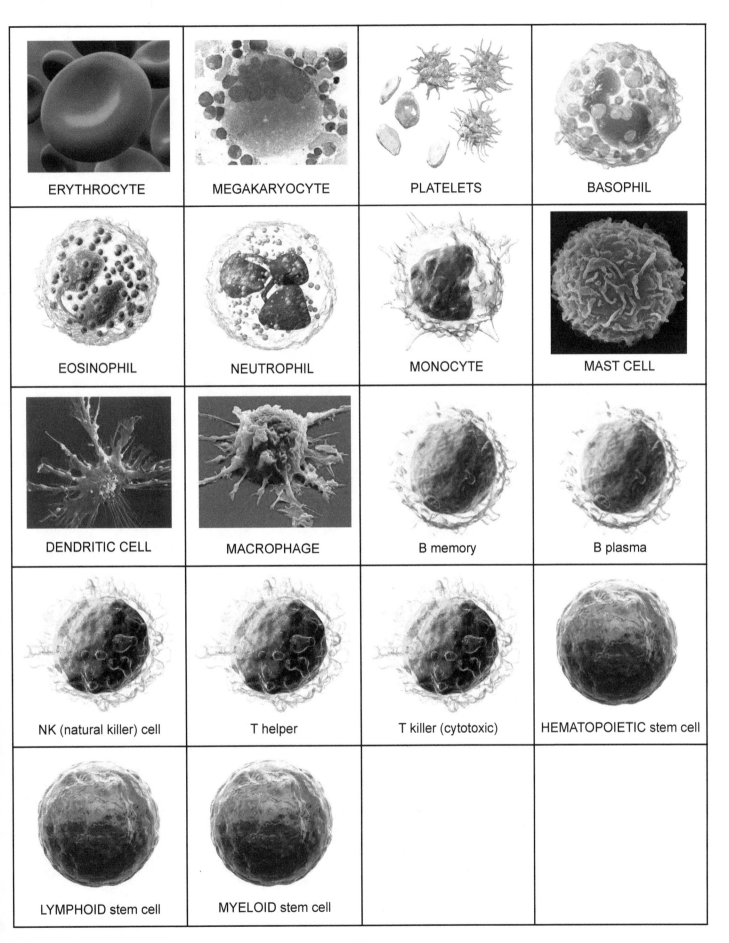

ERYTHROCYTE	MEGAKARYOCYTE	PLATELETS	BASOPHIL
EOSINOPHIL	NEUTROPHIL	MONOCYTE	MAST CELL
DENDRITIC CELL	MACROPHAGE	B memory	B plasma
NK (natural killer) cell	T helper	T killer (cytotoxic)	HEMATOPOIETIC stem cell
LYMPHOID stem cell	MYELOID stem cell		

BLOOD CELL BINGO ACTIVITY
These are either artists' renditions, or SEM electron micrographs. (This is NOT what you will see under an ordinary microscope. The other page shows compound microscope images.) This game does not intend to break copyright. Since it is not for sale, it should be considered "fair use" for educational purposes. The colored images are from Wikipedia, uploaded by BruceBlaus. Please consider making a donation to Wikipedia. :)

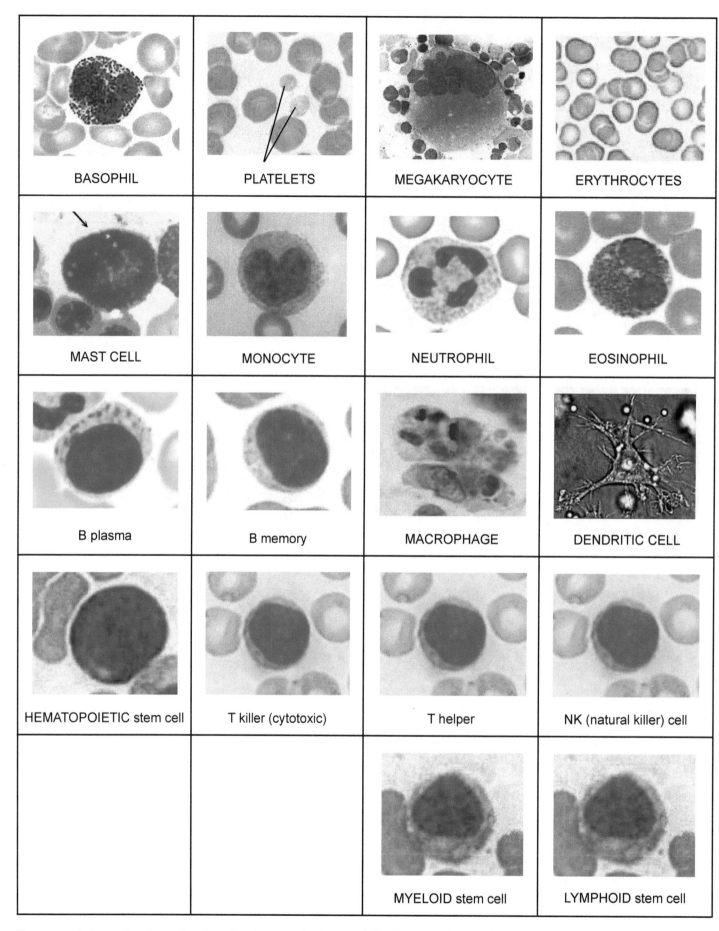

BASOPHIL	PLATELETS	MEGAKARYOCYTE	ERYTHROCYTES
MAST CELL	MONOCYTE	NEUTROPHIL	EOSINOPHIL
B plasma	B memory	MACROPHAGE	DENDRITIC CELL
HEMATOPOIETIC stem cell	T killer (cytotoxic)	T helper	NK (natural killer) cell
		MYELOID stem cell	LYMPHOID stem cell

These are photographs of samples viewed under a regular (compound) microscope (not an electron microscope). These images are easily found on the Internet, and you can view and download them yourself. No explicitly copyrighted images were used.
NOTE: Yes, the T cells are all the same photo. It is almost impossible to find pictures of separate cells because they look alike.

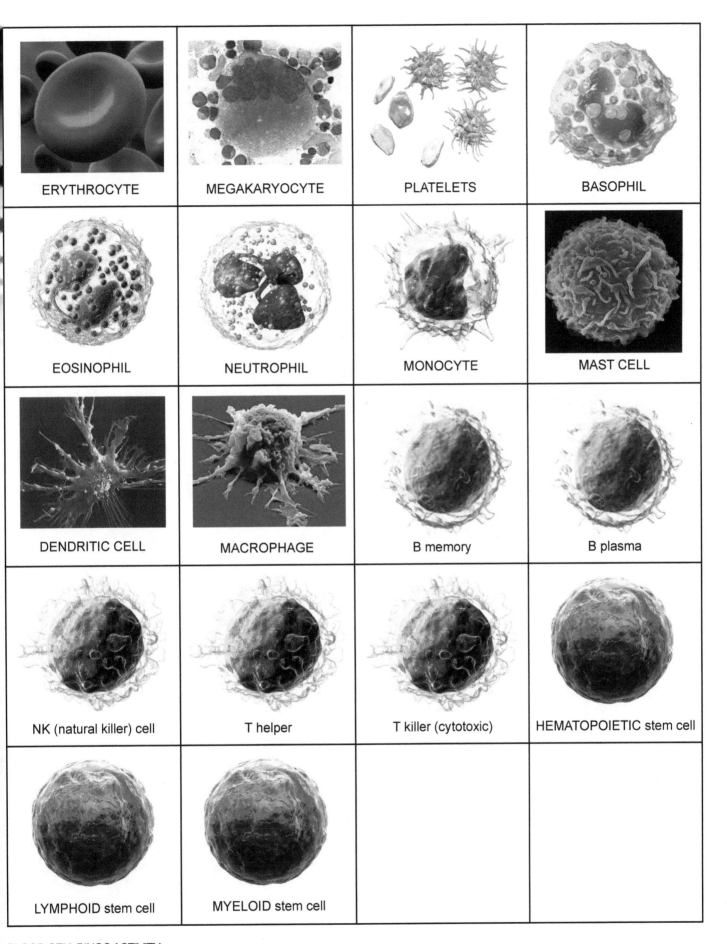

ERYTHROCYTE	MEGAKARYOCYTE	PLATELETS	BASOPHIL
EOSINOPHIL	NEUTROPHIL	MONOCYTE	MAST CELL
DENDRITIC CELL	MACROPHAGE	B memory	B plasma
NK (natural killer) cell	T helper	T killer (cytotoxic)	HEMATOPOIETIC stem cell
LYMPHOID stem cell	MYELOID stem cell		

BLOOD CELL BINGO ACTIVITY
These are either artists' renditions, or SEM electron micrographs. (This is NOT what you will see under an ordinary microscope. The other page shows compound microscope images.) This game does not intend to break copyright. Since it is not for sale, it should be considered "fair use" for educational purposes. The colored images are from Wikipedia, uploaded by BruceBlaus. Please consider making a donation to Wikipedia. :)

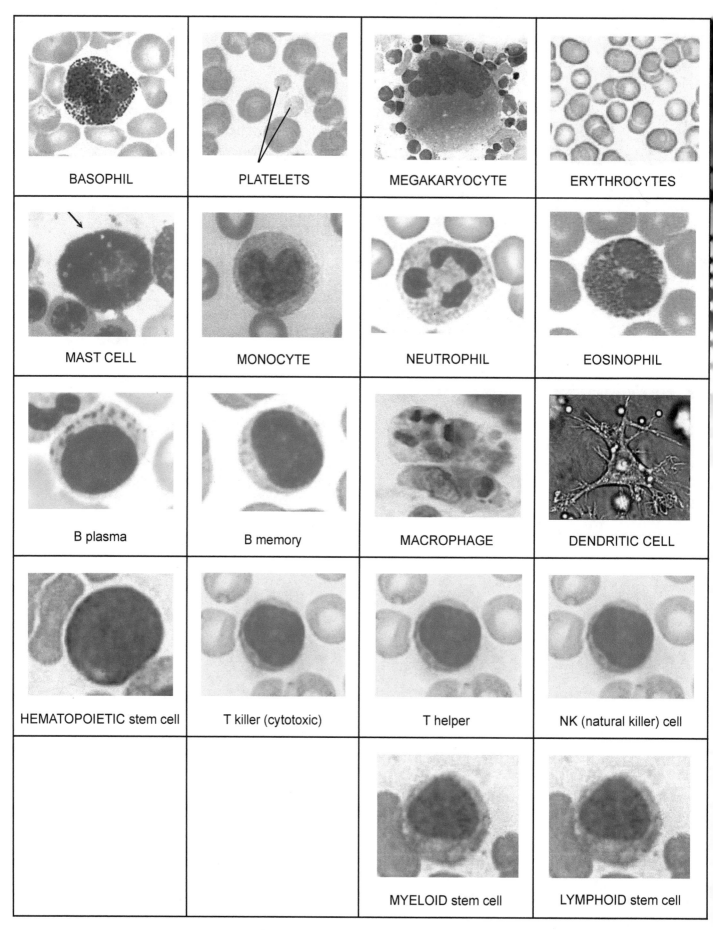

BASOPHIL	PLATELETS	MEGAKARYOCYTE	ERYTHROCYTES
MAST CELL	MONOCYTE	NEUTROPHIL	EOSINOPHIL
B plasma	B memory	MACROPHAGE	DENDRITIC CELL
HEMATOPOIETIC stem cell	T killer (cytotoxic)	T helper	NK (natural killer) cell
		MYELOID stem cell	LYMPHOID stem cell

These are photographs of samples viewed under a regular (compound) microscope (not an electron microscope). These images are easily found on the Internet, and you can view and download them yourself. No explicitly copyrighted images were used.
NOTE: Yes, the T cells are all the same photo. It is almost impossible to find pictures of separate cells because they look alike.

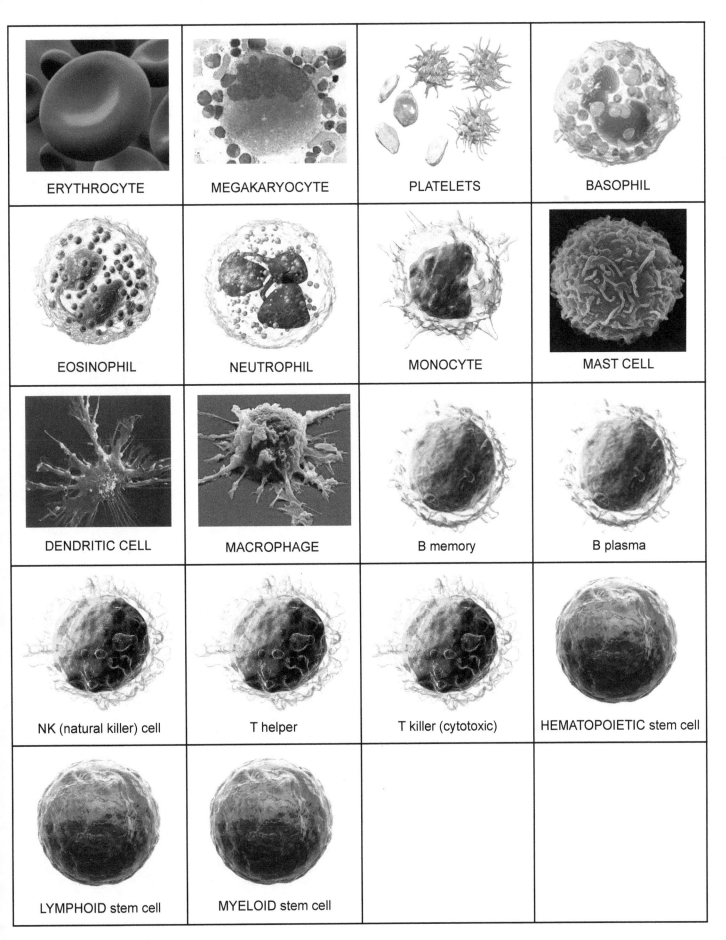

ERYTHROCYTE	MEGAKARYOCYTE	PLATELETS	BASOPHIL
EOSINOPHIL	NEUTROPHIL	MONOCYTE	MAST CELL
DENDRITIC CELL	MACROPHAGE	B memory	B plasma
NK (natural killer) cell	T helper	T killer (cytotoxic)	HEMATOPOIETIC stem cell
LYMPHOID stem cell	MYELOID stem cell		

BLOOD CELL BINGO ACTIVITY

These are either artists' renditions, or SEM electron micrographs. (This is NOT what you will see under an ordinary microscope. The other page shows compound microscope images.) This game does not intend to break copyright. Since it is not for sale, it should be considered "fair use" for educational purposes. The colored images are from Wikipedia, uploaded by BruceBlaus. Please consider making a donation to Wikipedia. :)

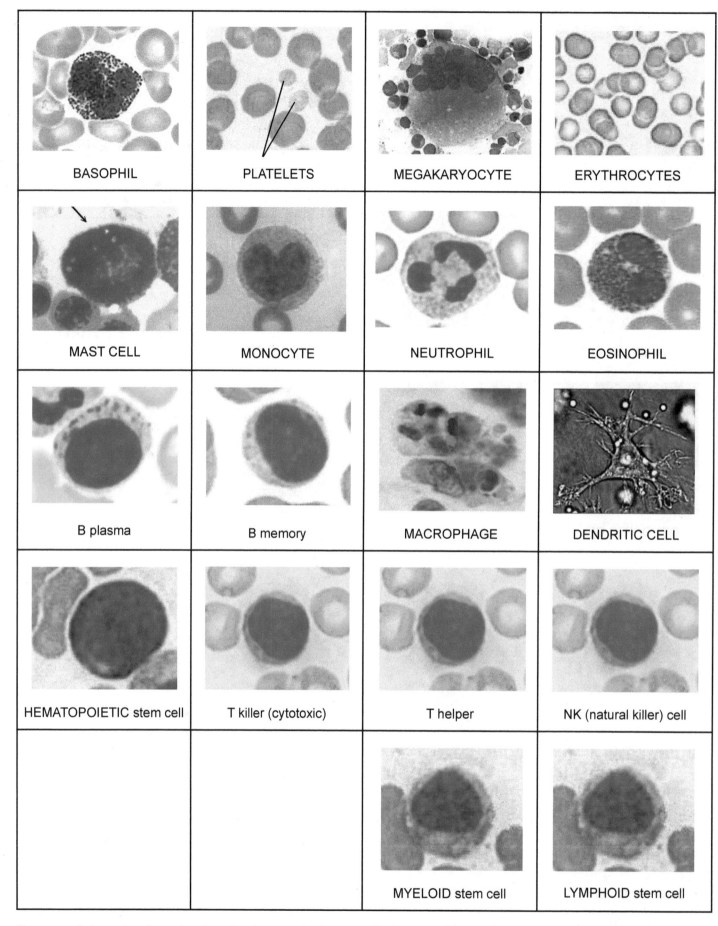

BASOPHIL	PLATELETS	MEGAKARYOCYTE	ERYTHROCYTES
MAST CELL	MONOCYTE	NEUTROPHIL	EOSINOPHIL
B plasma	B memory	MACROPHAGE	DENDRITIC CELL
HEMATOPOIETIC stem cell	T killer (cytotoxic)	T helper	NK (natural killer) cell
		MYELOID stem cell	LYMPHOID stem cell

These are photographs of samples viewed under a regular (compound) microscope (not an electron microscope). These images are easily found on the Internet, and you can view and download them yourself. No explicitly copyrighted images were used.
NOTE: Yes, the T cells are all the same photo. It is almost impossible to find pictures of separate cells because they look alike.